Glutamate Receptors in Peripheral Tissue: Excitatory Transmission Outside the CNS

Glutamate Receptors in Peripheral Tissue: Excitatory Transmission Outside the CNS

Edited by

SANTOKH GILL and OLGA PULIDO

Toxicology Research Division, HPFB, Health Canada, Ottawa, Canada

Kluwer Academic / Plenum Publishers
New York, Boston, Dordrecht, London, Moscow

Library of Congress Cataloging-in-Publication Data

Santokh Gill and Olga Pulido
    Glutamate receptors in peripheral tissue : excitatory
    transmission outside the CNS / edited by Santokh Gill and
    Olga Pulido.
        p. cm.
    Includes bibliographical references and index.
    ISBN 0-306-47973-7
      1. Glutamic acid—Receptors. 2. Nerves, Peripheral.
    I. Gill, Santokh.   II. Pulido, Olga. [DNLM: 1. Peripheral
    Nervous System—physiolgy. 2. Receptors, Glutamate. 3.
    Synaptic Transmission—physiology. WL 102.8 G5668 2004]

    QP364.7.G575 2004
    612.8'1—dc22
                                                    2003070342

ISBN: 0-306-47973-7 (hardback)
ISBN: 0-306-48644-X (ebook)

©2005 Kluwer Academic / Plenum Publishers, New York
233 Spring Street, New York, New York 10013

http://www.wkap.nl/

10 9 8 7 6 5 4 3 2 1

A C.I.P. record for this book is available from the Library of Congress

Permissions for books published in Europe: *permissions@wkap.nl*
Permissions for books published in the United States of America: *permissions@wkap.com*

Printed in the United States of America

To my daughters, Nitasha and Sonia, for putting my life into perspective.

Santokh Gill

To those who supported me and to those who opposed me, for each one has provided me with motivation and challenge in their own way.

Olga Pulido

Preface

Glutamate receptors (GluRs) in the central nervous system have been the subject of intense investigations for several decades, providing new avenues for the understanding of excitatory neurotransmission, excitotoxicity, mechanisms of injury, and therapeutics for several acute neurological conditions, such as brain trauma, and for neurodegenerative and neuropsychiatric disorders including addictions, Alzheimer disease, etc. Evidences of GluRs beyond the central nervous system were first reported in the early 1990s. When the idea of this book was conceived, the knowledge, specificity, and functional significance of GluRs in peripheral tissues was still in its embryonic stage. From our perspective, the idea of GluRs in peripheral tissues arose from our research on seafood toxins (see Chapter 1), and has now been reinforced by the results of other scientists working in similar areas. In this book, we have invited some of the leading authorities in the field to summarize their findings and to provide a framework for further investigations. The book is divided into three sections— Part I is on general concepts and concentrates on the distribution and cell-specific localization of glutamate receptors, their transporters, and the pharmacology in peripheral tissues and organs. Part II emphasizes the presence and implications of these receptors in specific target tissues, organs, and systems, including liver, lungs, endocrine tissues, bone, immune system, etc. Part III focuses on glutamate receptors in plants to illustrate their presence beyond the animal kingdom.

Recent advances in technology in molecular cell biology and the improvements in immunocytochemical and *in situ* hybridization methodologies have lead to a detailed knowledge of the distribution and characterization of GluRs in neural and nonneural cells and tissues. In spite of the anatomical, physiological, and pharmacological, evidence for GluRs in peripheral tissues, their functions remain elusive. The data presented shows that the molecular machinery required for glutamate signaling, including ionotropic GluRs, metabotropic GluRs, their transporters and vesicular glutamate transporters, are constitutively expressed in peripheral tissues. This suggests that the glutamatergic system plays fundamental roles in the maintenance of the functionality and integrity of these tissues. Furthermore, since the GluRs have a wide diversity, their distribution in specific cells and tissues open opportunities for the development of new pharmacological agents and compounds aiming to target selective sites for specific therapeutic effects. For this, dissection of the glutamate signaling pathways including glutamate receptors, their transporters, glutamate release mechanisms, ion channels, intracellular regulatory proteins and secondary signaling pathways, within all specific cell types or tissues is of primordial importance. The information in this book is intended to provide an overview on the current knowledge, and it is our hope that it will serve as a valuable source of practical information on glutamate receptors in peripheral tissues.

Santokh Gill
Olga Pulido

Contents

3. Expression of Non-Organelle Glutamate Transporters to Support Peripheral Tissue Function
James C. Matthews

4. Anticancer Effects of Glutamate Antagonists
Wojciech Rzeski, Lechoslaw Turski, and Chrysanthy Ikonomidou

5. Glutamate Receptors and their Role in Acute and Inflammatory Pain
Susan M. Carlton

Part II. Specific Target Tissues, Organs, and Systems

6. The Vertebrate Retina

Victoria P. Connaughton

7. Glutamate Receptors in Taste Receptor Cells

Albertino Bigiani

8. Glutamate Receptors in Endocrine Tissues

Tania F. Gendron and Paul Morley

9. Adrenal Glutamate Receptors: A Role in Stress and Drug Addiction?

Daniela Jezova and Marek Schwendt

10. Glutamate Receptors in the Stomach and their Implications

Li Hsueh Tsai and Jang-yen Wu

11. Glutamate Toxicity in Lung and Airway Disease

Sami I. Said

12. Glutamate: Teaching Old Bones New Tricks— Implications for Skeletal Biology

Gary J. Spencer, Ian S. Hitchcock, and Paul G. Genever

13. Expression and Function of Metabotrophic Glutamate Receptors in Liver

Marianna Storto, Maria Pia Vairetti, Francesc X. Sureda, Barbara Riozzi, Valeria Bruno, and Ferdinando Nicoletti

14. Neuroexcitatory Signaling in Immune Tissues
Helga S. Haas and Konrad Schauenstein

15. Platelet Glutamate Receptors as a Window into Psychiatric Disorders
Michael Berk

Part III. Non-Mammalian Organisms

16. Analysis of Glutamate Receptor Genes in Plants: Progress and Prospects
Joanna C. Chiu, Eric D. Brenner, Rob DeSalle,
Nora M. Barboza, and Gloria M. Coruzzi

Glutamate Receptors in Peripheral Tissue: Excitatory Transmission Outside the CNS

Part I

General Concepts

Glutamate Receptors in Peripheral Tissues: Distribution and Implications for Toxicology

Santokh Gill and Olga Pulido

1. Introduction

In 1987, a serious outbreak of food poisoning occurred in Canada following the consumption of mussels from Prince Edward Island. This incident resulted in several deaths and approximately 100 patients developing acute gastrointestinal and neurological symptoms. The excitatory neurotoxin domoic acid was subsequently identified the source responsible for the human intoxication (Thompson *et al.*, 1983; Iverson *et al.*, 1990; Perl *et al.*, 1990; Teitlebaum *et al.*, 1990; Tryphonas *et al.*, 1990; Peng *et al.*, 1994; Truelove *et al.*, 1996). Further toxicologic investigations of domoic acid revealed it as one of the most potent neurotoxins that can enter the food chain. Analysis of the human poisoning events together with the data obtained from toxicological studies in rodents and nonhuman primates provided the basis of the regulatory measures of 20 $\mu g/\mu l$ domoic acid within shellfish tissues. These measures are utilized both nationally and internationally to protect the population from exposure to this lethal seafood toxin. In addition to the acute effects, some of the survivors of this poisoning were left with severe residual memory deficits. This observation was the basis for naming the domoic acid intoxication "Amnesic Shellfish Poisoning." (Clark *et al.*, 1999; Whittle and Gallaher, 2000). Monitoring programs in Canada have been successful in preventing other incidents of human poisoning. However, recent reports of domoic acid contaminants of the coastal waters have made headlines in the news (*Seattle Post*, September 19, 2003). Hence, it is still a threat to human health, wildlife, and seafood industry. Therefore, the knowledge of mechanisms of action of domoic acid may open new avenues in the management and prevention of domoic acid toxicity and its analogues.

Glutamate (Glu) and aspartate (Asp) are naturally occurring amino acids found in the central nervous system (CNS) where they act as major excitatory neurotransmitters by stimulating and exciting the postsynaptic neurons (Olney, 1989, 1994; Gasic and Hollmann, 1992; Krogsgaard and Hansen, 1992; Dingledine and McBain, 1994; Asztely and Gustafasson,

Santokh Gill and Olga Pulido • Toxicology Research Division, HPFB, Health Canada, Banting Bldg, 2002D2, Tunney's Pasture, Ottawa, Ontario, Canada, K1A 0L2

Glutamate Receptors in Peripheral Tissue, edited by Santokh Gill and Olga Pulido. Kluwer Academic / Plenum Publishers, New York, 2005.

1996; Michaelis, 1998; Palmada and Centelles, 1998; Dingledine *et al.*, 1999; Moghaddam, 1999; Gallo and Ghiani, 2002; Herman, 2002; Moloney, 2002). These effects are mediated through the glutamate receptors (GluRs). Over excitation of these receptors by local and/or circulating concentrations of excitatory amino acids (EAAs) can lead to neural injury and cell death. Permanent brain lesions occur despite the protective mechanisms of the blood–brain barrier (BBB). Exogenous EAAs can access the brain through the circumventricular organs which are located outside the BBB (Price *et al.*, 1981; Bruni *et al.*, 1991). In the brain, GluRs mediate the neurotoxic effects of excitatory compounds such as domoic acid. Histopathology analysis of victims of intoxication and the brain of experimental animals exposed to domoic acid showed preferential damage to areas of the brain rich in GluRs. These areas include the hippocampus and other structures of the limbic system which are involved in the memory process (Tryphonas *et al.*, 1990; Peng *et al.*, 1994; Clark *et al.*, 1999). The main clinical manifestations of domoic acid intoxication were neurologic symptoms including severe seizures (Iverson *et al.*, 1990; Perl *et al.*, 1990; Teitelbaum *et al.*, 1990; Peng *et al.*, 1994). Other clinical symptoms such as gastrointestinal disturbances, cardiovascular collapse, and cardiac arrhythmia were also observed in some individuals (Iverson *et al.*, 1990; Gill *et al.*, 1998, 2000). Postmortem investigations of sea lions intoxicated with domoic acid off the coast of California in 1998 provided the first pathological evidence of severe toxic effects in peripheral tissues particularly the heart, kidney, and gastrointestinal tract (Gulland, 2000; Scholin *et al.*, 2000). Most investigations on domoic acid toxicity have focused on its effects on the CNS and less attention has been given to the potential toxicity to peripheral organs. Reports of cardiovascular disturbances and gastrointestinal distress associated with domoic acid intoxications and other excitatory compounds in foods such as monosodium glutamate (MSG), prompted our initial investigation of the heart as a possible target organ (Mueller *et al.*, 1996, 2003; Gill *et al.*, 1998, 1999, 2000; Gill and Pulido, 2001). Since then it has been established that Glu and its analogues interact with the GluRs in the CNS. As the first step in assessing the existence of a similar relationship we investigated the heart followed by other tissues and/or organs. The presence of GluRs in these tissues suggests they may also mediate the toxic effects of domoic acid in these sites. Furthermore, these findings might also offer an explanation for the idiosyncratic reactions to the ingestion of excitatory compounds in foods such as MSG, aspartame, and possibly sulfites in susceptible subsets of the population. The toxicity of each compound such as domoic acid and kainate varies according to the potency, chemical availability, rate of absorption, affinity to specific receptors, and the particular anatomical target site. In addition, considerations should be given to individual susceptibility determined by age, genetic predisposition, and health status as important factors in risk assessment.

In response to the increasing interest in GluRs and the safety of excitatory compounds in food, it seemed pertinent to compile current knowledge on GluRs in peripheral tissues to provide an overview for further investigation. Food toxicology is the science of evaluating the safety of chemicals that enter the human food chain either as natural compounds, contaminants and/or during processing. In order to assess chemical safety, tissues and/or organs are examined for structural, chemical, or functional alterations. These investigations help to establish the safety margins of such compounds for consumption by humans/animals as either food or therapeutic products. Therefore, product safety requires continual reassessment as new information becomes available with advances in technology.

In this opening chapter, we report on the anatomical distribution of the GluRs in peripheral tissues, an overview on the potential role of GluRs outside the CNS and the effects they can potentially mediate. In addition, we have tried to highlight the possible sources of the endogenous and exogenous ligands for these receptors. The following chapters will highlight the roles of GluRs and Glu transporters in specific tissues (immune organs and cells, liver, lungs, stomach) and their therapeutic potential in pain and cancer.

2. Excitatory Amino Acids

Glu and Asp are the most abundant dicarboxylic amino acids in the brain and were believed to be the primary neurotransmitters in the mammalian CNS (Olney, 1989, 1994; Gasic and Hollman, 1992; Krogsgaard and Hansen, 1992; Dingledine and McBain, 1994; Asztely and Gustafasson, 1996; Michaelis, 1998; Palmada and Centelles, 1998; Herman, 2002; Moloney, 2002). Although these amino acids are primarily involved in intermediary metabolism and other non-neuronal functions, their most important role is as neurotransmitters. It is estimated Glu mediates nearly 50% of all the synaptic transmissions in the CNS and its involvement is implicated in nearly all aspects of normal brain function including learning, memory, movement, cognition, and development (Dingledine and McBain, 1994; Schoepp,1994; Asztely and Gustafasson, 1996; Michaelis, 1998; Dingledine et al., 1999; Herman, 2002; Ozawa et al., 1998). Although Glu is found in various concentrations dependent on the brain region and serves a number of important metabolic functions in the CNS, it is surprising that systemic administration of Glu to infant mice and primates destroyed neurons in the retina and/or specific regions of the brain. At elevated concentrations, Glu acts as a neurotoxin capable of inducing severe neuronal damage. Glu can be considered to be a "two-edged sword" which undergoes a transition from a neurotransmitter to a neurotoxin.

In addition to the endogenous Glu, there are naturally occurring substances which have Glu-like excitatory properties and potentially excitotoxic effects. Glu and its structural analogues (Figure 1.1) may enter the food supply during preparation or processing as contaminants and/or additives (Zautcke et al., 1986; Olney, 1969; 1989, 1994; Iverson et al., 1990; Hayashi et al., 1996; Clark et al., 1999; La Bella and Picolli, 2000; Watters, 1995). These include MSG, L-aspartate (L-asp), L-cysteine (L-cys), related sulfur amino acids (i.e., homocysteate), B-N-oxalyamino-L-alanine (BOAA or ODAP), B-N-methyl-amino-L-alanine (BMAA), and the potent sea food toxin domoic acid (Zautcke et al., 1986; Olney, 1989, 1994; LaBella and Picolli, 2000). Ultrastructural studies localized the site of action of Glu and its analogues to the postsynaptic dendrosomal membranes. A plethora of findings in the past two decades has provided direct and circumstantial evidence for abnormal Glu transmission (and its analogues) in the etiology and pathophysiology of many neurological and psychiatric disorders such as epilepsy, schizophrenia, addiction, depression, anxiety, Alzheimer's, Huntington's, Parkinson's, and amyotrophic lateral sclerosis (Miller et al., 1996; Dingledine and McBain, 1994; Asztely and Gustafasson, 1996; Palmada and Centelles, 1998; Moghaddam, 1999; Herman, 2002).

The neurotoxic effects of the EAAs is dependent on the species, developmental stage of the animal, type of agonist, duration of exposure to the agonist, and the cellular expression of the GluR subtypes. Recent reports using rat brain slices when pre-exposed to domoic acid, show an age-dependent tolerance to the excitotoxic effects, suggesting that protective mechanisms are elicited in young but not in the aged animals (Kerr and Razak, 2000). There is also a growing body of evidence on astrocyte protective mechanisms in neurotoxicology and in domoic acid toxicity (Pentreath and Slamon, 2000; Ross et al., 2000). An array of GluRs are

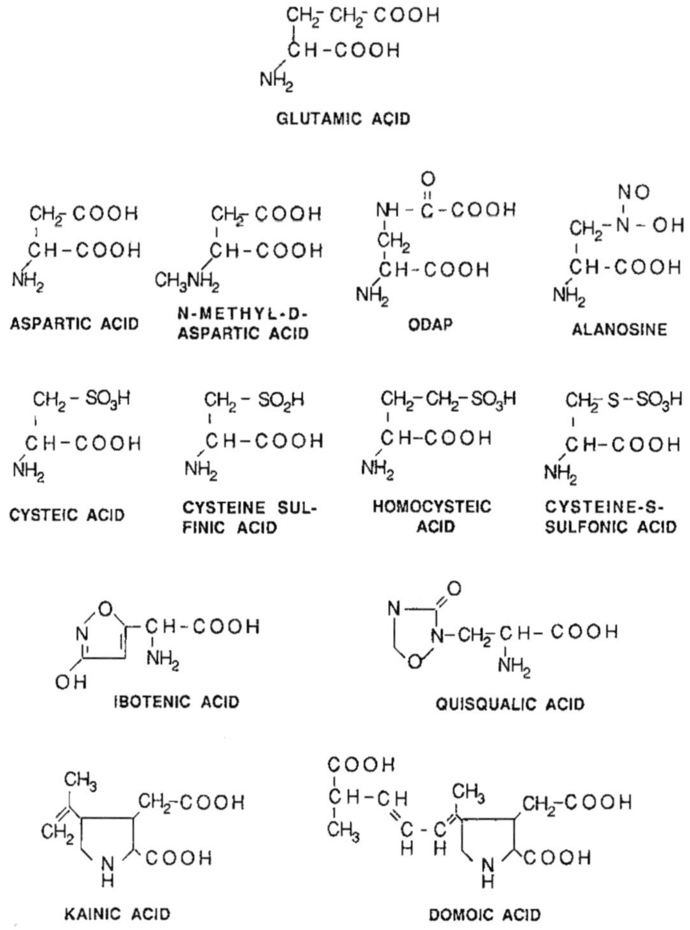

$$CH_2 - CH_2 - COOH$$
$$CH - COOH$$
$$NH_2$$

GLUTAMIC ACID

ASPARTIC ACID

N-METHYL-D-ASPARTIC ACID

ODAP

ALANOSINE

CYSTEIC ACID

CYSTEINE SULFINIC ACID

HOMOCYSTEIC ACID

CYSTEINE-S-SULFONIC ACID

IBOTENIC ACID

QUISQUALIC ACID

KAINIC ACID

DOMOIC ACID

Figure 1.1. Excitotoxic structural analogues of glutamate. Most of these compounds are known to mimic both the neuroexcitatory and neurotoxic properties of glutamate. ODAP is also *B-N*-Oxalylamino-L-Alanine (Olney, 1989).

known to be present on pre- and postsynaptic membranes which are used to transduce integrated signals using an increased ion flux and second messenger pathways (Figure 1.2) (Dingledine and McBain, 1994; Schoepp, 1994; Asztely and Gustafasson, 1996; Dingledine *et al.*, 1999; Moghaddam, 1999; Herman, 2002). GluRs are expressed both in neurons and astrocytes causing the excessive activation of these receptors and the interplay with protective mechanisms key to the neurotoxicity of domoic acid and other EAAs.

2.1. Possible Endogenous Sources of Excitatory Ligands

A potential site for endogenous synthesis of Glu could be the peripheral nervous tissue, most likely the autonomic nervous system including ganglia cells and nerve terminals. Glu synthesized in these locations could then be released from the axon buttons in close proximity with the target cells where it would bind to the specific receptors. Endogenous Glu release from nervous tissues could be increased under stressful conditions (Butcher *et al.*, 1987; Bertrand *et al.*, 1993). This is supported by the findings that phosphatase-activated

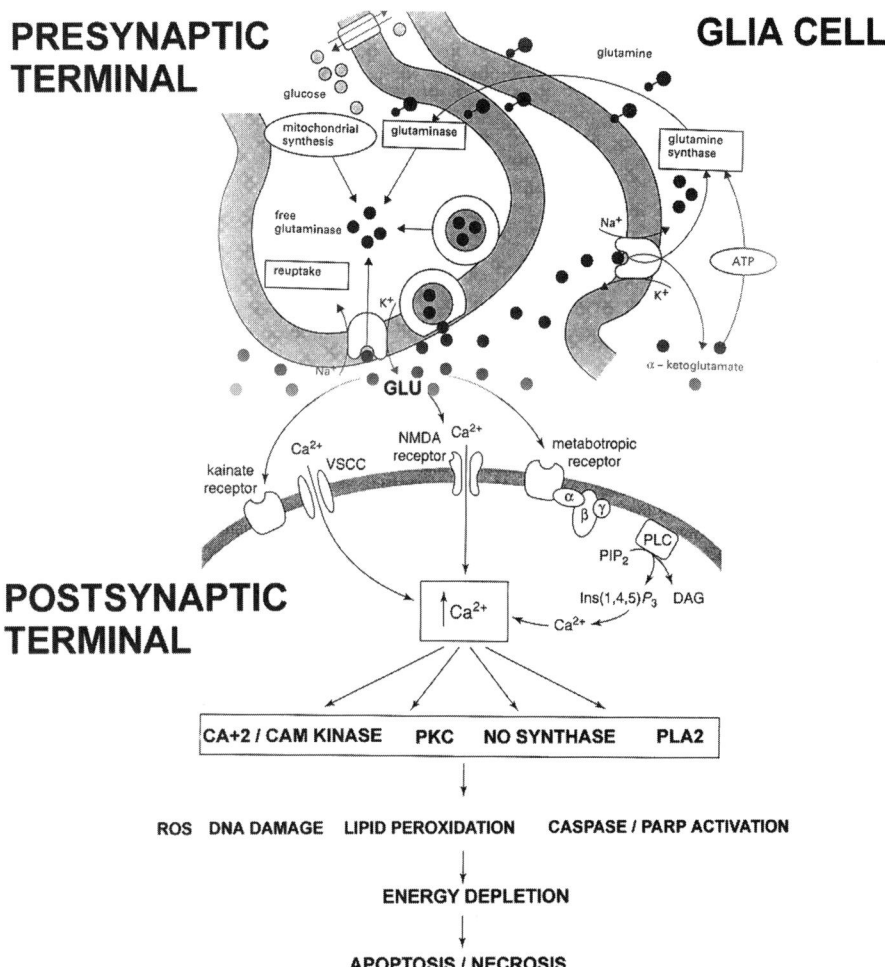

Figure 1.2. Illustration of the interaction between the presynaptic and the postsynaptic terminal. The figure shows vesicular release into the synaptic space, activation of the postsynaptic receptor systems, re-uptake into the presynaptic terminal, and surrounding glial cells. The excitatory amino acids (i.e., glutamate) activates the various glutamate receptors present in the postsynaptic membrane. This triggers the influx of Ca^{+2} from the extracellular environment to the synaptic cleft. The accumulation of Ca^{+2} is the crucial determinant of injury. This elevation of Ca^{+2} triggers the activation of several enzymes: calmodulin (CAM), protein kinase C (PKC), nitric oxide synthase (NO synthase), phospolipase A_2 (PLA2) and reactive oxygen species (ROS)—modified from G.J. Harry (1999) and Said (1990).

glutaminase, the enzyme involved in the synthesis of Glu, is present in the glucagon-secreting α-cells, the pancreatic ganglia, bone, and bone marrow (Weaver *et al.*, 1996; Chenu *et al.*, 1997; Genever *et al.*, 1999). The production of these EAAs could be triggered or stimulated by a variety of inflammatory mediators, such as arachidonate metabolites and reactive oxygen free radicals (Lipton and Rosenberg, 1994; Meldrum, 1994; Lipton and Gendelman, 1995; Demenes *et al.*, 1995; Said *et al.*, 1996, 2001; Said, 1999).

Although the D-amino acids may have a lower affinity for the GluRs than the L-form, they have been reported in various tissues and body fluids of chickens, rodents, and human plasma (Imai *et al.*, 1996; Snyder and Ferris, 2000; Wolosker *et al.*, 1999, 2002). D-aspartate (D-Asp) have been found in proteins of various human tissues such as teeth, normal and cataract eye lenses, myelin and myelin basic protein and both normal and Alzheimer's brains (Fisher *et al.*, 1991). The main free D-form amino acids found in vertebrate tissues are D-Asp and D-Serine (D-Ser). Others are found but only in smaller quantities (Imai *et al.*, 1996; Snyder and Ferris, 2000; Wolosker *et al.*, 2002). Experiments have shown that D-Ser and D-Ala (D-Alanine) have an affinity on the strychnine-insensitive glycine-binding site for the NMDAR 1 receptor. This further suggests a role for D-amino acids as intrinsic ligands of the receptor (allosteric effect). Also *N*-methyl-D-aspartate receptor (NMDAR) is a derivative of D-amino acid and some natural peptides, which perform important neurological functions. Kamatani *et al.* (1989) isolated a tetrapeptide, achatin-I from the subesophageal and cerebral ganglia of the African giant snail *Achatina fulica*. This endogenous D-amino acid derived peptide has neuroexcitatory properties by inducing a stereospecifically voltage-dependent inward current brought about by Na^+ in a neuron of the same species. It is significant that this endogenous peptide is neuroactive, thereby suggesting its role as a neurotransmitter or neurohormone in the CNS. Montecucchi *et al.* (1981) isolated a heptapeptide, dermorphin, with very potent opiate-like activity from the skin of a South African frog *Phyllomedusa sauvagei*. The amino acid sequence of this peptide is: H-Tyr-D-Ala-Phe-Gly-Try-Pro-Ser-NH_2. D-amino acids have also been known found in bacteria for more than 50 years but only recently they were identified in mammals. D-amino acids, particularly D-Asp, are also found in the testis, kidneys, pituitary, and pineal glands (Snyder and Ferris, 2000; Wolosker *et al.*, 2002). Their occurrence in mammals challenges the concept in which only L-amino acids were thought to play important roles. Recent discoveries by Wolosker *et al.* (2002) uncovered a role of endogenous D-Ser as a putative glial-derived transmitter that regulates glutamatergic neurotransmission in mammalian brain. Free D-Ser levels in the brain are about one third of L-serine values and its extracellular concentration is higher than many common L-amino acids. D-Ser also occurs in protoplasmic astrocytes. Biochemical and electrophysiological studies suggest that endogenous D-Ser is a physiological modulator at the coagonist site of the NMDA-type of GluRs. It has been shown that D-Ser is synthesized by a glial serine racemase, a novel enzyme converting L- to D-Ser in mammalian brain. Inhibitors of this enzyme have therapeutic implications for pathological processes in which overstimulation of NMDA receptors takes place such as stroke and neurodegenerative disease (Snyder and Ferris, 2000; Wolosker *et al.*, 2002).

2.2. Possible Exogenous Sources of Excitatory Ligands

Monosodium glutamate (MSG) and Glu are the most common food additive and enhancers, respectively, in commercial foods and in food condiments (Zautcke *et al.*, 1986; Allen *et al.*, 1987; Bertrand *et al.*, 1992). Free Glu is a natural constituent of many foods including meat, cheeses, and vegetables providing an excellent source for exogenous Glu. However, free Glu is absorbed slowly and is dependent on the presence or absence of other nutrients, in particular carbohydrates and probably glucose. Aspartame is a dipeptide of aspartic acid and phenylalanine which is widely used as a sweetener of foods, beverages, and drugs. It is hydrolyzed in the gut to release Asp which is efficiently absorbed. Hydrolyzed protein (HP) is also widely used by the food industry as a flavour additive. In addition to Glu, HP can also contain Asp and various sulfur-containing amino acid. L-Cys is a common sulfur containing amino acids. Although this compound is relatively weak, in the presence of physiological bicarbonate, L-Cys is transformed into a powerful EAA. L-Cys is unique in that it can penetrate the BBB. Group IV class of metabotropic receptors (mGluRs) are more efficiently activated by L-cysteine-sulfinic acid (L-CSA) rather than Glu (see introduction). Other Glu-like substances can also enter the human food supply and act on the same receptors with a possible potentiating effect. These include: BOAA—a component of chick pea, *Lathyrus sativus*, and BMAA that reacts with carbon dioxide to form a carbamate derivative. Domoic acid is manufactured by a sea phytoplankton, penna diatom, *Nitzchia pungens f. multiseries*, which is ingested and accumulated in the digestive system of seashells such as mussels. These mussels serve a dietary source for humans where their intake with high amounts of domoic acid causes severe neurological disorders. It is well documented that the excitotoxic effects of domoic acid and the above mentioned neurotoxins arises from their potent effect as Glu agonist. Since the GluRs appear to be omnipresent these circulating EAAs and analogues could potentially have a powerful effect on central and peripheral tissues by acting on the available receptor sites. Tissues not protected by a blood barrier are at a higher risk for over stimulation of the receptors as they are more available and susceptible to changes in circulating levels of these ligands. The circulating levels of Glu will vary with the ingestion and absorption of these excitatory compounds present in the food supply. The intensity of this effect will vary with the target tissues receptor availability, susceptibility, age, and the possible interaction or potentiating effect with ligands or their analogs.

3. Glutamate Receptors: An Overview

Numerous reviews are available on the GluRs in the CNS, their functional roles and implications on the pathobiology of neural injury, neuropsychiatric disorders, and addictions (Miller *et al.*, 1996; Dingledine and McBain, 1994; Hollmann and Heinemann, 1994; Schoepp, 1994; Asztely and Gustafasson, 1996; Conn and Pin, 1997; Dingledine *et al.*, 1999; Herman, 2002). GluRs have been individually characterized by their sensitivity to specific Glu analogues and by the features of the Glu-elicited current. GluR agonists and antagonists are structurally similar to Glu which allows them to bind onto the same receptors. Recently, it has been shown that these Glu antagonist exhibit anticancer properties *in vitro* (Rzeski *et al.*, 2002; Wojciech *et al.*, Chapter 4, this volume).

Two classes of GluRs have been characterized based on the studies in the CNS: ionotropic (iGluRs) and metabotropic (mGluRs). Their cloning has revealed the molecular diversity of the gene families encoding various receptor types which are responsible for the pharmacological and functional heterogeneity in the brain.

3.1. Ionotropic Glutamate Receptors

The iGluRs contain integral cationic channels associated with ligand-binding sites and they are known to mediate rapid synaptic transmission. They are classified into three major subtypes according to their sequence similarities, their electro-physiological properties and their affinity to selective agonists: *N*-methyl-D-aspartate (NMDA), α-amino-3-hydroxy-5-methyl-4-isoxazole propionic acid (AMPA), and kainate (Ka) receptors (Miller *et al.*, 1996; Dingledine and McBain, 1994; Hollman and Heinemann, 1994; Schoepp, 1994; Asztely and Gustafasson, 1996; Conn and Pin, 1997; Dingledine *et al.*, 1999; Herman, 2002). The membrane channels associated with these receptors exhibit varied pharmacological and electrophysiological properties including ionic channel selectivity to sodium (Na^+), potassium (K^+), and calcium (Ca^{+2}) (Miller *et al.*, 1996; Dingledine and McBain, 1994; Hollman and Heinemann, 1994; Schoepp, 1994; Asztely and Gustafasson, 1996; Conn and Pin, 1997; Dingledine *et al.*, 1999; Herman, 2002). The non-NMDA receptors control a nonselective cationic channel permeable to Na^+ and K^+ whereas NMDA is more permeable to Ca^{+2} ions than either AMPA or Ka (Miller *et al.*, 1996; Dingledine and McBain, 1994; Hollman and Heinemann, 1994; Schoepp, 1994; Asztely and Gustafasson, 1996; Conn and Pin, 1997; Dingledine *et al.*, 1999; Herman, 2002). Recombinant technology has identified several gene families encoding iGluRs: (a) AMPA family—composed of GluR 1–4; (b) Ka family—composed of GluR 5–7 and Ka 1–2, and (c) NMDA family—composed of NMDAR 1 and NMDAR 2A–D. NMDAR 1 is the most tightly regulated neurotransmitter receptor by forming the channel where other subunits (NMDA 2A–D) are involved in the receptor modulation. It is the most intensively studied and complex receptor that is linked to the Na^+/Ca^{+2} ion channel which has five distinct binding sites for endogenous ligands that influence its opening (Dingledine and McBain, 1994; Hollman and Heinemann, 1994; Schoepp, 1994; Asztely and Gustafasson, 1996; Conn and Pin, 1997; Dingledine *et al.*, 1999; Herman, 2002). These include two different agonist recognition sites (one for Glu and the other for glycine or D-Seri) and a polyamine regulatory site that promotes receptor activation. The remaining sites separate recognition sites for Mg^{+2}, Zn^{+2}, and phencyclidine (PCP) which inhibit the ion flux through agonist-bound receptors (Dingledine and McBain, 1994; Hollman and Heinemann, 1994; Asztely and Gustafasson, 1996; Conn and Pin, 1997; Dingledine *et al.*, 1999; Snyder and Ferris, 2000; Herman, 2002). The complexity of these GluR families is further increased by alternate splicing, RNA editing and post-translational modifications such as phosphorylation, glycosylation, and palmitoylation (Holmann and Heinemann, 1994; Hampson and Malano, 1998; Michaelis, 1998; Dingledine *et al.*, 1999). Each of these modifications is important in the regulation of channel functions. Within each family, GluRs subunits can also form homo-oligomeric or hetero-oligomeric channels that exhibit different functional properties depending on the subunit composition. For example the presence of the GluR 2 subunit decreases Ca^{+2} permeability of AMPA channels.

3.2. Metabotropic Glutamate Receptors

The mGluRs exert their effects either on the second messengers or ion channels via the activation of the of GTP-binding proteins which regulate the synthesis of different intracellular second messengers such as IP_3, cAMP, or cGMP (Dingledine and McBain, 1994; Hollman and Heinemann, 1994; Schoepp, 1994; Asztely and Gustafasson, 1996; Conn and Pin, 1997; Michaelis, 1998; Dingledine *et al.*, 1999; Herman, 2002). This process is predominantly used for the long term aspects of cellular control. A single mGluR protein can cross-talk with multiple second messengers in the same cell. As with iGluRs, the mGluRs are also classified into four groups based on amino acid sequence similarities, agonist pharmacology and the signal transduction pathways to which they are coupled. Group I (mGluR 1 and 5) stimulates inositol phosphate metabolism and mobilization of intracellular Ca^{+2}. Group II (mGluR 2 and 3) and group III (mGLuR 4,6–8) are coupled to adenyl cyclase (Miller *et al.*, 1996; Dingledine and McBain, 1994; Schoepp, 1994; Conn and Pin, 1997, Dingledine *et al.*, 1999). Group IV is coupled to the activation of phospholipase D (PLD). The latter class is more efficiently activated by L-CSA rather than Glu, which suggests that L-CSA may serve as an endogenous agonist of this receptor. As with the iGluRs, the mGluRs also have a unique distribution in the

CNS and retina which reflects a diversity of function in normal and pathological processes. These receptors have been shown to exert a wide variety of modulating effects on both excitatory and inhibitory synaptic transmission. This is expected if a receptor activation is coupled to multiple effector enzymes (Miller *et al.*, 1992; Dingledine and McBain, 1994; Herman, 2002; Hollman and Heinemann, 1994; Schoepp, 1994; Asztely and Gustafasson, 1996; Conn and Pin, 1997; Dingledine *et al.*, 1999).

The mGluRs have certain features that distinguish them from the iGluRs. First, the mGluRs modulate the activity of neurons rather than mediate fast synaptic neurotransmission. Second, the distribution of the mGluRs is highly diverse and heterogenous. Different sub-classes are localized uniquely at both the anatomical and cellular levels. For example, mGluR 2 and 3 are found in high density in the cerebral cortex mGluR 4 is found in high density in the thalamus but not in the cortex and mGluR 6 is almost exclusively found in the retina.

Further characterization of the GluRs in the CNS and the peripheral tissues may help to clarify the role of these molecules in the different processes and may stimulate the development of therapeutic compounds or food products (Gill *et al.*, 1998, 1999, Dingledine *et al.*, 1999; 2000; Morley *et al.*, 2000).

4. Glutamate Receptors in Peripheral Tissues

A recent surge of publications (and this book) supports the presence, importance, and functionality of GluRs outside the CNS with unique distributions within various tissues and species. These tissues include adrenal medulla (Yoneda and Ogita, 1986; Watanabe *et al.*, 1994; Hinoi *et al.*, 2002a), peripheral nerves both myelinated and unmyelinated (Aas *et al.*, 1989; Coggeshall and Carlton, 1998), bone (Chenu *et al.*, 1997; Bhanga *et al.*, 2001; Itzstein *et al.*, 2001; Chenu, 2002; Gu *et al.*, 2002; Hinoi *et al.*, 2002), bone marrow (Genever *et al.*, 1999), bronchial smooth muscle, endocrine pancreas (Betrand *et al.*, 1992, 1993; Gonoi *et al.*, 1994; Inagaki *et al.*, 1995; Molnar *et al.*, 1995; Weaver *et al.*, 1996; Liu *et al.*, 1997; Gill *et al.*, 2000; Morley, chapter 8, this volume), gut (Moroni *et al.*, 1986; Shannon and Sawyer,1989; Burns *et al.*, 1994; Tsai *et al.*, 1994, 1999), esophagus (Burns *et al.*, 1994), duodenum, ileum, and descending colon (Burns *et al.*, 1994), hepatocytes (Sureda *et al.*, 1997; Storto *et al.*, 2000; Gill and Pulido, unpublished), heart (Rockhold *et al.*, 1989; Morhenn *et al.*, 1994; Winter and Baker, 1996; Gill *et al.*, 1998, 1999; Leung *et al.*, 2002; Mueller *et al.*, 2003), taste buds (Chaudhari *et al.*, 1996; Hayashi *et al.*, 1996), keratinocytes (Morhenn *et al.*, 1994), kidney (Gill *et al.*, 2000; Deng *et al.*, 2002; Leung *et al.*, 2002), lungs (Said *et al.*, 1999, 2001; Gill *et al.*, 2000), pituitary (Kiyama *et al.*, 1993; Villalobos *et al.*, 1996), pineal gland (Mick, 1995), retina (Lucas and Newhouse, 1957; Grunder *et al.*, 2002; Xue *et al.*, 2002), autonomic and sensory ganglia, rat glaborous skin (Carleton *et al.*, 1995), vagus and other cholinergic nerves (Aas *et al.*, 1989), tachykinin containing sensory nerves, and vestibular tissues (Demenes *et al.*, 1995). In our laboratory, we have conducted a thorough analysis on the distribution of the GluRs in peripheral tissues of the rat and monkey (Gill *et al.*, 1998, 1999, 2000; Gill and Pulido 2001; Mueller *et al.*, 2003). Different subtypes of GluRs were observed in the heart, spleen, lungs, liver, male and female reproductive systems, adrenals, and kidney (Gill *et al.*, 2000; Tsibris *et al.*, 2002). These findings supports the view that GluRs are widely present in peripheral tissues with a specific cellular distribution. Some of the GluRs from peripheral tissues have been cloned and sequenced (Gill *et al.*, 1998) and their sequences correspond with those GluRs that have been cloned in the CNS. Further physiological and pharmacological experiments support that GluRs in the periphery have similar properties to those in the CNS or those expressed in host cells transfected with cloned subunits (Moroni *et al.*, 1986; Rockhold *et al.*, 1989; Shannon and Sawyer, 1989; Gonoi *et al.*, 1994; Hardy *et al.*, 1994; Mick, 1995; Inagaki *et al.*, 1995; Weaver *et al.*, 1996). For example, the AMPA receptors in the rat pancreas respond to L-Glu, AMPA, and kainate. These receptors were blocked by competitive antagonists, 6-cyano-7-nitroquinoxaline (CNQX) and potentiated by cyclothiazide (Shannon and Sawyer, 1989). These properties are also shared by neuronal AMPA receptors. In addition, the stimulation of cultured rat myocardial cells by L-Glu leads to an increase in the intracellular Ca^{+2} oscillation frequency (Winter and Baker, 1996). Similar physiological studies with agonists and antagonists of the NMDAR 1 receptor in the pig ileum have shown that these receptors are similar to those characterized in the CNS (Shannon and Sawyer, 1989). NMDAR 1 receptors in the pig ileum were also blocked by Mg^{+2} ions and competitively antagonized by DL-2-amino-5-phosphonovaleric acid.

4.1. GluRs in the Heart and Cardiac Arrhythmia

The reports of cardiovascular disturbances associated with domoic acid intoxications and other excitatory compounds in foods prompted our initial investigation of the heart as a possible target organ (Peng *et al.*, 1994; Winter and Baker, 1996; Gill *et al.*, 1998, 1999). The pathogenesis of these myocardial dysfunction could be explained either on the basis of central cardiovascular–neuronal control or by the direct effect of the chemical on the target organ (Rockhold *et al.*, 1989; DiMicco and Monroe, 1996; Gill *et al.*, 1998, 1999). Visualizing the cardiac conducting system and the neural structures of the heart was the first step in the evaluation of the cardiologic effects of these compounds. PGP 9.5 and neurofilaments (NF) are among the biomarkers that have been used to visualize the conducting system and neural structures of the heart (Vitadello *et al.*, 1990, 1996; Schofield *et al.*, 1995). PGP 9.5 is a neuron specific cytoplasmic protein and its abundance in the brain (1%–5% of soluble protein) makes it a useful marker for neurons and the diffuse neuroendocrine system. It has also been used to study the pattern of distribution of the conducting system in the heart of guinea pigs, bovine, rat and monkey (Figure 1.3a). The NF is a cytoskeleton neural protein that has been used to identify the neural structures in the heart (Vitadello *et al.*, 1990, 1996; Schofield *et al.*, 1995).

Recent studies in rodents have shown the differential distribution of NMDAR 1 in the atrium and ventricles but only visible in the pulmonary artery and descending aorta. NMDAR 2 subunits were not found in the adult heart (Seeber *et al.*, 2000). Hence it can be speculated that the NMDA receptor consists of homoligomeric NMDA 1 subunits in the heart. Studies conducted in our laboratory showed a preferential localization of several subtypes of GluRs within the conducting system, nerve terminals, and intramural ganglia cells of the rat and monkey heart (Figures 1.3b–f, Mueller *et al.*, 2003). However, this distribution was more defined in the monkey heart. It is our opinion that this probably reflects the higher level of anatomical differentiation of the conducting system in nonhuman primates vs the rodents or may represent species specific differences. We have demonstrated the differential distribution in the monkey and rat heart of the following GluRs: GluR 1, GluR 2/3, GluR 5/6/7, mGluR 2/3, mGluR 5, and NMDAR 1. The antibodies for mGluR 2/3, mGluR 5, and NMDAR 1 showed the strongest reactivity in the monkey and rat (Figure 1.3b–f). However, the distribution of the mGluR 5 is different between these two species. In the rat, mGluR 5 was present in the intercalated discs whereas it was absent in the discs of the monkey heart

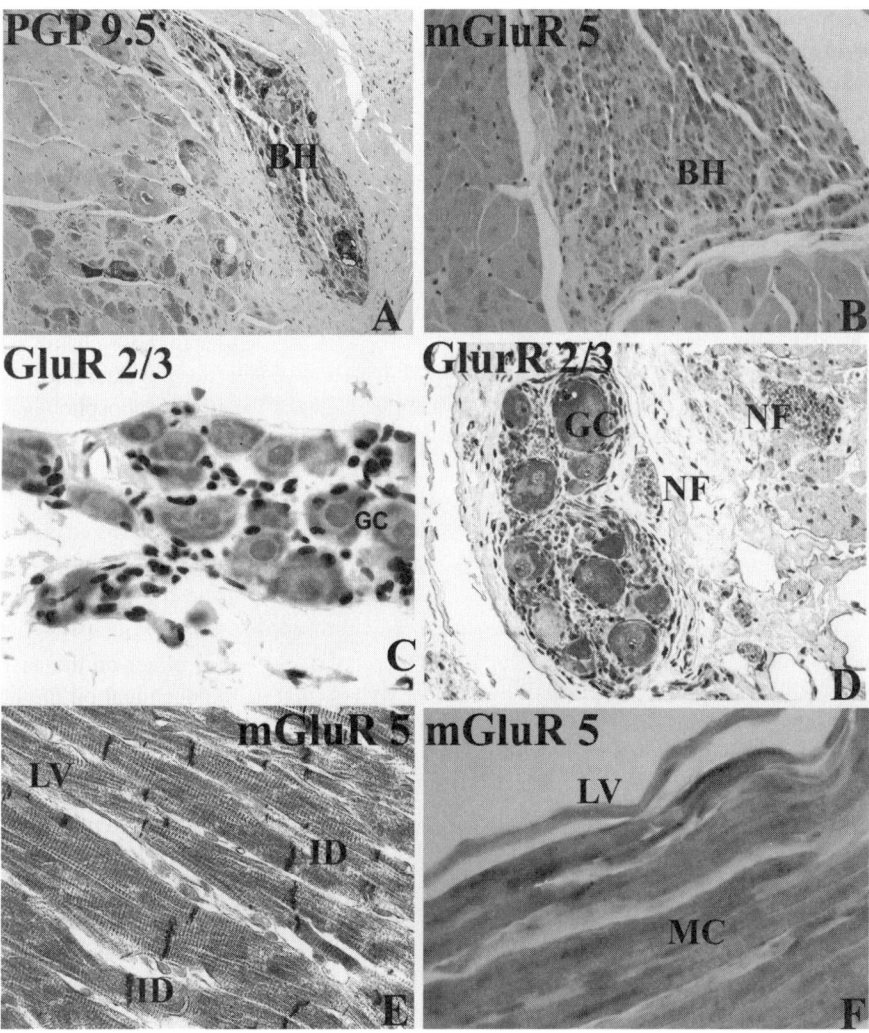

Figure 1.3. Photomicrographs of the heart perfused with 4% paraformaldehyde and processed for immunohistochemistry. Microwaved paraffin sections were immunostained with specific antibodies to GluRs using the LAB/avidin biotin method and DAB as chromogen. (A) Distribution of the neural biomarkers protein gene product (PGP 9. 5) in section of the monkey heart through the atrioventricular junction and the left ventricle. (B) Anti- mGluR 5 in the Bundle of His (BH) in the monkey heart in the same region as in (A) but of a higher magnification (C) rat heart ganglion cells (GC) showing cytoplasmic distribution of the anti-GluR 2/3 immunostain. (D) Immunostaining of the monkey atrium with the antibody GluR 2/3 and aminoethylcarbazole (AEC) as chromogen. Clumps of strong cytoplasmic stain are seen in the GC with preferential distribution in the periphery of the cells. The stain is less intense within the nerve fibers (NF). (E) Anti-mGluR 5 showing preferential stain in the intercalated discs (ID) of the rat heart. (F) In contrast to (E) intercalated discs of the monkey heart lack preferential affinity for mGluR 5.

(Figures 1.3e–f). Absorption controls for the antibodies GluR 2/3, mGluR 2/3, NMDAR 1, and mGluR 5 blocked the corresponding immunoreactivity thus confirming the specificity of the immunohistochemical reactions. The intrinsic cardiac ganglia are typically associated with interconnecting nerves that form a ganglionated plexus. They are consistently identified in the atria and atrio-ventricular regions. In humans, intrinsic cardiac ganglia are connected to the cranially located extracardiac ganglia via cardio-pulmonary nerves adjacent to the larger vessels (Armour *et al.*, 1997). The presence of GluRs in the intrinsic cardiac ganglia and interconnecting plexus suggests that cardiac function could be modified by the amino acid Glu or its analogues. Their localization in areas specifically involved in the conduction of impulses suggests their involvement in the control of heart rhythm and could explain some of the clinical symptoms described in susceptible individuals with MSG and domoic acid ingestion (Zautcke *et al.*, 1986; Winter and Baker, 1996). Symptoms such as burning, facial pressure, palpitations, and chest pains associated with MSG consumption are known as the "Chinese distress syndrome" (CDS—Zautcke *et al.*, 1986; Winter and Baker, 1996). It is associated with the ingestion of Chinese food and usually occurs in individuals who cannot degrade MSG. Winter and Baker (1996) have shown that L-Glu increases the frequency of Ca^{+2} oscillations, an effect positively correlated with increased contraction frequency in myocardial cells, which could lead to reduced cardiac filling, hypoxia, and angina-like chest pains. In addition, NMDA, AMPA, and Ka all have been shown to increase the contraction on cultured myocardial cells. Therefore GluRs can potentially play a role in cardiac function and cardiotoxicity of excitotoxins. This is supported by clinical findings of cardiovascular collapse and cardiovascular symptoms in individuals intoxicated with domoic acid and ingesting MSG respectively. Further, Gulland (2000) and Scholin *et al.* (2000) reported severe cardiac lesions in sea lions that died of domoic acid intoxication off the coast of California in 1998. This report is the first to describe direct structural cardiac changes associated with domoic acid poisoning.

Since the intramural ganglia cells and cardiac nerve fibers are known to be components of the peripheral autonomic nervous system, we were prompted to expand our investigations of GluRs in other tissues.

4.2. GluRs in Kidney and Electrolyte/Water Homeostasis

We have shown a wide distribution of NMDAR 1 and the presence of mGluR 2/3 and GluR 2/3 in the juxtaglomerular apparatus (JGA) and proximal tubules (Figures 1.4a, b, c, and d). This suggests that these receptors may be involved in electrolytes and water home-ostasis. The strong immunoreactivity with anti-mGluR 2/3 and anti-GluR 2/3 within the granular cells of the afferent arteriole suggests a potential involvement in the control of renin release (Jackson *et al.*, 1985; Figures 1.4b and 1.4c). The renin–angiotensin system is a major hormonal system involved in the regulation of electrolyte, fluid balance, and blood pressure (Jackson *et al.*, 1985). Recently, Leung *et al.* (2002) and Deng *et al.* (2002) used Western blotting to show the presence of NMDAR 1 and NMDAR 2C in the renal cortex and medulla. The other NR2 subunits were not expressed. Furthermore, they showed that expression of the NMDAR1 subunit increased with renal development. Miller *et al.* (1992) and Heymen *et al.* (1992) have shown that kidney tubules are sensitive to strychnine, glycine ACPC, D-alanine, and D-Ser all of which are specific agonists of the NMDA receptor found in the CNS.

4.3. GluRs in Sex Organs and Reproduction

We have shown the differential distribution of GluRs in reproductive organs of the male (Figures 1.5a–e) and female rats and monkeys (Figures 1.5f–i). In the testis, these receptors have a specific affinity for different structures. There is intense anti-mGluR 2/3 immuno-labelling of the head of the mature spermatids/spermatozoa, interstitial, and myoid cells.

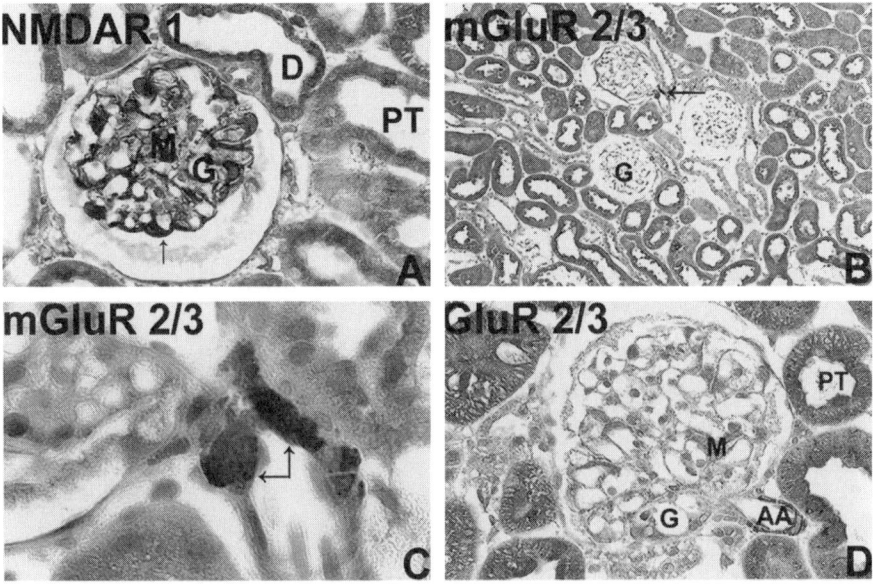

Figure 1.4. Photomicrographs of the rat kidney processed for immunohistochemistry. (A) Immunostaining with anti-NMDAR 1 is seen in the distal tubule, proximal tubule and glomeruli, particularly in the mesangium and podocytes (shown by an arrow). (B) Anti-mGluR 2/3 strong immunostain in the convoluted proximal tubules and the JGA (arrow). (C) Higher magnification of the JGA, showing anti-mGluR 2/3 dark cytoplasmic staining of the granular cells in the wall of the afferent arteriole. (D) Anti-GluR 2/3 stain distribution is similar to the anti-mGluR 2/3 (Gill *et al.*, 2000).

Anti-NMDAR 1 has a strong affinity for the germinal epithelium, particularly the spermatogonia adjacent to the basal lamina and the more mature spermatids near the lumen (Figure 1.5c), whereas the anti-GluR 2/3 immunostain is limited to the cells in the interstitial spaces (Figures 1.5a and 1.5b). This differential distribution suggests that GluRs may be involved in spermatogenesis, spermatozoa motility, and testicular development each linked to a specific receptor. Previously it has been shown that specific binding sites for [^{3}H]-TCP, a ligand that labels a binding site within the NMDA receptor ion channel, has been demonstrated on membranes of mammalian spermatozoa. Morever Cl$^-$ independent [^{3}H]-Glu binding, which could be partially displaced by NMDA and AP5, has also been detected in seminal vesicles (Erdo, 1990, 1991). Further, the results of Lara and Bastos-Ramos (1988) suggest the noradrenergic neurons innervating the rat vas deferens are controlled by a Glu-dependent excitatory process in the ganglia. They showed that a single dose of kainate to the ganglia induced a decrease in the norepinephrine content of the vas deferens. Binding studies using [^{3}H]-Glu to the membrane fraction showed that the binding was saturable. This binding was inhibited by different analogues according to different potencies: L-Glu > kainate >quisqualate >> NMDA. Thus, it is likely that the vas deferens has a glutamatergic excitatory mechanism for the control of their activity and these are responsible for the depolarizing potential on the noradrenergic neurons. This mechanism might be important for the contractile activity of the vas deferens and hence in the control of seminal fluid (Lara and Bastos-Ramos, 1988).

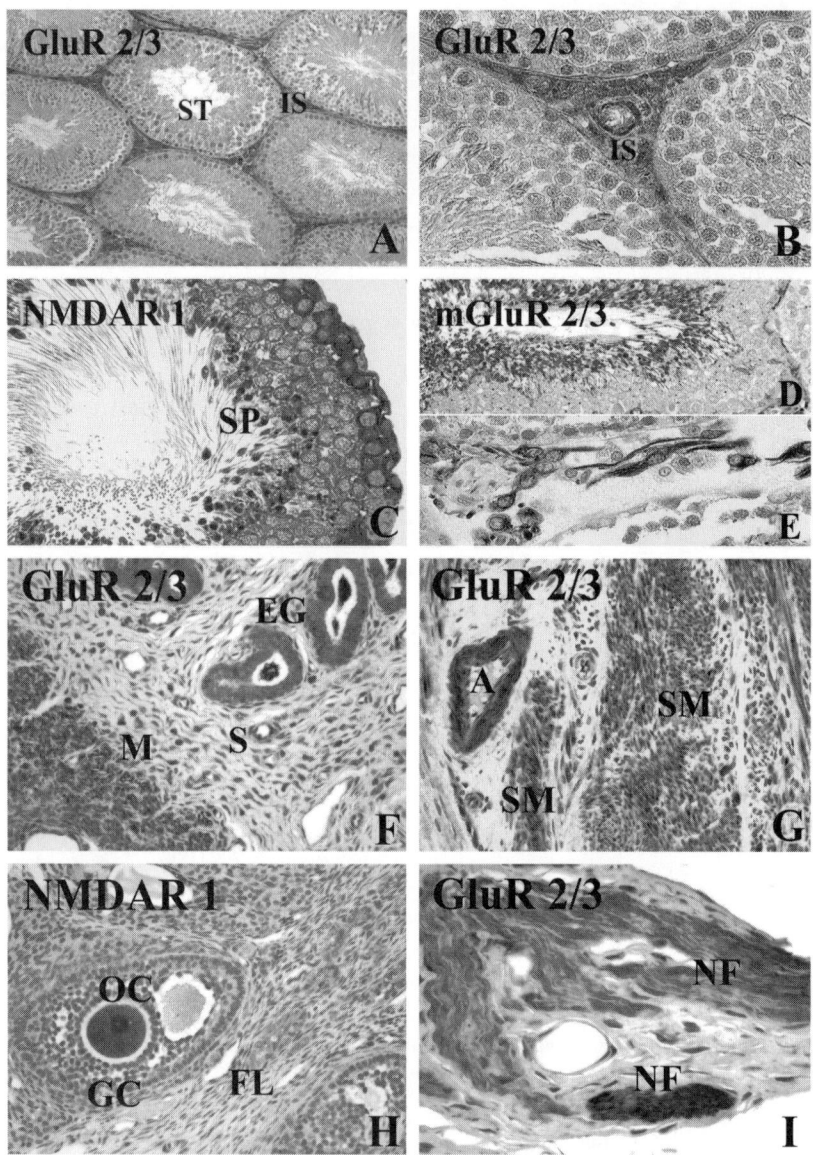

Figure 1.5. Photomicrographs of the rat testis (A–E) perfused with 4% PFA and processed for immunohistochemistry. (A) & (B) Anti-GluR 2/3 immunostain is observed in the interstitial spaces (IS), but not in the seminiferous tubules (ST). (C) Anti-NMDAR 1 strong affinity for the germinal epithelium, particularly the spermatogonia (SP) adjacent to the basal lamina and the more mature spermatids near the lumen. (D) Intense anti-mGluR2/3 immunolabeling of the head of the mature spermatids/spermatozoa, and (E) Interstitial and myoid cells.

F–I: Photomicrographs of the female reproductive tissues (rat—F, H, and I) and monkey (G). (F) Anti-Glu R 2/3 of the rat uterus showing strong affinity for the endometrial glands (EG) and myometrium (M). Stain is observed to a lesser extent in the stroma (S); (G) Anti-GluR 2/3 showing strong affinity for smooth muscles (SM) within the arteriole (A) in the cervical wall of the monkey; (H) Rat ovary: Anti-NMDAR 1 has strong selective affinity for the oocyte (OC) but is absent from the focciles (FL), or granulosa cells (GC) and (I) Anti-GluR 2/3 staining of the nerve fiber (NF) within the suspensory ligament in the ovaries.

In the female reproductive system of the rat and monkey, GluRs also have a unique distribution within each organ (Figures 1.5f–i). Each antibody has a differential affinity to specific structures in the ovaries, fallopian tubes, cervix, myometrium, and endometrium (Figures 1.5f–i). In the ovary (Figures 1.5h and i), the distribution of GluRs within the follicles varies at different stages of their maturation. In the rat, anti-NMDAR 1 and to some extent anti-GluR 2/3 and anti-mGluR 2/3 have a remarkable selective affinity for the oocyte. GluR 2/3 and mGluR 2/3 but not NMDAR 1 are visualized in the corpus luteum. In the uterus (Figure 1.5f), anti-GluR 2/3 showed strong affinity for the ciliated epithelium of the fallopian tubes, smooth muscle of the myometrium, and endometrial glands whereas anti-NMDAR 1 is more selective to the endometrial glands and the exocervical epithelium. This suggests that these receptors may be involved in ovulation, fertilization, implantation of the ovum, and excitability of the uterus. In order to examine if similar preferential distributions exist in higher mammals, we have tested the location of GluRs in the sex organs of nonhuman primates using ovary, uterus, and fallopian tubes of *Macaca fascicularis*. In this species, the corpus luteum and oocytes display intense, selective immunolabeling with anti-NMDAR 1 and anti-GluR 2/3 (manuscript in preparation). The GluRs were also present in the uterus of monkey. This observation, together with the report of uterine rupture in pregnant sea lions that died from intoxication of domoic acid off the coast of California (Gulland, 2000), suggests a cause–effect relationship. More recently, Tsibris *et al.* (2002) have shown that GluR 2 was upregulated in the leiomyomata relative to myometrium.

The presence of the GluRs within the reproductive organs and the known functional inhibitory/excitatory effects of GABA receptors (Erdo, 1990, 1991) suggest that similar excitatory/inhibitory neurotransmission interplay may also be present using GluRs as mediators. Therefore, reproductive functions such as gonadal maturation, steroidal sex hormone regulation, maturation/motility/excitability of the spermatozoa, ovulation, fertilization, excitability of the fallopian tubes, implantation of the ovum, and excitability of the myometrium may all be affected. This warrants testing since it has important therapeutic and toxicological implications.

4.4. GluRs in Other Tissues

Hass and Schaustein (2004, Chapter 4) describe the distribution and possible functions of various GluRs and Glu transporters in the immune tissues. It has been shown that mGluRs 1, 2, 3, and 5 are present in the thymic stromal cell line TC1S and the thymocyte. Using FACS analysis, majority of the unfractionated thymocytes (70%) showed the presence of mGluR 5 whereas 50% of the cells expressed mGluR 3 and only 15% expressed mGluR 1. In contrast, isolated $CD4^-/CD8^-$, double negative thymocyte precursors expressed mGluR 3 (45%) and mGluR 1 (40%) whereas mGluR 5 was barely detectable. Therefore, it was hypothesized that changes in mGluR subtype expressions maybe related to T-cell maturation stages (Grazia *et al.*, 1999; Storto *et al.*, 2000). GluRs are also expressed in other lymphoid tissues and inflammatory infiltrates.

In addition to the above-discussed tissues, the GluRs and Glu transporters are present in other tissues such as liver, lungs, the immune tissues (different cell types), and endocrine tissues (this book). The reader is encouraged to read the following chapters for further details.

5. Glutamate Receptors as Possible Mediators of Cell and Tissue Injury and Pathology

The view that Glu or related EAAs can cause neuronal injury as a result of overexcitation is called excitotoxicity. The GluRs are known to act as mediators of inflammation and cellular injury through a common injury pathway (Beal, 1992; Farooqui and Horrock, 1994; Lipton, 1993; Lipton and Rosenberg, 1994; Meldrum, 1994; Lipton and Gendelman, 1995; Said, 1999; Figure 1.2). In the CNS, the stimulation of GluRs triggers an excessive influx of calcium into neurons through the ion channels which mediate neural injury (Dingledine and McBain, 1994; Asztely and Gustafasson et al., 1996; Herman, 2002). Since the iGluRs are ion-gated channels selective to Na^+, K^+, and Ca^{+2} any sustained stimulation of the GluRs results in osmotic damage due to the entry of excessive ions/water. This increases the intracellular Ca^{+2} concentration which is crucial to the determinant of injury. The high concentration of Ca^{+2} triggers the activation of several enzyme pathways and signaling cascades including phospholipases, protein kinase C, proteases, protein phosphatases, nitric acid synthases, and the generation of free radicals (Beal, 1992; Lipton, 1993; Lipton and Rosenberg, 1994; Meldrum, 1994; Lipton and Gendelman, 1995; Said, 1999). The destabilization of Ca^{+2} homeostasis also causes the translocation of protein kinase C (PKC) from the cytoplasm to the membrane. This leads to the phosphorylation of the membrane proteins promoting the destabilization of the regulatory mechanisms for Ca^{+2} homeostasis which mediates toxicity (Choi, 1992; Dingledine and McBain, 1994; Dingledine et al., 1999). The activation of phospholipase A_2 generates arachidonic acid with its metabolites and platelet-activating factors. Platelet-activating factors increase the neuronal calcium levels by stimulating the release of Glu. Arachidonic acid potentiates NMDA-evoked currents and inhibits the reabsorption of Glu into astrocytes and neurons. This further exacerbates the situation by a positive feedback mechanism where free radicals are formed during arachidonic acid metabolism leading to further phospholipase A_2 activation. This results in an increased concentration of extracellular Glu which contributes to the sustained activation of the GluRs (Choi, 1992; Dingledine and McBain, 1994; Dingledine et al., 1999; Farooqui and Horrock, 1994). As a consequence, Cys transport is inhibited causing a decrease of intracellular reducing sulfhydryls and the generation of oxygen radicals which results in cell death by contributing to DNA fragmentation. In addition to enzymes of the cell cytosol, the nuclear enzymes are also activated by increases of Ca^{+2}. This increased concentration of Ca^{+2} raises the nitric oxide via the calmodulin activation of nitric oxide synthetases which generate free oxygen radicals. Nitric oxide has been observed in peripheral tissues, ganglion cells, nerve fibers, cardiocytes, and myocytes in the heart of pig and rat. Our work has demonstrated the presence of the GluRs in the same structures. Liu et al. (1997) showed the co-localization of nitric oxide and AMPA receptors in ganglion cells of the pancreas. Since the nitric oxide is calcium- and calmodulin-dependent and the AMPA receptors are Ca^{+2} permeable, it is possible that nitric oxide is activated through the AMPA receptors. We propose that the mechanism of injury in the CNS may be a basic process of injury in all tissues. This is supported by the finding that excessive activation of the NMDAR 1 in the lungs induces acute edema and injury as seen in "adult respiratory distress syndrome" (Said et al., 1996, 2001; Said, 1999). This injury can further be modulated by blockage of one of three critical steps: NMDAR 1 binding, inhibition of NO synthesis or activation of poly(ADP-ribose) polymerase (Said et al., 1996, 2001; Said, 1999). Our results showed immunolabeling for various GluRs in bronchial epithelium, blood vessels, mast cells, and inflammatory cells within the lung. This supports the view GluRs play a role in airway responses to injury and inflammation. All the antibodies tested for the subtypes of GluRs showed affinity to mast cells in all tissues analyzed, particularly in the lungs, gastrointestinal tract, and female reproductive tissues (unpublished observations). Their presence in the airway structures such as the larynx, esophagus, and mast cells also implicate the GluRs in the mediation of asthmatic episodes (Allen *et al.*, 1987; Said *et al.*, 1996, 2001; Said, 1999). The excitation of GluRs in the air passages therefore may be important in airway inflammation and hyper-reactivity observed in bronchial asthma (Haxhiu *et al.*, 1997; Said, 1999). Their presence also could explain the enhancement of acute asthmatic attacks by Glu-containing foods (Allen *et al.*, 1987; Haxhiu *et al.*, 1997; Said, 1999). Researchers did blind placebo-controlled experiments where subjects with asthma received MSG in tablet form. Their studies showed that MSG provoke asthma which in some cases was severe and life threatening (Allen *et al.*, 1987).

Mast cells are known to be found in connective tissues throughout the body, most abundantly in the submucosa tissues and the dermis. Purcell *et al.* (1996) showed that spermidine-induced release from the mast cells are dependent upon the presence of Ca^{+2} in the external mileu. The influx of Ca^{+2} is known to be accompanied by NMDAR 1 activation. This increased intracellular Ca^{+2} concentration initiates the exocytotic degranulation process in mast cells. Spermidine is a natural polyamine and the opening of the ion channel associated with NMDAR receptors is facilitated by its binding sites. In neuronal tissue, polyamines triggers histamine secretion through interaction with a polyamine site associated with an NMDAR 1 macro-complex. Therefore, spermidine can modulate activation of the macro-complex at polyamine-binding sites in the lung or other sites. MK801, the antagonist of NMDAR 1, blocked the release of histamine secretion which was induced by the spermidine. If the NMDAR 1 is present (Purcell *et al.*, 1996; Said, 1999; Gill *et al.*, 2000), then it opens up the possibility that EAAs can also influence allergic reactions.

Glu and related EAA agonists induce neurotoxic damage in the CNS, under conditions of hypoxia or ischemia due to the increased intracellular Ca^{+2} concentration (Dingledine and McBain, 1994; Asztely and Gustafasson *et al.*, 1996; Herman, 2002). Intestinal mucosa damage has also been reported after hypoxia and ischemia (Otamri, 1988). Ulcerations of the mucosa of the gastrointestinal tract have been observed by our laboratory and others in experimental animals exposed to domoic acid (unpublished observations, Gulland, 2000).

6. Considerations On the Phylogeny of Glutamate Receptors

Chen *et al.* (1999) recently showed the presence of Glu-like receptors in bacteria. They showed existence of a K^+ selective GluR that binds Glu and forms K^+ selective ion channels. It showed the highest sequence similarity to the rat delta (δ) 1 GluR followed by a putative GluR from *Arabidopsis thaliana*. The δ receptors (1 and 2) have recently been identified and share low homology with the other three classes of iGluRs and no clear response or ligand binding has been obtained from cells transfected alone or in combination with other ionotropic receptors. As a result of these observations, it has been proposed that GluRs of eukaryotes arose from a primordial prokaryotic protein (Chen *et al.*, 1999; Chiu *et al.*, 1999; Daveport, 2002). Furthermore, iGluRs and Glu transporters are now known to be present in both monocotyledons and dicotyledons based on Northern and heterologous Southern blot analysis (Lam *et al.*, 1998; Kang and Turano, 2003). Preliminary data indicates that the GluRs are involved in light signal transduction. Therefore, it is possible that signaling between cells by EAAs may have evolved from primitive mechanisms before the divergence of plants and animals (Chen *et al.*, 1999; Daveport, 2002). GluRs and other similar signaling systems are actually ancestral methods of communication, common to plants and animals alike. This is supported by the fact that these receptors are present in mollusk (Stumer *et al.*, 1996), leech (Dierkes *et al.*, 1996), freshwater fish *Oreochromis* sp. (Wu *et al.*, 1996), blue-green algae *C. elegans* (Maricq *et al.*, 1996), and insects (Chiang *et al.*, 2002). In plants, the Glu-like receptors respond to the same antagonist as the GluRs in the CNS. The DNA sequence also shows variable (60%–16%) homology to the GluRs characterized in the mammalian system. At the present, too little is known about the plant Glu-like receptors to assign or predict any physiological role. This will require extensive characterization of tissues and membrane localization as well as determination of ion selectivity and ligand sensitivities of homomeric and heteromeric GluRs. Transgenic plants and mutants should help to identify both gene functions and physiological role.

7. Conclusions and Future Research Considerations

In summary, GluRs have a wide and unique distribution in peripheral tissues. These receptors are pharmacologically similar to their counterparts in the CNS, although the possibility that there are subtle distinctions such as glycosylation cannot be ignored (Gonoi *et al.*, 1994). The presence of the GluRs in peripheral tissues may provide explanations for the autonomic disturbances (GI, salivation, cardiovascular, vomiting) that have been reported in human intoxications and in animals dosed with potent excitotoxins such as domoic acid (Thompson *et al.*, 1983; Iverson *et al.*, 1990; Perl *et al.*, 1990; Teitlebaum *et al.*, 1990; Tryphonas *et al.*, 1990; Peng *et al.*, 1994; Truelove *et al.*, 1996). Since the EAA excitotoxicity is intimately associated with the GluRs, the toxic effects may be more generalized than initially assumed, particularly in the light that GluRs are widely present in peripheral tissues which are not protected by a blood barrier (Price *et al.*, 1981; Bruni *et al.*, 1991). Excitotoxicity depends on the intracellular Na^+ and Cl^- ions and in the influx of Ca^{+2}. It is the Ca^{+2} influx which is thought to be the ultimate trigger of the toxic effects (Choi, 1992; Dingledine and McBain, 1994; Asztely and Gustafasson *et al.*, 1996; Herman, 2002). Based on the pattern of anatomical distribution, we suggest that these receptors may be important for the mediation of functions such as hormone regulation, heart rhythm, blood pressure, circulation, and reproduction. Some of the noteworthy locations are the heart, kidney, lungs, ovary, testis, and endocrine cells suggesting they play a role on cardiorespiratory, endocrine, and reproductive functions. Furthermore, the tissues expressing GluRs could also be target sites for the toxic effect of Glu-like products. Many of these are known toxins that contaminate food and others are used as food additives or enhancers during processing. Currently, evidence is lacking to suggest the reassessment of the regulated safety levels for these products in food since little is known about how these receptors work in each organ. However, it has been shown that MSG can trigger severe asthma (Aas *et al.*, 1989). The relatively high concentrations of the endogenous EAAs in the blood and other tissue fluids (10–200 μM) suggest that peripheral GluRs may be constantly saturated and therefore would argue against a physiological role (Villalobos *et al.*, 1996). The true nature of the GluRs/ligand interaction in each tissue is not known. It is possible that some ligands have more potentiating ability than others. From the toxicological point of view it is known that various EAAs contaminants have different potencies. Some of these would have the ability to replace the weaker ligand. In addition to the potentiating ability of the compound of interest, the true local concentrations of endogenous EAAs at peripheral tissues are not known. Currently, more research will be needed to assess the extent that these receptors participate in normal functions and/or in the development of diseases and how they mediate the toxic effects of EAAs.

In addition to food safety, a growing area of interest is the development of therapeutic products specifically designed to interact with synaptic transmission of the GluRs in the CNS (Cunningham *et al.*, 1994; Michaelis, 1998; Dingledine *et al.*, 1999; Moghaddam, 1999; Gallo and Ghiani, 2000). It is established that a myriad of pre- and postsynaptic mechanisms exist by which iGluRs and mGluRs could modulate cell functions in the CNS. Selective agonists and antagonists could be used to modulate glutamatergic neuronal transmissions in very select areas of the CNS (Moroni *et al.*, 1986). For example, it has been suggested that NMDAR 1 antagonists may be useful in preventing tolerance to opiate analgesia and helping control withdrawal symptoms from addictive drugs. Also, the overactivation of NMDAR 1 has been suggested as one of the factors to cause chronic diseases such as Huntington's, Alzheimer's, Parkinson's, HIV-related neuronal injury, and amyotrophic lateral sclerosis. Therefore, antagonists of NMDAR 1 receptor function are expected to be useful in the treatment of some of these diseases. Research demonstrating the presence of the GluR, dopamine, and GABA receptors in peripheral tissues suggests that these antagonists might have modulating functions in the peripheral organs and tissues. GluRs in peripheral tissues could also be targets for pharmacological manipulations. For example, GluRs in pancreatic islets offers a potential target for therapeutic intervention to fine tune insulin or glucagon secretion (Inagaki *et al.*, 1995; Weaver *et al.*, 1996). The presence of these receptors in osteoblasts and the demonstration that NMDAR activation is effective in inhibiting bone reabsorption *in vitro* may contribute to the development of new therapeutics for osteoporosis (Chenu *et al.*, 1997; Chenu, 2002). Of all GluRs, NMDA is the most widely studied because of its involvement in neuronal injury and death in acute conditions such as head injury, strokes, and epileptic seizures and in chronic neurodegenerative disease as Alzheimer, Huntington, Parkinson, and ALS. The subunit NMDAR 1 is the most widely distributed in the CNS (Burns *et al.*, 1994). Our data also supports that the NMDAR 1 has also a wide distribution in peripheral tissues (Gill *et al.*, 2000; Gill and Pulido, 2001). This supports a possible role in cell injury outside the CNS for this receptor.

In conclusion, it is evident that GluRs have a cell specific distribution in neural and non-neural tissues. In these locations, they may play a pathophysiological role or as target-effector sites for excitatory compounds in foods or therapeutic products. Further research is required to assess the significance and the role of the GluRs and their impact in various fields.

Acknowledgments

We are grateful to Dr. A. Arnold for the use of the monkey tissues. We would like to thank Peter Smyth, Ian Greer, Mike Barker, and Joann Clausen for their support in pathology and histotechnology. We are also indebted to Dr. R. Mueller for the photomicrographs in Figure 1.3. We are grateful to Meghan Kavanagh for editing and assisting in the preparation of this manuscript.

References

Aas, P., R. Tanso, and F. Fonnum (1989). Stimulation of peripheral cholinergic nerves by glutamate indicates a new peripheral glutamate receptor. *Eur. J. Pharamacol.* **164**, 93–102.

Allen, D.H., H. Delohery and G. Baker (1987). Monosodium L-glutamate-induced asthma. *J. Allergy Clin. Immunol.* **80**, 530–537.

Armour, J.A., D.A. Murphy, B.X. Yuan, S. MacDonald, and D.A. Hopkins (1997). Gross microscopic anatomy of the human intrinsic cardiac nervous system. *Anatom. Rec.* **247**, 289–298.

Asztely, F. and B. Gustafasson (1996). Ionotropic glutamate receptors: Their possible role in the expression of hippocampal synaptic plasticity. *Mol. Neurobiol.* **12**, 1–11.

Beal, M.F. (1992). Mechanism of excitotoxicity in neurologic disease. *FASEB J.* **6**, 3338–3344.

Bhanga, P.S., P.G. Genever, G.J. Spencer, T.S. Grewal, and T.M. Skerry (2001). Evidence for targeted vesicular glutamate exocytosis in osteoblasts. *Bone* **29**, 16–23.

Bertrand, G., R. Gross, R. Puech, M.M. Loubatieres-Mariana, and J. Bockaert (1992). Evidence for a glutamate receptor of the AMPA subtype which mediates insulin release from rat perfused pancreas. *Br. J. Pharmacol.* **106**, 354–359.

Bertrand, G., R. Gross, R. Puech, M.M. Loubatieres-Mariana, and J. Bockaert (1993). Glutamate stimulates glucagon secretion via an excitatory amino acid receptor of the AMPA subtype in rat pancreas. *Br. J. Pharmacol.* **237**, 45–50.

Bruni, J.E., R. Bose, C. Pinsky, and G. Gavin (1991). Circumventricular organ origin of domoic acid-induced neuropathology and toxicology. *Brain Res. Bull.* **26**, 419–424.

Burns, G.A., K.E. Stephens, and J.A. Benson (1994). Expression of mRNA for N-methyl-D-aspartate (NMDAR 1) receptor by the enteric neurons of the rat. *Neurosci. Lett.* **170**, 87–90.

Butcher, S.P., M. Sandberg, H. Hagberg, and A. Hamberger (1987). Cellular origins of endogenous amino acids released into the extracellular fluid of the rat striatum during severe insulin-induced hypoglycemia. *J. Neurochem.* **48**, 722–723.

Carlton, S.M., G.L. Hargett, and R.E. Coggeshall (1995). Localization and activation of glutamate receptors in unmyelinated axons of rat glaborous skin. *Neurosci. Lett.* 197, 25–28.

Chaudhari, N., H. Yang, C. Lamp, E. Delay, C. Cartford, T. Than *et al.* (1996). The taste of monosodium glutamate: Membrane receptors in taste buds. *J. Neurosci.* **16**, 3817–3826.

Chenu, C. (2002). Glutamatergic innervation in bone. *Microsc. Res. Tech.* **58**, 70–6.

Chenu, C., C.M. Serre, C. Raynal, B. Burt-Pichat, and P.D. Delmas (1997). Glutamate receptors are expressed by bone cells and are involved in bone reabsorption. *Bone* **22**, 295–299.

Chen, G.-Q., C. Cui, M.L. Mayer, and E. Gouaux (1999). Functional characterization of a potassium-selective prokaryotic glutamate receptor. *Nature* **402**, 817–819

Chiang, A.S., Wy. Lin, Pszczolkowski, T.F. Fu, Sl. Chiu, and G. Holbrook (2002). Insect NMDA receptors mediate juvenile hormone biosynthesis. *PNAS* **99**, 37–42.

Chiu, J., R. DeSalle, H.M. Lam, L. Meisel, and G. Coruzzi (1999). Molecular evolution of glutamate receptors: A primitive signaling mechanisms that existed before plants and animals diverged. *Mol. Biol. Evol.* **16**, 826–838.

Choi, D.W. (1992). Excitotoxic Cell Death. *J. Neurobiol.* **23**, 1261–1276.

Clark, R.F., S.R. Williams, S.P. Nordt, A.S. Manoguerra (1999). A review of selected seafood poisoning. *Undersea Hyperb Med.* **26**, 175–184.

Coggeshall, R.E. and S.M. Carlton (1998). Ultrastructural analysis of NMDA, AMPA, and kainate receptors on myelinated and unmyelinated axons in the periphery. *J. Comp. Neurol.* **391**, 78–86

Conn, P.J., and J.P. Pin (1997). Pharmacology and functions of metabotropic receptors. *Annu. Rev. Pharmacol. Toxicol.* **37**, 205–237.

Cunningham, M.D., J.W. Ferkany, and S.J. Enna (1994). Excitatory amino acid receptors:A gallery of new targets for pharmacological intervention. *Life Sci.* **54**, 135–148.

Daveport, R. (2002). Glutamate receptors in plants. *Ann. Bot.* **90**, 49–557.

Demenes, D., A. Lleixa, and C.J. Dechesne (1995). Cellular and subcellular localization of AMPA-selective glutamate receptors in the mammalian peripheral vestibular system. *Brain Res.* **671**, 83–94.

Deng, A., J.M. Valdivielso, K.A. Munger, R.C. Blantz, and S.C. Thomson (2002). Vasodilatory N-methyl-D-aspartate receptors are constitutively expressed in rat kidney. *J. Am. Soc. Nephrol.* **13**, 1381–1384.

Dierkes, P.W., P. Hochstrate, and W.R. Schlue (1996). Distribution and functional properties of glutamate receptors in the leech central nervous system. *J. Neurophysiol.* **75**, 2312–2321.

DiMicco, J. and A.J. Monroe (1996). Stimulation of metabotropic glutamate receptors in the dorsal-medial hypothalamus elevates heart rate in rat. *Am. J. Physiol.* **270**, 115–1121.

Dingledine, R., K. Borges, D. Bowie, and S.F. Traynelis (1999). The glutamate receptor ion channels. *Pharmacol. Rev.* **51**, 7–61.

Dingledine, R. and C.J. McBain (1994). Excitatory amino acids transmitters. In G.J. Siegal, R.W. Agronoff, B.W. Albers, and P.B. Molinof (eds), *Basic Neurochemistry*. Raven Press, NY, pp. 367–387.

Erdo, S.L. (1990). The GABAergic system in the human female genital organs. In S.L. Erdo (ed.), *GABA: Outside the CNS*. Springer-Verlag, NY, pp. 183–197.

Erdo, S.L. (1991). Excitatory amino acid receptors in the mammalian periphery. *TIBS* **121**, 426–429.

Farooqui, A.A. and L.A. Horrocks (1994). Involvement of glutamate receptors, lipases, and phospholipases in long-term potentiation and neurodegeneration. *J. Neurosci. Res.* **38**, 6–11.

Fisher, G.H., A. Aniello, A. Vetere, L. Padula, G.P. Cusano, and E.H. Man (1991). Free D-Amino acid and D-Alanine in normal and Alzheimer brain. *Brain Res. Bull.* **26**, 983–985.

Gallo, V. and Ghiani (2002). Glutamate receptors in glia: New cells, new inputs and new functions. *TIPS* **21**, 252–258.

Gasic, G.P. and M. Hollmann (1992). Molecular neurobiology of glutamate receptors. *Ann. Rev. Physiol.* **54**, 507–536.

Genever, P.G., D.J.P. Wilkinson, A.J. Patton, N.M. Peet, Y. Hong, A. Mathur *et al.* (1999). Expression of a functional N-methyl-D-aspartate-type glutamate receptor by bone marrow megakaryocytes. *Blood* **93**, 2876–2883.

Gill, S.S. and O.M. Pulido (2001).Glutamate receptors in peripheral tissues: Current knowledge, future research and implications for toxicology. *Toxicol. Pathol.* **29**, 208–223.

Gill, S.S., O.M. Pulido, R.W. Mueller, and P.F. McGuire (1998). Molecular and immunological characterization of the ionotropic glutamate receptors in the rat heart. *Brain Res. Bull.* **46**, 429–435.

Gill, S.S, O.M. Pulido, R.W. Mueller, and P.F. McGuire (1999). Immunological characterization of the metabotropic glutamate receptors in the rat heart. *Brain Res. Bull.* **48**, 143–146.

Gill, S.S., O.M. Pulido, R.W. Mueller, and P.F. McGuire (2000). Potential target sites in peripheral tissues for excitatory neurotransmission and excitotoxicity. *Toxicol. Pathol.* **28**, 277–284.

Gonoi, T., N. Mizuno, N. Inagaki, H. Kuromi, Y. Seino, J. Miyazaki *et al.* (1994). Functional neuronal ionotropic glutamate receptors are expressed in the non-neuronal cell line MIN6. *J. Biol. Chem.* **269**, 16989–16992.

Grazia, U., M. Storto, G. Battaglia, M.P. Felli, M. Maroder, A. Gulino *et al.* (1999). Evidence for the expression of metabotropic receptors in the thymic cells. 29th Annual Meeting Miama Beach, Fla. Oct. 23–28. Society of Neuroscience. P.449. Ab.177.16.

Grunder, T., K. Kohler, A. Kaletta, and E. Guenther (2002). The distribution and developmental regulation of NMDA receptor subunit proteins in the outer and inner retina of the *rat. J Neurobiol.* **44**; 333–342.

Gu., Y., P.G. Genever, T.M. Skerry, and S.J. Publicover (2002). The NMDA type glutamate receptors expressed by primary rat osteoblasts have the same electrophysiological characteristics as neuronal receptors. *Calcif. Tissue Int.* **70**, 194–203.

Gulland, F. (2000). Domoic acid toxicity in California sea lion (*Zalophus californianus*) stranded along the central California coast, May–October 1998. *Report to the National Marine Fisheries Service Working Group on Unusual Marine Mammal Mortality Events*. U.S. Dep. Commer., MOAA Tech. Memo, NMFS-OPR-17; pp. 1–45.

Hampson, D.R. and J.L. Manalo (1998). The activation of the glutamate receptors by kainic acid and domoic acid. *Nat. Toxins* **6**, 153–158.

Hardy, M., D. Younkin, Tang C.M., J. Pleasure, Q.Y. Shi, M. Williams *et al.* (1994). Expression of non-NMDA glutamate receptor channel genes by clonal human neurons. *J. Neurochem.* **63**, 482–489.

Harry, G.J. (1999). Basic principles of disturbed CNS and PNS functions. In R.J.M. Niesink, R.M.A. Jaspers, L.M.W. Kornet, J.M. van Ree, and H.A. Tislosn (eds), *Introduction to Neurobehavioral Toxicology: Food and Environment*. CRC Press, Washington, DC, pp. 115–162.

Hayashi, Y., M.M. Zviman, J.G. Brand, J.H. Teeter, and D. Restrepo (1996). Measurement of membrane potential and [Ca^{2+}] in cell ensembles: Application to the study of glutamate taste in mice. *Biophys. J.* **71**, 1057–1070.

Haxhiu, M.A., B. Erokwu, and I.A. Dreshaj (1997). The role of excitatory amino acids in airway reflex responses in anaesthetized dogs. *J. Auton. Nerv. Syst.* **67**, 192–199.

Herman, B. (ed.) (2002). *Glutamate and Addiction*. Humana Press.

Heyman, S., K. Spokes, S. Rosen, and F.H. Epstein (1992). Mechanism of glycine protection in hypoxia: Analogies with glycine receptor. *Kidney Int.* **42**, 41–45.

Hinoi, E. and Y. Yoneda (2001). Expression of GluR 6/7 subunits of kainate receptors in rat adenohypophysis. *Neurochem. Inte.* **38**, 539–547.

Hinoi, E., S. Fujimori, T. Takarada, H. Taniura, and Y. Yoneda (2001). Group III metabotropic glutamate receptors in rat cultured calvarial osteoblasts. *Biochem. Biophys. Res. Commun.* **281**, 341–346.

Hinoi, E., S. Fujimori, M. Yoneyama, and Y. Yoneda (2002a). Blockade by N-methyl-D-aspartate of elevation of activator protein-1 binding after stress in rat adrenal gland. *J. Neurosci. Res.* **70**, 161–171.

Hinoi, E., S. Fujimori, A. Takemori, H. Kurabayashi, Y. Nakamura, and Y. Yoneda (2002b). Demonstration of expression of mRNA for particular AMPA and kainate receptor subunits in immature and mature cultured rat calvarial osteoblasts. *Brain Res.* **943**, 112–116.

Hinoi, E., S. Fujimori, T. Takarada, H. Taniura, and Y. Yoneda (2002c). Facilitation of glutamate release by ionotropic glutamate receptors in osteoblasts. *Biochem. Biophys. Res. Commun.* **297**, 452–458.

Hinoi E., S. Fujimori, Y. Nakamura, V.J. Balcar, K. Kubo, K. Ogita *et al.* (2002d). Constitutive expression of heterologous N-methyl-D-aspartate receptors subunits in rat adrenal medulla. *J. Neurosci. Res.* **68**, 36–45.

Hollmann, M. and S. Heinemann (1994). Cloned glutamate receptors. *Annu. Rev. Neurosci.* **17**, 31–108.

Imai, K., T. Fukushima, T. Sante, H. Homma, Y. Huang, K. Sakai *et al.* (1996). Distribution of free D-amino acids in tissues and body fluids of vetebrates. *Enantiomer* **2**, 143–144.

Inagaki, N., H. Kuromi, T. Gonoi, Y. Okamoto, H. Ishida, Y. Seino *et al.* (1995). Expression and role of ionotropic glutamate receptors in pancreatic islet cells. *FASEB J.* **9**, 686–691.

Iverson, F., J. Truelove, L. Tryphonas, and E.A. Nera (1990). The toxicology of domoic acid administered systemically to rodents and primates. *Can. Dis. Wkly. Rep.* **16**(Suppl.1E), 15–19.

Itzstein, C., H. Cheynel, B. Burt-Pichat, B. Merle, L. Espinosa, P.D. Delmas *et al.* (2001). Molecular identification of NMDA glutamate receptors expressed in bone cells. *J. Cell Biochem.* **82**, 134—44.

Jackson, E.K., R.A. Branch, H.S. Margoius, and J.A. Oates (1985). Physiological functions of the renal prostaglandin, renin and kallikrein systems. In D.W. Seldin, and G. Giebisch (eds), *The Kidney—Physiology and Pathology*. Raven Press, NY, pp. 613–644.

Kamatani, Y., H. Minakata, P.T.M. Kenny, T. Iwashita, K. Watanabe, K. Funase *et al.* (1989). Achatin-I, an endogenous neuroexcitatory tetrapeptide from *Achatina fulica* ferussac containing a D-amino acid residue. *Biochem. Biophys. Res. Commun.* **160**, 1015–1020.

Kang, J. and Turano (2003). The putative glutamate receptors 1.1 (atGluR1.1) functions as a regulator of carbon and nitrogen metabolism in *Arabidopsis thaliana*. *PNAS* **100**, 6872–6877.

Kerr, D.S. and A.N. Razak (2000). Domoic acid induces tolerance to domoic acid in hippocampal slices. *Int. J. Neurosci.* **109**, 186

Kiyama, H., K. Sato, and M. Tohyama (1993). Characteristic localization of non-NMDA type glutamate receptor subunits in the rat pituitary gland. *Mol. Br. Res.* **19**, 262–268.

Krogsgaard-Larsen, P. and J.J. Hansen (1992). Naturally occurring excitatory amino acids as neurotoxins and leads in drug design. *Toxicol. Lett.* **64/65**, 409–416.

La Bella, V. and Piccoli (2000). Differential effect of B-N-oxalylamino-L-alanine, the *Lathyrus sativus*, neurotoxin, and (+_)-a-amino-3-hydroxy-5-methylisoxazole-4-propionate on the excitatory amino acid and taurine levels in the brain of freely moving rats. *Neurochem. Int.* **36**, 523–530.

Lam, H.M., J. Chiu, M.H. Hsieh, L. Meisel, I.C. Oliveira, M. Shin *et al.* (1998). Glutamate receptor genes in plants. *Nature* **396**, 125–126.

Lara, H. and W. Bastos-Ramos (1988). Glutamate and kainate effects on the noradrenergic neurons innervating rat vas deferens. *J. Neurosci. Res.* **19**, 239–244.

Leung, J.C., B.R. Travis, J.W. Verlander, S.K. Sandhu, S.G. Yang, A.H. Zea *et al.* (2002). Expression and developmental regulation of the NMDA receptor subunits in the kidney and cardiovascular system. *Am. J. Physiol. Regul. Integr. Comp. Physiol.* **283**, R964–R971.

Liu, P.H., S.S.W. Tay, and S.K. Leong (1997). Localization of glutamate receptors subunits of the a-amino-3-hydroxy-5-methyl-4-isoxazolepropionate (AMPA) type in the pancreas of newborn guinea pig. *Pancreas* **14**, 360–368.

Lipton, S.A. (1993). Prospects for clinically tolerated NMDA antagonists: Open-channel blockers and alternative redox states of nitric oxide. *Trends Neuorsci.* **16**, 527–532.

Lipton, S.A. and H.E. Gendelman (1995). Dementia associated with the acquired immunodeficiency syndrome. *N. Engl. J. Med.* **332**, 934–940.

Lipton, S.A. and P.A. Rosenberg (1994). Excitatory amino acids as a final common pathway for neurologic disorders. *N. Engl. J. Med.* **330**, 613–622.

Lucas, D.R. and J.P. Newhouse (1957). The toxic effect of sodium L-glutamate on the inner layers of the retina. *AMA Arch Ophthalmol.* **58**, 193–201.

Maricq, A.V., E. Peckol, M. Driscoll, C.I. Bargmann (1996). Mechanosensory signalling in *C. elegans* mediated by the GLR-1 glutamate receptor. *Nature* **379**, 749–781.

Meldrum, B.S. (1994). The role of glutamate in epilepsy and other central nervous disorders. *Neurology* **44**, 14–23.

Mick, G. (1995). Non-N-methyl-D-aspartate glutamate receptors in glial cells and neurons of the pineal gland in a higher primate. *Neuroendocrinology* **61**, 256–264.

Michaelis, E. (1998). Molecular biology of glutamate receptors in the central nervous system and their role in excitotoxicity, oxidative stress and aging. *Prog. Neurobiol.* **54**, 369–415.

Miller, G.W., E.A. Lock, and R.G. Schnellman (1992). Strychnine and glycine protect renal proximal tubules from various nephrotoxicants and act in the late phase of necrotic cell injury. *Toxicol. App. Pharmacol.* **125**, 192–197.

Miller, S., J.P. Kesslak, C. Romano, and C.W. Cotman (1996). Roles of metabotropic receptors in brain plasticity and pathology. *Ann New York Academy of Sciences.* **757**, 460–474.

Moghaddam, B. (1999). Glutamate and schizophrenia. *Sci. Med.* March/April, 22–31.

Montecucchi, P.C., R. de Castiglione, S. Pianio, L. Gozzini, and V. Erspamier (1981). Amino acid composition and sequence of Dermorphin, a novel opiate-like peptide from skin of *Phyllomedusa sauvagei. Int. J. Peptide Protein Res.* **17**, 275–283.

Molnar, E., A. Varadi, R.A.J. McIlhinney, and S.J.H. Ashcroft (1995). Identification of functional ionotropic glutamate receptor proteins in the pancreatic B-cells and in the islets of Langerhans. *FEBS Lett.* **371**, 253–257.

Moloney, K.G. (2002). Excitatory amino acids. *Nat. Prod. Rep.* **19**, 597–616.

Morhenn V.B., N.S. Waleh, J.N. Mansbridge, D. Unson, A. Zolotorev, P. Cline *et al.* (1994). Evidence for an NMDA receptor subunit in human keratinocytes and rat cardiocytes. *Eur. J. Pharmacol.* **268**, 409–414.

Moroni, F., S. Luzzi, S.F. Micheli, and L. Zilleti (1986). The presence of N-methyl-D-aspartate type receptors for glutamic acid in the guinea pig myenteric plexus. *Neurosci. Lett.* **68**, 57–62.

Mueller, R., S. Gill, O. Pulido, K. Kapal, and P. Smyth (1996). Demonstration and differential localization of glutamate receptors in the rat and monkey (*Macaca fascicularis*). *FASEB J.* **9**, LB146.

Mueller, R., S. Gill, and O. Pulido (2003). The monkey (*Macaca Fascicularis*) heart neural structures and conducting system: An immunochemical study of selected neural biomarkers and glutamate receptors. *Toxicol. Pathol.* **31**, 227–234.

Olney, J.W. (1969). Brain lesions, obesity and other disturbances in the mice treated with monosodium glutamate. *Science* **164**, 719–721.

Olney, J.W. (1989). Excitotoxicity: An overview. *Can. Dis. Wkly. Rep.* **16**(Suppl.1E), 49–58.

Olney, J.W. (1994). Excitotoxins in foods. *NeuroToxicology* **15**, 535–544.

Otamiri, T. (1988). Quinacrine prevention of intestinal ischaemic mucosal damage is partly mediated through inhibition of intraluminal phospholipase A_2. *Agents Actions* **25**, 378–384.

Ozawa, S., H. Kamiya, and K. Tsuzuki (1998). Glutamate receptors in the mammalian central nervous system. *Prog. Neurobiol.* **54**, 581–618.

Palmada, M. and Centelles (1998). Excitatory amino acid neurotransmission: Pathways for metabolism, storage and reuptake of glutamate in brain. *Front. in Biosci.* D701–718.

Peng, Y.G., T.B. Taylor, R.E. Finch, R.C. Switzer, and J.S. Ramsdell (1994). Neuroexcitatory and neurotoxic actions of the amnesic shellfish poison, domoic acid. *Neuroreport* **5**, 981–985.

Pentreath, V.W. and N.D. Slamon (2000). Astrocyte phenotype and prevention against oxidative damage in neurotoxicity. *Hum. Exp. Toxicol.* **1**, 641–649.

Perl, T.M., L. Bedard, T. Kosatsky, J.C. Hockin, E.C.D. Todd *et al.* (1990). An outbreak of toxic encephalopathy caused by eating mussels contaminated with domoic acid. *N. Engl. J. Med.* **322**, 1775–1780.

Price, M.D., J.W. Olney, O.H. Lowry, and S. Buchsbaum (1981). Uptake of exogenous glutamate and aspartate by circumventricular organs but not other regions of brain. *J. Neurochem.* **36**, 1734–1780.

Purcell, W.M., K.M. Doyle, C. Westgate, and C.K. Atterwill (1996). Characterization of a functional polyamine site on rat mast cells:Association with a NMDA receptor macrocomplex. *J. Neuroimmunol.* **65**, 49–53.

Rockhold, R.W., C.G. Acuff, and B.R. Clower (1989). Excitotoxin-induced myocardial necrosis. *Eur. J. Pharmacol.* **166**, 571–576.

Ross, I.A., W. Johnson, P.P. Sapienza, C.S. Kim (2000). Effects of the seafood toxin domoic acid on glutamate uptake by rat astrocytes. *Food Chem. Toxicol.* **38**, 1005–1011.

Rzeski, W., C. Ikonomidou, L. Turski (2002). Glutamate antagonists limit tumor growth. *Biochem. Pharmacol.* **64**, 1195–200.

Said, S.I. (1999). Glutamate receptors and asthmatic airway disease. *TIBS.* **20**, 132–135.

Said, S.I., H.I. Berisha, and H. Pakbaz (1996). Excitotoxicity in the lung: N-methyl-D-aspartate induced, nitric oxide-dependent, pulmonary edema is attenuated by vasoactive intestinal peptide and by inhibitors of poly (ADP-ribose) polymerase. *Proc. Natl. Acad. Sci. USA* **93**, 4688–4692.

Said, S.I., R.D. Dey, and K. Dickman (2001). Glutamate signalling in lung. *Trends Pharmacol. Sci.* **22**, 344–345.

Schoepp, D.D. (1994). Novel function for subtypes of metabotropic glutamate receptors. *Neurochem. Int.* **24**, 439–449.

Schofield, J.N., I.N. Day, R.J. Thompson, and Y.H. Edwards (1995). PGP 9.5, a ubiquitin C-terminal hydrolase; pattern of mRNA and protein expression during neural development in the mouse. *Dev. Brain Res.* **85**, 229–238.

Seeber, S., K. Becker, T. Rau, T. Eschenhagen, C.M. Becker, and M. Herkert (2000). Transient expression of NMDA receptor subunit NR2B in the developing rat heart. *J. Neurochem.* **75**, 2472–2477.

Scholin, C.A., F. Gulland, G.J. Doucette, S. Benson, M. Busman, F.P. Chavez *et al.* (2000). Mortality of sea lions along the central coast linked to a toxic diatom bloom. *Nature* **403**, 80–84.

Shannon, H.E. and B.D. Sawyer (1989). Glutamate receptors of the N-methyl-D-aspartate subtype in the myentric plexus of the guinea pig ileum. *J. Pharmacol. Exp. Ther.* **251**, 518–523.

Snyder, S.H. and C.D. Ferris (2000). Novel transmitters and their neuropsychiatric relevance. *Am. J. Psychiatry* **157**, 1738–1751.

Storto, M., U.D. Grazia, T. Knoppel, P.L. Canonica, A. Copani, P. Richelmi *et al.* (2000). Selective Blockade of mGluR 5 metabotropic glutamate receptors protects rat hepatocytes against hypoxic damage. *Hepatology* **31**, 649–655.

Stumer, T., M. Amar, R.J. Harvey, I. Bermudez, J.V. Minnen, and M.G. Darlison (1996). Structure and pharmacological properties of a molluscan glutamate-gated cation channel and its likely role in the feeding behavior *J. Neurosci.* **16**, 2869–2880.

Sureda, F., A. Copani, V. Bruno, T. Knopel, G. Meltzger, and F. Nicoletti (1997). Metabotropic glutamate receptor agonists stimulate polyphosphoinositide hydrolysis in primary cultures of rat hepatocytes. *Eur. J. Pharmacol.* **338**, R1–R2.

Tryphonas, L., J. Truelove, F. Iverson, E.C.D. Todd, and E.A. Nera (1990). Neuropathology of experimental domoic acid poisoning in non-human primates and rats. *Can. Dis. Wkly. Rep.* **16**(Suppl.1E), 75–81.

Teitelbaum, J., R.S. Zatorre, S. Carpenter, D. Gendron, A.C. Evans, A. Gjedde *et al.* (1990). Neurologic sequelae of domoic acid intoxication due to the ingestion of contaminated mussels. *N. Engl. J. Med.* **322**, 1781–1787.

Thompson, R.J., J.F. Doran, P. Jackson, A.P. Dhillon, and J. Rode (1983). PGP 9.5—a new marker for vertebrate neurons and neuroendocrine cells. *Brain Res.* **278**, 224–228.

Truelove, J., R. Mueller, O. Pulido, and F. Iverson (1996). Subchronic toxicity study of domoic acid in the rat. *Food Chem. Toxicol.* **34**, 525–529.

Tsai, L.H., Y.J. Lee, and J.Y. Wu (1994). Effect of L-glutamate acid on acid secretion and immunohistochemical localization of glutamatergic neurons in the rat stomach. *J. Neurosci. Res.* **38**, 188–195.

Tsai, L.H., Y.J. Lee, and J.Y. Wu (1999). Effect of excitatory amino acid neurotransmitters on acid secretion in the rat stomach. *J. Biomed. Sci.* **6**, 36–44.

Tsibris, J.C., J. Segars, D. Coppola, S. Mane, G.D. Wilbanks, W.F. O'Brien *et al.* (2002). Insights from gene arrays on the development and growth regulation of uterine leiomyomata. *Fertility and Sterility.* **78**, 114–121.

Vitadello, M., M. Matteoli, and L. Gorza (1990). Neurofilament proteins are co-expressed with desmin in heart conduction system myocytes. *J. Cell Sci.* **97**, 11–21.

Vitadello, M., S. Vettore, E. Lamar, K.R. Chien, and L. Gorza (1996). Neurofilament M mRNA is expressed in conduction system myocytes of the developing and adult rabbit heart. *J. Mol. Cell Cardiol.* **28**, 1833–44.

Villalobos, C., L. Nunez, and J. Garcia-Sancho (1996). Functional glutamate receptors in a subpopulation of anterior pituitary cell. *FASEB J.* **10**, 654–660.

Watanabe, M., M. Mishina, and Y. Inoue (1994). Distinct gene expression of the N-methyl-D-aspartate receptor channel subunit in peripheral neurons of the mouse sensory ganglia and adrenal gland. *Neurosci. Lett.* **165**, 183–186.

Watters, M.R. (1995). Organic neurotoxins in seafoods. *Clin. Neurol. Neurosurg.* **97**, 119–124.

Weaver, C.D., T.L. Yao, A.C. Powers, and T.A. Verdoorn (1996). Differential expression of glutamate receptor subtypes in rat pancreatic islets. *J. Biol. Chem.* **271**, 12977–12984.

Whittle, K. and S. Gallaher (2000). Marine Toxins. *Br. Med. Bull.* **56**, 236–253.

Winter, C.R. and R.C. Baker (1996). L-Glutamate induced changes in intracellular calcium oscillation frequency through non-classical glutamate receptor binding in cultured rat myocardial cells. *Life Sci.* **57**, 1925–1934.

Wolosker, H., R. Panizzutti, and J. Miranda (2002). Neurobiology through the looking glass: D-Serine as a new glial neurotransmitter. *Neurochem. Int.* **41**, 327–332.

Wolosker, H., S. Blackshaw, and S.H. Snyder (1999). Serine Racemase: A glial enzyme synthesizing D-Serine to regulate glutamate-N-methyl-D-asparatate neurotransmission. *PNAS.* **9**, 13409–13414.

Wu, Y.M., S.S. Kung, and W.C. Chow (1996). Determination of relative abundance of splicing variants of *Oreochromis* glutamate receptors by quantitative reverse-transcriptase PCR. *FEBS Letters* **390**, 157–160.

Xue, J., G. Li, E. Bharucha and N.G. Cooper (2002). Developmentally regulated expression of CaMKII and iGluRs in the rat retina. *Brain Res. Dev. Brain Res.* **138**, 61–70.

Yoneda, Y. and K. Ogita (1986). Localization of [^{3}H-] glutamate binding sites in rat adrenal medulla. *Brain Res.* **38**, 387–391.

Zautcke, J.L., J.A. Schwartz, and E.J. Mueller (1986). Chinese restaurant syndrome: A Review. *Ann. Emerg. Med.* **15**, 1210–1213.

2

Glutamate Receptor Pharmacology: Lessons Learned from the Last Decade of Stroke Trials

Daniel L. Small and Joseph S. Tauskela

1. Introduction

Interest in the pharmacology of glutamate receptors has stemmed primarily from the identification of glutamate as the principal excitatory neurotransmitter in the brain and, more recently, as a potential neurotoxin. The therapeutic potential of glutamate receptor antagonists were being explored in animal models for the indication of acute stroke in the late 1980s (Park *et al.*, 1988) but even before that, almost two decades earlier, the Chinese Restaurant Syndrome (Schaumburg *et al.*, 1969) and the neurotoxicity associated with the ingestion of the food preservative monosodium glutamate (MSG) was described in detail (Arees and Mayer, 1970). These reports foreshadowed the importance of the tremendous concentration gradient that exists between plasma and the cerebrospinal fluid (CSF). Given that glutamate is the most abundant amino acid in the diet, it is not surprising that plasma glutamate levels are 30–100 μM. What is remarkable, perhaps, is that concentrations in the CSF are maintained at levels below 1 μM and that although glutamate is the principal excitatory neurotransmitter in the brain, exposure of the neurons to the levels found in plasma, even for periods as brief as 5 min, results in neuronal death (Choi *et al.*, 1987). This largely accounts for why the peripheral effects of glutamate and glutamate receptor antagonists have largely been ignored until recently. With such high levels of glutamate in peripheral tissues and the pronounced sensitivity to glutamate in the central nervous system (CNS), the focus of glutamate receptor pharmacology has been restricted to CNS. It was not until the late 1990s that attention was devoted to target organs other than the brain, in spite of the fact that in the late 1970s an association was established between MSG and the ventricular tachycardia experienced by a patient presenting with Chinese Restaurant Syndrome (Gann, 1977).

Now with more than a decade of effort in the clinic to develop glutamate receptor antagonists, and more recently glutamate receptor agonists, peripheral glutamate receptor pharmacology and the systemic effects of glutamate receptor pharmacophores must be revisited

Daniel L. Small • Lexicon Genetics, 8800 Technology Forest Place, The Woodlands, Texas. **Joseph S. Tauskela** • Institute for Biological Sciences, National Research Council of Canada, Ottawa, Canada.

Glutamate Receptors in Peripheral Tissue, edited by Santokh Gill and Olga Pulido.
Kluwer Academic / Plenum Publishers, New York, 2005.

and scrutinized. Such studies are not only necessary to tap the therapeutic potential of peripherally acting glutamate receptor pharmacophores, but will also serve to predict undesirable side effects due to peripheral receptor activity.

1.1. Glutamate Receptor Pharmacology: Early Stages

Ironically, an association between glutamate and "excitotoxicity" (Hayashi, 1952) predates the demonstration of the excitatory role of glutamate in the mammalian brain and spinal cord (Curtis *et al.*, 1959; Curtis and Watkins, 1960). Spanning the late 1960s to early 1970s, further studies with L-glutamate and L-aspartate in the thalamus and spinal cord demonstrated the existence of subsets of neurons with different pharmacology between the two neurotransmitter candidates (McLennan *et al.*, 1968; Duggan, 1974). As a result of early medicinal chemistry studies, substitution of a methyl on the amino group of D-aspartate yielded N-methyl-D-aspartate as a potent excitant (Curtis and Watkins, 1963). One of the active ingredients from the poisonous mushroom *Amanita muscaria* that causes hallucinations, ibotenic acid, was identified (Eugster *et al.*, 1965; Sirakawa *et al.*, 1966)—shortly thereafter, the site of action of this endogenous glutamate receptor agonist was identified in insect muscle fibers (Lea and Usherwood, 1973), strengthening the hypothesis for pharmacologically distinct populations of glutamate receptors. During the early 1970s, several other natural products, such as quisqualate and kainate (KA), were identified as highly potent neuronal excitants in the mammalian CNS (Shinozaki and Konishi, 1970; Johnston *et al.*, 1974; Shinozaki and Shibuya, 1974; Biscoe *et al.*, 1976). Potency ratios of such compounds varied between cells, further suggesting receptor subtypes.

Final acceptance of the idea of pharmacologically distinct subtypes of glutamate receptors came with the emergence of early antagonists such as D-alpha-amino-adipate (DAA), gamma-glutamyl-amino-methyl-sulphonate (GAMS), and glutamate diethyl ester (GDEE; Haldeman and McLennan, 1972; Bisco *et al.*, 1977; Evans *et al.*, 1978; Lodge *et al.*, 1978; Davies and Watkins, 1979, 1985; McLennan and Lodge, 1979). These three agents proved to be weak and somewhat nonselective antagonists of NMDA, KA, and quisqualate, respectively. NMDA excitation was also blocked by increasing levels of extracellular Mg (Davies and Watkins, 1977; Evans *et al.*, 1977; Ault *et al.*, 1980) and by 3-amino-l-hydroxy-2-pyrrolidone (HA-966; Haldeman *et al.*, 1972; Davies and Watkins, 1973; Biscoe *et al.*, 1977; Evans *et al.*, 1978). Three separate classes of antagonists of NMDA excitation (DAA, Mg, and HA-966) also supported the concept of the NMDA receptor as a separate pharmacological entity. The pharmacological separation of the non-NMDA glutamate receptors came in 1980 with the emergence of alpha-amino-3-hydroxy-5-methyl-4-isoxazole proprionic acid (AMPA), as a potent GDEE-sensitive excitatory amino acid agonist clearly distinguishable from KA (Krogsgaard-Larsen *et al.*, 1980).

2. Glutamate Receptor Subtypes

The holy grail to understanding the pharmacological distinction of glutamate receptor subtypes resulted from application of molecular biological techniques to identify the individual molecular subunits. The subunit complement defined the pharmacological specificity of each of the subtypes of glutamate receptors. The explosion in the number of studies utilizing molecular biological approaches to study glutamate receptors, following the cloning and expression of the first glutamate receptor (GluR1) by Hollmann, Heinemann, and colleagues in 1989, led to the realization that the presence or absence of particular subunits within the heteromeric glutamate receptor complex dictated the pharmacological and biophysical properties of that receptor subtype. In addition to the NMDA and non-NMDA receptor subtypes that make up the ionotropic family of glutamate receptors, a family of metabotropic glutamate receptors (mGluRs) was also identified in the late 1980s that are G-protein linked. In discussing glutamate receptor subtypes, we have attempted to cover the development of the various classes of receptor antagonists, as well as to examine the pharmacokinetics and dynamics of these compounds in the context of safety and tolerability in clinical applications. Emphasis will be focused on the NMDA receptor subtype primarily because the experience gained with antagonists for this type of glutamate receptor in clinical trials outnumbers antagonists for other glutamate receptor subtypes.

2.1. NMDA

Within the NMDA receptor subtypes, there are numerous subdivisions based on (a) binding affinity, (b) the binding site on the receptor (in addition to the ligand-binding site, there are several allosteric sites on the receptor; Figure 2.1), and (c) the specific subunit to which the antagonist binds (Table 2.1). More types or classes of NMDA receptor antagonists have been developed than for any other glutamate receptor subtype, reflecting more than a decade of intensive interest in potential therapeutic application. As alluded to earlier, the driving force behind the NMDA receptor pharmacology was compelling evidence obtained in the late 1980s from animal models, which suggested that the NMDA receptor was a viable therapeutic target for the intervention of acute CNS injury.

2.1.1. Competitive Antagonists

Competitive antagonists bind to the ligand-binding site of the ionotropic receptor complex. Although the subunit composition *in vivo* which accounts for the functional diversity of NMDA receptors is not well understood, the NR1 subunit can couple with the NR2A, -B, -C, or -D subunits, or certain combinations (Wafford *et al.*, 1993; Sheng *et al.*, 1994; Didier *et al.*, 1995). Combinations of recently identified NR3A and NR3B subunits with NR1 form relatively impermeable cation channels resistant to Mg^{2+}, MK-801, memantine, and competitive antagonists (Chatterton *et al.*, 2002). Two lines of evidence support the idea that each of the NMDA receptor subunit subtypes possesses a glutamate-binding site. From molecular biology studies, two domains, S1 and S2 (Figure 2.2), located on the N-terminal as it enters the

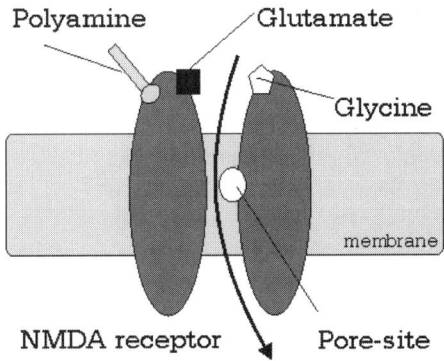

Figure 2.1. Schematic illustration of an NMDA receptor showing sites of pharmacological modulation.

Table 2.1. Pharmacological Classification of NMDA Receptor Subtypes

NMDA receptor antagonists **Class**	**Compound (Company) (K_i μM)**
Competitive	CGS 19755/Selfotel (Ciba) (0.067) MDL 100 453 (Merrel Dow) ATA (Eli Lilly) dCPP-ene (SDZ EAA 494; Sandoz) CGP 40116 (Ciba-Geigy) NPC 12626 (Nova Pharmaceuticals Corp.) NPC 17742 (Nova Pharmaceuticals Corp.) AP5 AP7
Noncompetitive, high affinity	MK-801/Dizoclipine (Merck) (0.014) Phencyclidine (0.05) CNS 1102/Atiganel HCl (Cambridge Neuroscience) (0.041) FR115427 (Fujisawa Pharmaceutical Co.) MDL 27266 (Merrell Dow)
Intermediate affinity	Dextrorphan/Ro 01–6794/706 (Hoffmann LaRoche) (0.22)
Low affinity	Ketamine (0.60) Kyurenate Dextromethorphan (Hoffmann LaRoche) (1.5) Memantine (Merz) (0.54) Amantadine (20.4) Magnesium Remacemide (Astra) (60.0) ARR15896 (Astra) (1.3)
Glycine site	C-W 554/Felbamate (Carter-Wallace) (269) ACEA 1021/Licostinel (CoCensys) GV 196771 (Glaxo Welcome) GV 150526 (Glaxo Welcome)
Partial agonist	HA-966 L 687,414 (Hoffmann LaRoche) ACPC/SYM 2030 (Symphony)
Polyamine site	SL 82–0715/Eliprodil (Synthelabo/Lorex) Ifenprodil (Synthelabo/Lorex) CP-101,606 (Pfizer) Ro-25–6981 (Hoffmann LaRoche) Co-101244/PD 174494

membrane and TM3 as it leaves the membrane, are thought to form the glutamate-binding domains due to their homology with QBP1 and QBP2 (bacterial glutamate-binding protein; O'Hara *et al.*, 1993; Kuryatov *et al.*, 1994). Also, in spite of the fact that homomeric recombinant NMDA receptors composed of NR2 subunits produce current responses of orders of magnitude smaller than those produced by homomeric NR1 subunits, each of the subunits can elicit a response (Monyer *et al.*, 1992).

The earliest competitive NMDA receptor antagonists were water-soluble compounds that did not cross the blood brain barrier, rendering them unsuitable for clinical use. More lipophilic drugs were subsequently developed, the prototype compounds being 3-(2-carboxypiperazin-4-yl) propyl-1-phosphonic acid (CPP), with its derivative d-CPPene and

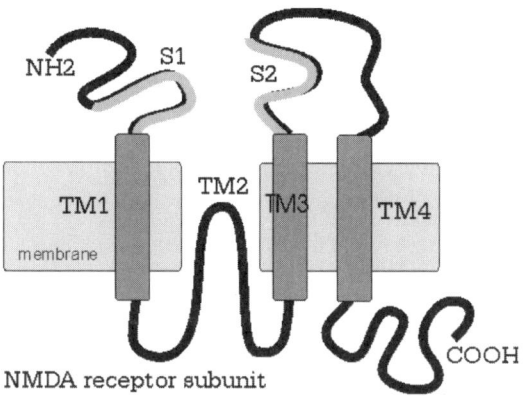

Figure 2.2. Schematic illustration of an NMDA receptor secondary structure and membrane topology.

CGS19755 (Selfotel, Novartis, formerly Ciba-Geigy). Reported use of CPP in humans has been limited to intrathecal doses for a patient with chronic pain. Maximal analgesia was evident at 100 nM levels of CPP, and delayed CNS side effects were evident with severe anxiety, agitation, hyperacusis, and nightmares for several nights (Kristensen *et al.*, 1992). CPPene was given orally to patients with refractory epilepsy in two small trials (Taylor, 1994). All eight patients reported sedation, ataxia, and confusion. Systemic effects of CPPene in humans were unremarkable.

CGS19755 has been extensively investigated in humans. In normal healthy volunteers, doses up to 1 mg/kg did not produce signs or symptoms. Doses of 2 and 3 mg/kg (i.v. bolus) produced transient CNS effects such as anxiety, sedation, altered sensory perception, decreased consciousness, or a feeling of light-headedness. All subjects with adverse experiences completely returned to normal when discontinuing treatment, with no significant effects on vital signs (Grotta *et al.*, 1995). Sedated head-injured patients tolerated higher doses (6 and 12 mg/kg) well but more than half of patients undergoing elective neurosurgical procedures (12 of 21) reported agitation, dizziness, hallucinations, confusion, ataxia, and paranoia after 2 mg/kg (Steinberg *et al.*, 1994). Although the predominant adverse effects of CGS 19755 were CNS related, significant non-CNS adverse experiences were also noted. In a small safety and tolerability study of twelve patients, three patients experienced vomiting, two experienced nausea, and one experienced hypertension (Grotta *et al.*, 1995). In a much larger randomized phase 3 study of 567 patients treated with 1.5 mg/kg CGS 19755, the incidence of nausea or vomiting did not significantly increase, but the incidence of atrial fibrillation and hypertension did. Interestingly, in a small safety study which preceded the larger phase 3 study, although the blood concentrations of CGS 19755 were proportional to the given dose with a mean blood half-life of 2–3.3 hr, the duration of the CNS adverse effects lasted for an average of 24 hr, suggesting neuronal binding of CGS 19755 for a duration longer than the plasma half-life. This implied that a single dose of the drug might have a biological effect at the NMDA receptor for 24 hr or longer, long after extracellular glutamate concentrations have returned to normal, based on animal studies using micro-dialysis. The phase 3 trial of CGS 19755 was terminated early in December of 1995 by Ciba-Geigy based on the recommendation from an independent data and safety monitoring board reporting that the benefit-to-risk ratio was not significantly high enough to warrant continuation. A non-significant increase in deaths in the CGS 19755 treated patients was reported. It is not clear whether more promising results would have been achieved with administration of CGS 19755 earlier than 6 hr after onset of acute stroke. At the time of the trial, a 3-hr treatment window may not even have been feasible.

2.1.2. Noncompetitive Antagonists

Competitive antagonists were not the first antagonists synthesized, although realization of this role was only understood later. Although not recognized at the time as antagonist of the NMDA receptor, phencyclidine (PCP) was developed as an anesthetic in the 1950s. PCP is the archetypal noncompetitive NMDA receptor antagonist. One of the first NMDA receptor antagonists, HA-966, recognized as an (noncompetitive) antagonist of the NMDA receptor was a partial agonist. In contrast to HA-966, PCP and similar compounds bound preferentially to the open channel within the pore of the channel. Kornhuber *et al.* (1993) proposed the term "uncompetitive antagonist" as a subclass of noncompetitive antagonists: uncompetitive blockers such as PCP do not act at a site that is completely independent of the endogenous agonist, unlike the partial agonist HA966.

There were two main reasons why NMDA receptor antagonists that bound to the pore site of the channel were the focus of the pharmacological discovery efforts. Following CNS injury, extracellular glutamate levels transiently increase (Benveniste *et al.*, 1984), and findings of this type formed the basis for the concept of excitotoxicity. A competitive antagonist becomes less efficacious at inhibiting calcium influx through the channel pore as the concentration of glutamate rises. An ideal pharmacological agent for use in CNS injury is a use-dependent antagonist like MK-801 (dizoclipine maleate), which was developed by Merck in the early 1980s. Another reason why the pore-site antagonists were ideal for treating CNS injury was that the pore channel blockers were much more likely to cross the blood brain barrier than molecules designed to interact with the sites that were normally occupied by highly polar amino acids like glutamate. The molecules that are cationic channel blockers commonly have two general characteristics, a cationic region centered on a protonatable nitrogen and a hydrophobic region. Compounds of this type proved to rapidly penetrate the blood brain barrier, an essential feature of a therapeutic for acute CNS injury. The problem, as we will describe, was that the earlier uncompetitive antagonists were very high-affinity compounds and required a long time to reach steady state block, unlike the low-affinity compounds.

In the 1950s when PCP was used as an anesthetic, patients reported marked agitation, disinhibition, and occasionally hallucinations and paranoia long after the analgesia had subsided (Greifenstein *et al.*, 1958; Johnston *et al.*, 1959). This was an omen to the problems that would accompany other high-affinity uncompetitive NMDA receptor antagonists like MK-801 and CNS 1102 (aptiganel; Cerestat, Cambridge Neurosciences). MK-801 (dizoclipine), one of the uncompetitive antagonists with the highest affinity (0.014 μM; Palmer, 2001), was demonstrated in rodent models of cerebral ischemia in the late 1980s to be potently neuroprotective (Park *et al.*, 1988). It was subsequently shown to be consistently neuroprotective in several models of excitotoxicity. However, its clinical development was short-lived because of the CNS effects normally associated with NMDA receptor antagonists, like agitation and hallucinations. Shortly after a phase II investigation was initiated, Barnes (1987) and Olney *et al.* (1989) published provocative articles showing that MK-801 and PCP produced transient vacuoles in cingulated and retrosplenal cortical regions of the rat brain. Numerous studies followed in which the focus was the neurotoxicity of MK-801 and PCP.

CNS-1102, another high-affinity, uncompetitive NMDA receptor antagonist (0.028 μM; Kirk *et al.*, 1994) was examined in the mid-1990s with high expectations in spite of the bad press that MK-801 received in the late 1980s. The project team at Cambridge Neuroscience was optimistic that the properties of the compound were suitable for an adequate test of the hypothesis that an uncompetitive NMDA receptor antagonist could reduce brain damage following stroke and traumatic brain injury. The optimism for CNS-1102 was generated in large measure due to its pharmacokinetics. It distributed rapidly into a large volume (<10 min following i.v.; ~30 min following s.c.) with a plasma half-life of 60–90 min, its metabolic products were inactive, and it loaded very quickly into brain tissue (6 min after administration of [14C]-CNS-1102, the level of radioactivity in the brain was ~12 times that in plasma). Moreover, Olney *et al.* (1989) had demonstrated reversible neuronal vacuolation with MK-801 but at doses much higher (>1 mg/kg, s.c.; >2.5 mg/kg i.v.) than doses at which neuroprotection in animal models was demonstrated (0.25 mg/kg bolus i.v. followed by 0.17 mg/kg/hr; McBurney, 1997). The affinity of CNS-1102 for the NMDA receptor was less than half that of MK-801, so the expectation was for good efficacy combined with eradication of the problems plaguing MK-801. However, CNS side effects were still noted at plasma concentrations (~10 ng/ml) of CNS-1102 associated with neuroprotection in rat models. As with the other high-affinity uncompetitive antagonists, ataxia and sedation were associated with lower doses of CNS-1102 while paranoid ideation and hallucinations were associated with higher doses. CNS-1102 also caused a dose-dependent hypertension (rises in mean arterial pressure of up to 30 mmHg) and an increase in heart rate in healthy volunteers (Grosset *et al.*, 1995) but not in stroke patients (Fisher, 1994). In the end, the therapeutic window or index was too narrow, so that a large phase III trial was discontinued 2 years following initiation. In a subsequent analysis, some potential therapeutic benefit was identified in a subset of stroke patients. Cambridge Neuroscience announced in 1998 that they planned to further investigate CNS-1102 with Boehringer Ingelheim, but by 2001, CNS had changed their name to CeNeS Pharmaceuticals and had refocused on pain and contract research and selling their ion-channel focused chemical library to Scion Pharmaceuticals Inc.

Dextrorphan is an intermediate affinity uncompetitive NMDA receptor antagonist (0.22 μM Ki). Dextrorphan is the O-methylated metabolite of dextromethorphan, an antitussive agent used for years before its recognition as a low-affinity NMDA receptor antagonist (1.5 μM K_i). Both dextrorotary opioid derivatives exhibit inhibitory effects on L-type voltage-sensitive calcium channels (VSCC; Carpenter *et al.*, 1988) and dextromethorphan has additional affinity for sigma receptors (Tortella *et al.*, 1989). In a multi-center, ascending dose phase II trial of dextrorphan in 67 patients, 60–260 mg/hr were infused over 1 hr followed by ascending maintenance doses for 11 or 23 hr reaching total cumulative doses of up to 1,280 and 2,140 mg over 12 and 24 hr, respectively (Albers *et al.*, 1990). Adverse side effects associated with these doses of dextrorphan were nystagmus (53% of patients), somnolence (45%), agitation (53%), hallucinations (43%), confusion (51%), and nausea or vomiting (28%). The dose-limiting effect during bolus infusion was a highly significant and clinically concerning hypotension (up to 50 mmHg drop) that occurred in 11 of 51 patients. During the maintenance infusion over the 11 to 23 hr, hypertension was recorded in more than 40% of patients. Further development of dextrorphan has been abandoned.

Among the low-affinity uncompetitive NMDA receptor antagonists, memantine has enjoyed the most publicity and success of late. Memantine is not a new drug though. It was first synthesized by researchers at Eli Lilly as an agent to lower blood sugar levels (Gerzon *et al.*, 1963), but it was completely devoid of such activity. Merz and Co. filed for a patent for memantine for the potential treatment of Parkinson's disease, spasticity, and cerebral disorders including cerebrovascular disorders in 1972 and was granted the patents in 1975 (Germany) and 1978 (USA). Memantine was first used clinically in the early 1980s as an anti-Parkinsonian agent and as an anti-spastic agent in chronic neurodegenerative diseases, long before electrophysiological studies revealed its NMDA receptor blocking properties (Bormann, 1989). Prior to evidence of NMDA receptor antagonism, it was speculated that memantine had direct and indirect dopaminergic activity as well as effects on serotonergic and noradrenergic systems (Wesemann et al., 1983). It has since been well characterized electrophysiologically and shown to be an uncompetitive NMDA receptor antagonist with an IC_{50} of ~2 μM. The therapeutic index is much better than the high-affinity uncompetitive antagonist MK-801. The therapeutic ratio of memantine vs MK-801 is 3.12 and 5.4 in vitro and in vivo, respectively (Parsons et al., 1999). The length of time memantine has been in the clinic is testimony to its safety. It is estimated that in more than 15 years in the clinic, memantine has been provided to more than 200,000 patients and in that long history it has demonstrated good tolerability (Parsons et al., 1999). Psychotomimetic effects, normally associated with other uncompetitive NMDA receptor antagonists, appear only if the recommended dosing range from 5 to 20 mg over the 3–4 weeks is skipped or when memantine is combined with dopaminergic therapies. The half-life in humans is surprisingly much longer than in rats (up to 100 hr vs 3–5 hr) making it much easier to maintain a steady-state concentration in humans over extended periods with chronic dosing, such as is the case for treatment paradigms for dementia. Although memantine has demonstrated improved safety over other uncompetitive NMDA receptor antagonists, at doses higher than those considered therapeutic, memantine in animals does produce a number of side effects, like ataxia, myorelaxation, and amnesia, which are typical of uncompetitive NMDA receptor antagonists. Memantine has not been reported to induce the infamous transient neuronal vacuolization seen with high-affinity uncompetitive NMDA receptor antagonists, first reported by Olney (Olney et al., 1989). At doses of the high-affinity uncompetitive antagonists higher than those associated with the transient vacuoles, neuronal death and apoptosis occurs, while at doses lower than those required to induce vacuolization, a heat shock protein, HSP70, is induced in the posterior cingulated, retrosplenial cortex and dentate gyrus (Sharp et al., 1994). At doses much higher than are therapeutic, 25, 50, and 75 mg/kg i.p., memantine does induce HSP70 (Tomitaka et al., 1996, 1997), suggesting that memantine is capable of neuronal toxicity associated with other uncompetitive NMDA receptor antagonists, although not to the same degree as the affinity compounds.

In addition to the recent success in treating dementia, memantine has been proposed for other therapeutic indications such as for AIDS, glaucoma, hepatic encephalopathy, multiple sclerosis, tinnitus, Parkinson's disease, tardive dyskinesia, chronic pain, addiction, epilepsy, spasticity, depression, and anxiety (Parsons et al., 1999).

Another NMDA receptor antagonist in this class, AR-R 15896AR, a product of previous work with remacemide and its principal metabolite remacemide desglycine (FPL 12495), demonstrated very good tolerability in both phase I (Palmer et al., 1999) and phase II studies (Lees et al., 2001). In the phase II dose ranging study, plasma concentrations of 1,524 +/− 536 ng/ml at the end of the 250-mg loading infusion and 1,847 +/− 478 ng/ml at steady state after the nine maintenance doses of 120 mg were achieved (Lees et al., 2001). Adverse effects associated with other uncompetitive NMDA receptor antagonists were reported including dizziness, vomiting, nausea, stupor, and some agitation/hallucination. Unfortunately, in spite of achieving putative neuroprotective concentrations of 1,240 ng/ml (based on data from animal models), no significant difference in outcome was observed between groups (Lees *et al.*, 2001). No further development of AR-R15896AR has been reported.

Magnesium is yet another low-affinity uncompetitive NMDA receptor antagonist that has the attraction of being cheap, widely available and with an established safety profile, having been used for many years as standard therapy for preeclampsia in North America. It is extremely well tolerated. In a small dose tolerating study, 25 stroke patients were randomized to receive placebo or one of three intravenous $MgSO_4$ infusions: a loading infusion of 8, 12, or 16 mmol, followed by 65 mmol over 24 hr (Muir and Lees, 1998). No effects of magnesium on heart rate, blood pressure, or blood glucose were evident. A multicenter, double-blind, randomized placebo-controlled trial of i.v. magnesium sulfate, delivered in a 16-mmol bolus infused over 15 min, followed by a 65-mmol maintenance dose infused over 24 hr is underway with 2,102 patients randomized as of March 2002 and an expected enrollment of 2,700 patients at 100 centers. One concern overshadowing the optimism for success in this trial is the fact that Mg is known to produce hyperglycemia, which has been suggested to underlie the failure of Mg in animal models of cerebral ischemia (Muir and Lees, 1998).

2.1.3. Polyamine Site NMDA Receptor Antagonists

Antagonists which act at the polyamine site of the NMDA receptor, such as eliprodil (SL 82-0715) and ifenprodil, have fared poorly in clinical trials (Small *et al.*, 1999a). Administration of either of these compounds induces unacceptable cardiovascular side affects, believed to emanate from interactions with additional Na^+ and Ca^{2+} channels (Fisher, 1995; Synthelabo Press Release, 1996). At reduced doses which minimized these side effects, eliprodil was not efficacious against stroke and phase III trials were abandoned in 1997 (Synthelabo Press Release, 1996). A potentially important feature of the polyamine site antagonists is a relatively specific block of the NR2B receptor subunit of NMDA receptors (Chenard and Menniti, 1999). The affinity of eliprodil for NR2B subunits is more than 100-fold greater than that for NR2A or NR2C receptor subunits. Other NR2B subunit selective antagonists synthesized include Ro 25–6981 and CP-101,606, which are both high-affinity NR2B-selective antagonists with very slow kinetics (Fischer *et al.*, 1997; Chenard and Menniti, 1999). CP-101,606 is neuroprotective in animal models of cerebral ischemia, but the slow kinetics may limit the rate at which neuroprotective doses may be achieved in human stroke. The lack of psychotomimetic effects of CP-101,606 in phase II trials for moderate head injury and hemorrhagic stroke, combined with good clinical tolerance, suggest that CP-101,606 may be effective in occlusive stroke (Merchant *et al.*, 1999). Another high-affinity compound, Co 101244/PD 174494, may be suitable for testing against stroke, due to a much lower affinity for $\alpha1$ adrenergic receptors, as well as less interference with potassium channels, problems that plague most other high-affinity compounds (Zhou *et al.*, 1999). A high-affinity compound, isoxsuprine, possesses faster kinetics than CP-101,606, potentially avoiding too slow a rate of achievement of a neuroprotective dose (Zhou *et al.*, 1999). Future testing of these compounds may be re-directed from stroke research, though as a result of an altered corporate landscape. Compounds which were being developed under the collaborative efforts of Parke-Davis and Cocensys, including Co 101244/PD 174494 and Cl-1041 (PD 0196860/Co200461), may be targeted to pain (Taniguchi *et al.*, 1997; Fillhard *et al.*, 2000), and Parkinson's disease (Wright *et al.*, 1999, 2000; Schelkun *et al.*, 2000), rather than against stroke. Pfizer recently merged with Warner-Lambert, which included Parke-Davis as a division of Warner-Lambert. Now, with a more recent merger between Pfizer and Pharmacia, the direction adopted for newer NR2B specific antagonists is not yet apparent.

2.1.4. Glycine Site Antagonists

Glycine site antagonists generally exhibit better safety profiles than antagonists that bind to other sites of the NMDA receptor (Muir and Lees, 1995). Much clinical attention of glycine site antagonists has been directed toward ACEA 1021 (5-nitro-6,7-dichloro-2,3-quinoxaline-dione, known as Licostinel) and GV150526. ACEA 1021 belongs to the kappagem family of compounds. ACEA-1021 was neuroprotective in animal models of focal and global ischemia (Warner *et al.*, 1995), exhibiting minimal adverse CNS or cardiovascular side effects (Kretschmer *et al.*, 1997; Albers *et al.*, 1999). The compound, originally obtained by ACEA, was then developed by CoCensys for Novartis, but Novartis halted its participation after crystals of ACEA-1021 were found in the urine of some participants in a phase I study (Drugs RD, 1999). CoCensys retains the rights to the drug and is continuing to evaluate it in phase II trials. To address the problems of excretion, ACEA-1021 has been applied in combination with probenecid to nonselectively inhibit secretion of anionic compounds (Xue *et al.*, 1999). This combination resulted in significantly smaller infarct in animal models of cerebral ischemia compared to ACEA 1021 alone. This improvement suggested that inclusion of probenecid may allow achievement of higher steady-state levels of ACEA 1021 (Xue *et al.*, 1999).

GV150526 is significantly neuroprotective in animal models of cerebral ischemia and, as for ACEA-1021, has shown good tolerability with minimal CNS side effects in the Glycine Antagonist in Neuroprotection (GAIN) phase I and II trials (The GAIN Investigators) (Dyker and Lees, 1999). Minor abnormalities in liver function were noted with higher maintenance doses, but these changes were asymptomatic and resolved within 10 days (Dyker and Lees, 1999). The results of the dose-escalation phase II clinical trial for GV150526 (GAIN-1) were reported in 2001 and were followed with the GAIN Americas and the GAIN International phase III trials. Approximately 1,600 patients were recruited into randomized double-blinded placebo controlled trials. Two sub-studies in each trial were planned, to measure lesion volume by Magnetic Resonance Imaging-Diffusion Weighted Imaging (MRI-DWI) and to measure health-related quality of life outcomes. The results of the GAIN International were presented at the 25th International American Heart Association meeting and did not meet expectations, demonstrating no improvement over untreated patients.

2.2. AMPA and KA Receptors

AMPA receptors are composed of GluR1-4 subunits, with the heterogeneous combination of tetramers or pentamers determining the functional properties of the receptors. The presence of a GluR2 subunit renders an AMPA channel impermeable to potentially neurotoxic Ca^{2+}. Activation of AMPA receptors following ischemia allows Na^+ influx which contributes to the depolarization of the neuron and to the influx of Ca^{2+} through other Ca^{2+} channels such as NMDA receptors. Fewer binding sites have been characterized on the AMPA receptor compared to the NMDA receptor. Most AMPA antagonists are members of the following chemical families: (a) tricyclic derivatives, (b) substituted quinoxalinediones, (c) quinolones, (d) isatin oximes, and (e) benzodiazepine derivatives. Types of AMPA receptors include competitive antagonists, noncompetitive antagonists such as GYKI [1-(4-aminophenyl)-4-methyl-7,8-methylenedioxy-5H-2,3-benzodiazepine HCl, also known as GYKI 52466,], site-binding antagonists, and uncompetitive antagonists (Gill and Lodge, 1997). Pore blocking uncompetitive antagonists have been isolated primarily from spider toxins like JSTX and are not appropriate for clinical development. Although fewer in number, AMPA receptor antagonists may offer greater protection against cerebral ischemia (Xue *et al.*, 1994; Small *et al.*, 1999b) than NMDA receptor antagonists. Several competitive AMPA receptor antagonists (NBQX, PNQX, YM90K, and YM872) and noncompetitive AMPA receptor modulators (GYKI 52466, EGIS-9637) have shown robust neuroprotective efficacy in experimental focal and global models (Gill and Lodge, 1997; Levay, 1999; Small *et al.*, 1999b). Some quinoxalinediones act as antagonists at the NMDA receptor glycine site, making resolution of the relative importance of the AMPA vs NMDA receptor blocking capability different to determine. Moreover, almost all quinoxalinediones are relatively insoluble in water, resulting in nephrotoxicity.

Much less experience has been gained using AMPA and KA compared to NMDA receptor antagonists to treat acute stroke. One of the more common CNS effects of AMPA receptor antagonists is sedation. All of the AMPA receptor antagonists examined in clinical trials for the treatment of acute stroke have been competitive antagonists based on the quinoxalinediones like NBQX. NBQX and other quinoxalinediones are neuroprotective in animal models of ischemia, even when administered up to 12 hr following reperfusion (Li and Buchan, 1993). Problems encountered with quinoxaline-diones in trials have been nephrotoxicity, resulting from the poor solubility of these compounds, and sedation (Xue *et al.*, 1994). One AMPA receptor antagonist, ZK-200775, progressed to phase IIa but these trials were discontinued in 1998 due to excessive sedation. Two of the competitive AMPA receptor antagonists that Yamanouchi Pharmaceuticals had demonstrated were neuroprotective in animal models, and also had improved solubility (Kawasaki-Yatsugi *et al.*, 1998, 2000; Takahashi *et al.*, 1998), progressed to clinical trials. In phase I trials, YM90K was well tolerated, and in spite of a high rate of urinary excretion, only mild changes were observed in kidney function markers with a single i.v. dose of 36 mg and repeated doses of 24 mg over 3 hr (Umemura *et al.*, 1997). Following promising preclinical studies, these phase I trials were followed by phase II trials using the AMPA receptor antagonist, YM872, which has a solubility 800-fold greater than NBQX or YM90K at pH 7 (Kohara *et al.*, 1996). Phase IIa trials are ongoing. The only adverse effect reported in the YM872 phase I study was euphoria in some patients at the higher dose levels. The challenge with YM872 will be overcoming the short half-life of the drug to maintain the elevated plasma levels necessary to effect neuroprotection.

In addition to the problems of solubility and safety which have plagued most AMPA receptor antagonists in efforts to develop stroke therapeutics, the problem facing all stroke therapeutics remains; the economics of drug development suggest that even those compounds appearing promising as therapeutics for stroke may first be directed toward other, seemingly less challenging diseases than stroke, such as pain (Simmons *et al.*, 1998), epilepsy (Anderson *et al.*, 1999; Pratt *et al.*, 2000), or other neurodegenerative diseases like Parkinson's disease (Konitsiotis *et al.*, 2000), or ALS (Hurko and Walsh, 2000).

Evidence suggests that activation of KA receptors does not substantially contribute to ischemic cell death, although overactivation of these receptors can kill neurons. The neuroprotective quinoxalinediones have a 100-fold selectivity for AMPA receptors over KA receptors (Kohara *et al.*, 1996). The relatively ischemia resistant CA3 region of the hippocampus has an abundance of KA receptors and neuroprotection. Finally, blocking KA receptors alone does not protect against ischemia (Monaghan and Cotman, 1982).

2.3. Metabotropic Glutamate Receptors

Although mGluRs are not ion channels, these will be included to complete the review of glutamate receptor pharmacology. Interest in the role of mGluRs in ischemia has grown recently. The mGluRs subtypes, mGluR1–8, are grouped into three main families based on sequence similarities, signal transduction systems, and pharmacology. Group I receptors (mGluR1 and -5) are linked to phospholipase C, generally producing excitatory effects, while groups II (mGluR2 and -3) and III (mGluR4, -6–8) are negatively linked to adenylate cyclase (Conn and Pin, 1997). Although mGluRs agonists and antagonists have not yet entered clinical trials, interest in developing these as neuroprotective therapeutics is growing. Activation of the Group I mGluRs may exacerbate neuronal damage, whereas activation of Group II and III mGluRs are neuroprotective (Miller et al., 1992; Nicoletti et al., 1996). The arsenal of pharmacological agents targeting individual receptor subtypes has traditionally been small but is now growing rapidly. Pharmacological activation of Group 1 mGluRs may either facilitate or attenuate excitotoxic-based death depending on the functional state of these receptors (Nicoletti et al., 1999). mGluR1 antagonists are effective against excitotoxic-based insults in vitro and may also be appropriate as antiepileptic agents (Greenwood et al., 2000), possibly by enhancing release of the inhibitory neurotransmitter, GABA (Battaglia et al., 2001). However, ataxia may be a serious side effect (Sillevis Smitt et al., 2000). mGluR5 receptors may positively modulate NMDA receptors, likely accounting for the potent anti-excitotoxic activity. Activation of mGluR5 receptors inhibits apoptosis, suggesting that the clinical effectiveness of mGluR5 antagonists may be highest in CNS disorders with a higher ratio of excitotoxic to apoptotic contributions (Bruno et al., 2001). Relatively selective mGluR1 and -5 antagonists have been developed to test these and other hypotheses. CPCCOEt exhibits selectivity for mGluR1 over mGluR5 (Toms et al., 1996), which may be an attractive property, given that activation of mGluR5 activation/mGluR1 inhibition may be beneficial (Nicoletti et al., 1996).

Compounds exhibiting activity at both Group I and II receptor subtypes like (1S, 3R)-ACPD have yielded mixed results in vivo (Chiamulera et al., 1992; Toms et al., 1996). DCG IV is a Group II selective agonist but has had limited usefulness in studies of cerebral ischemia. As an orally active Group II selective agonist, LY 354740 represented a feasible alternative to test compared to DCG-IV (Toms et al., 1996). In a comparison with diazepam for use as an anxiolytic LY354740 did not produce sedation, cause deficits in neuromuscular coordination, interact with CNS depressants, produce memory impairment, or change convulsive thresholds at doses 100–1,000-fold the efficacious doses in animal models of anxiety (Helton et al., 1998). However, LY354740 failed to protect against either focal (Lam et al., 1998) or global (Bond et al., 1998) ischemia, but is in a clinical trial for anxiety. A Group I antagonist/Group II agonist, [(S)-4C3HPG], however, was significantly neuroprotective in both focal (Rauca et al., 1998) and global ischemia (Henrich-Noack et al., 1998). Group II receptor agonists inhibit glutamate release and induce production of neurotrophic factors by astrocytes, raising the interesting possibility of extension to most acute and chronic neurodegenerative disorders, without concerns of poor blood–brain barrier (Bruno et al., 2001).

The role of Group III mGluRs in cerebral ischemia is unclear. The development of Group III selective agonists is in its infancy, due to the requirement for an omega-phosphonic group in agonist molecules such as L-AP4, and the acidic moiety limits blood–brain permeability. This may be related to the difficulty of functionally differentiating Group II and III mGluRs and the fact that two of the four Group III mGluRs, mGluR6 and -8, are found in abundance only in the retina and olfactory bulb. L-AP4 and CPPG exhibit selectivity for Group III mGluRs (Toms *et al.*, 1996) but *in vivo* data is not yet available. Trans-MCG-I is an interesting agent because it exhibits selectivity for mGluR2 over mGluR3 (Nicoletti *et al.*, 1996). This may be a useful trait in selectively targeting the hippocampus after transient forebrain ischemia since there are high levels of mGluR2 expression relative to mGluR3 in the hippocampus (Iversen *et al.*, 1994). Mechanistically, mGluR4/7/8 receptor agonists are neuroprotective by limiting glutamate release, while mGluR4/7 receptor agonists exert anticonvulsant activity. In summary, it seems clear that further development of mGluR agents is imminent, driven mainly by the possibility of potentially high clinical efficacy applicability to a broad array of CNS disorders.

3. Conclusion

Much of the *in vivo* pharmacology discussed in this chapter has been drawn from clinical stroke trials. Treatment durations with glutamate receptor antagonists for acute stroke has been transient, not lasting more than a day in the longest cases. More recent CNS indications for which NMDA receptor antagonists are being targeted, like pain, Parkinson's disease, and Alzheimer's disease imply a different treatment paradigm: patient exposure to these agents will be chronic. Except in isolated cases, though, a dearth of knowledge exists regarding the effects of chronic administration of many of the NMDA receptor antagonists. Although not highlighted in this chapter, CNS side effects such as behavioral changes resulting from chronic administration of glutamate receptor antagonists may have further downstream consequences in peripheral tissues. For instance, alterations in feeding activity may in turn cause changes in parameters such as bone formation and resorption (Skerry and Genever, 2001).

The fact that memantine was first used in the early 1980s as an anti-Parkinsonian agent means we do have some information about the safety of this compound administered chronically. This compound was well tolerated but it is one of the safest NMDA receptor antagonists in its class. Other uncompetitive antagonists, like MK-801, have not been given to humans chronically. In experimental rodent models, chronic administration of MK-801 causes sensitization of low dose effects like hyper-kinetic behavior and desensitization of high dose effects like ataxia (Hesselink *et al.*, 1999). In nonhuman primates, tolerance to specific behavioral alterations induced by chronic (2-year) exposure of remacemide and MK-801 developed only in the latter case (Popke *et al.*, 2001a,b). A more recent study performed by this same group, using a broad range of toxicology data and tests of cognitive function to measure peripherally mediated effects, indicated both drugs were well tolerated (Popke *et al.*, 2002), in contrast to their earlier reports of effects on complex operant behaviors. Another uncompetitive NMDA receptor antagonist we have some experience with chronically is PCP (angel dust). Following chronic administration, there is a desensitization of the CNS effects motivating abusers of this drug to self administer higher doses to achieve the sought after high. Unfortunately, the cardiovascular system does not adapt or desensitize as readily as the CNS and most addicts die from cardiac failure. The presence of glutamate receptors throughout the heart as described in an earlier chapter warrant closer examination of these receptors following chronic administration of therapeutics targeting glutamate receptors. Clearly there is precedence for examining effects in the cardiovascular system from the cases of tachycardia associated with Chinese Restaurant Syndrome and the victims of the shell fish poisoning presenting with cardiovascular symptoms.

Much of the discussion in this chapter on glutamate receptor antagonists has dealt with negative side effects. It was not our intention to necessarily portray glutamate receptor compounds in a negative light, but to indicate potential sources of problems in the development of therapeutics in peripheral tissues. A potentially key strength of the voluminous work already performed in the CNS field is application of the libraries of compounds already developed by pharmaceutical companies to peripheral disorders. Groundwork research on the numerous organs and peripheral systems that have been shown to possess functional glutamate receptors described throughout this book strongly suggest that glutamate receptor agonists and antagonists will have therapeutic value in the periphery. Numerous possibilities already exist for the therapeutic use of glutamate-receptor-based compounds in diabetes, asthma, and osteoporosis by virtue of the receptors shown in the pancreas, lung, and bone in both osteoclasts and osteoblasts. There are numerous other organs that have been shown to possess glutamate receptors and some of these reports have gone on further to demonstrate functional receptors. Undoubtedly, further sites will be identified. The effects of antagonists, particularly *in vivo*, have not yet been well studied. This is an important evolving field which will advance as the spotlight perhaps dims on the CNS stage, following the failure of so many agents in clinical stroke trials, and emphasis is re-directed toward peripheral targets.

Acknowledgments

JST was supported in part by a grant from the Heart & Stroke Foundation of Ontario (#T-4475).

References

Albers, G.W., R. Saenz, J. Moses, Jr., and D.W. Choi (1990). Pilot study of oral dextromethorphan in patients at risk for brain ischemia. *Neurology* **40**(Suppl 1), 193.

Albers, G.W., W.M. Clark, R.P. Atkinson, K. Madden, J.L. Data, and M.J. Whitehouse, (1999). Dose escalation study of the NMDA glycine-site antagonist licostinel in acute ischemic stroke. *Stroke* **30**, 508–513.

Anderson, B.A., N.K. Harn, M.M. Hansen, A.R. Harkness, D. Lodge, and J.D. Leander, (1999). Synthesis and anticonvulsant activity of 3-aryl-5H-2,3-benzodiazephine AMPA antagonists. *Bioorg. Med. Chem. Lett.* **9**, 1953–1956.

Arees, E.A. and J. Mayer (1970). Monosodium glutamate-induced brain lesions: Electron microscopic examination. *Science* **170**, 549–550.

Ault, B., R.H. Evans, A.A. Francis, D.J. Oakes, and J.C. Watkins (1980). Selective depression of excitatory amino acid induced depolarizations by magnesium ions in isolated spinal cord preparations. *J. Physiol.* **307**, 413–428.

Barnes, D.M. (1987). Drug may protect brains of heart attack victims. *Science* **235**, 632–633.

Battaglia, G., V. Bruno, A. Pisani, D. Centonze, M.V. Catania, P. Calabresi *et al.* (2001). Selective blockade of type-1 metabotropic glutamate receptors induces neuroprotection by enhancing Gabaergic transmission. *Mol. Cell. Neurosci.* **17**, 1071–1083.

Benveniste, H., J. Drejer, A. Schousboe, and N.H. Diemer (1984). Elevation of the extracellular concentrations of glutamate and aspartate in rat hippocampus during transient cerebral ischemia monitored by intracerebral microdialysis. *J. Neurochem.* **43**, 1369–1374.

Biscoe, T.J., R.H. Evans, P.M. Headley, M.R. Martin, and J.C. Watkins (1976). Structure-activity relations of excitatory amino acids on frog and rat spinal neurones. *Br. J. Pharmacol.* **58**, 373–382.

Biscoe, T.J., R.H. Evans, A.A. Francis, M.R. Martin, J.C. Watkins, J. Davies *et al.* (1977). D-alpha-Aminoadipate as a selective antagonist of amino acid-induced and synaptic excitation of mammalian spinal neurones. *Nature* **270**, 743–745.

Bond, A., M.J. O'Neill, C.A. Hicks, J.A. Monn, and D. Lodge (1998). Neuroprotective effects of a systemically active group II metabotropic glutamate receptor agonist LY354740 in a gerbil model of global ischaemia. *Neuroreport* **9**, 1191–1193.

Bormann, J. (1989). Memantine is a potent blocker of N-methyl-D-aspartate (NMDA) receptor channels. *Eur. J. Pharmacol.* **166**, 591–592.

Bruno, V., G. Battaglia, A. Copani, M. D'Onofrio, P. Di Iorio, A. De Blasi *et al.* (2001). Metabotropic glutamate receptor subtypes as targets for neuroprotective drugs. *J. Cereb. Blood Flow Metab.* **21**, 1013–1033.

Carpenter, C.L., S.S. Marks, D.L. Watson, and D.A. Greenberg (1988). Dextromethorphan and dextrorphan as calcium channel antagonists. *Brain Res.* **439**, 372–375.

Chatterton, J.E., M. Awobuluyi, L.S. Premkumar, H. Takahashi, M. Talantova, Y. Shin *et al.* (2002). Excitatory glycine receptors containing the NR3 family of NMDA receptor subunits. *Nature* **415**, 793–798.

Chenard, B.L. and F.S. Menniti (1999). Antagonists selective for NMDA receptors containing the NR2B subunit. *Curr. Pharm. Des.* **5**, 381–404.

Chiamulera, C., P. Albertini, E. Valerio, and A. Reggiani (1992). Activation of metabotropic receptors has a neuroprotective effect in a rodent model of focal ischaemia. *Eur. J. Pharmacol.* **216**, 335–336.

Choi, D.W., M. Maulucci-Gedde, and A.R. Kriegstein (1987). Glutamate neurotoxicity in cortical cell culture. *J. Neurosci.* **7**, 357–368.

Conn, P.J. and J.P. Pin (1997). Pharmacology and functions of metabotropic glutamate receptors. *Annu. Rev. Pharmacol. Toxicol.* **37**, 205–237.

Curtis, D.R., J.W. Phillis, and J.C. Watkins (1959). Chemical excitation of spinal neurons. *Nature* **183**, 611–612.

Curtis, D.R. and J.C. Watkins (1960). The excitation and depression of spinal neurons by structurally related amino acids. *J. Neurochem.* **6**, 117–141.

Curtis, D.R. and J.C. Watkins (1963). Acidic amino acids with strong excitatory actions on mammalian neurons. *J. Physiol.* **166**, 1–14.

Davies, J. and J.C. Watkins (1973). Antagonism of synaptic and amino acid induced excitation in the cuneate nucleus of the cat by HA-966. *Neuropharmacology* **12**, 637–640.

Davies, J. and J.C. Watkins (1977). Effect of magnesium ions on the responses of spinal neurones to excitatory amino acids and acetylcholine. *Brain Res.* **130**, 364–368.

Davies, J. and J.C. Watkins (1979). Selective antagonism of amino acid-induced and synaptic excitation in the cat spinal cord. *J. Physiol.* **297**, 621–635.

Davies, J. and J.C. Watkins (1985). Depressant actions of gamma-D-glutamylaminomethyl sulfonate (GAMS) on amino acid-induced and synaptic excitation in the cat spinal cord. *Brain Res.* **327**, 113–120.

Didier, M., M. Xu, S.A. Berman, and S. Bursztajn (1995). Differential expression and co-assembly of NMDA zeta 1 and epsilon subunits in the mouse cerebellum during postnatal development. *Neuroreport* **6**, 2255–2259.

Drugs, R.D. (1999). Licostinel. *ACEA 1021* **1**, 27–28.

Duggan, A.W. (1974). The differential sensitivity to L-glutamate and L-aspartate of spinal interneurones and Renshaw cells. *Exp. Brain Res.* **19**, 522–528.

Dyker, A.G. and K.R. Lees (1999). Safety and tolerability of GV150526 (a glycine site antagonist at the N-methyl-D-aspartate receptor) in patients with acute stroke. *Stroke* **30**, 986–992.

Eugster, C.H., G.F. Muller, and R. Good (1965). The active ingredients from Amanita muscaria: Ibotenic acid and muscazone. *Tetrahedron Lett.* **23**, 1813–1815.

Evans, R.H., A.A. Francis, and J.C. Watkins (1977). Selective antagonism by Mg^{2+} of amino acid-induced depolarization of spinal neurones. *Experientia* **33**, 489–491.

Evans, R.H., A.A. Francis, and J.C. Watkins (1978). Mg^{2+}-like selective antagonism of excitatory amino acid-induced responses by alpha, epsilon-diaminopimelic acid, D-alpha-aminoadipate and HA-966 in isolated spinal cord of frog and immature rat. *Brain Res.* **148**, 536–542.

Fillhard, J.A., K.A. Serpa, J.J. Kinsora, and L.T. Meltzer (2000). The effects of Cl-1041 in two tests of analgesia: Acetic acid-induced writhing test and formalin foot pad test. *Soc. Neurosci. Abstr.* **26**, 617.4.

Fisher, M. (1994). Cerestat (CNS 1102), a non-competitive NMDA antagonist, in ischemic stroke patients: Dose-escalating safety study. *Cerebrovasc. Dis.* **4**, 245.

Fisher, M. (1995). Potentially effective therapies for acute ischemic stroke. *Eur. Neurol.* **35**, 3–7.

Fischer, G., V. Mutel, G. Trube, P. Malherbe, J.N. Kew, E. Mohacsi *et al.* (1997). Ro 25-6981, a highly potent and selective blocker of N-methyl-D-aspartate receptors containing the NR2B subunit. Characterization in vitro. *J. Pharmacol. Exp. Ther.* **283**, 1285–1292.

Gann, D. (1977). Ventricular tachycardia in a patient with the "Chinese restaurant syndrome." *South Med. J.* **70**, 879–881.

Gerzon, K., E.V. Krumkalns, R.L. Brindle, F.J. Marshall, and M.A. Root (1963). The adamantyl group in medicinal agents. I. Hypoglycemic N-arylsulfonyl-N′-adamantylureas. *J. Med. Chem.* **6**, 760–763.

Gill, R. and D. Lodge (1997). Pharmacology of AMPA antagonists. *Int. Rev. Neurobiol.* **40**, 197–232.

Greenwood, R.S., Z. Fan, R. McHugh, and R. Meeker (2000). Inhibition of hippocampal kindling by metabotropic glutamate receptor antisense oligonucleotides. *Mol. Cell. Neurosci.* **16**, 233–243.

Greifenstein, F.E., M. DeVault, J. Yoshitake, and J.E. Gajewski (1958). A study of 1-aryl cyclohexylamine for anesthesia. *Anesth. Analg.* **37**, 283–294.

Grosset, D.G., K.W. Muir, and K.R. Lees (1995). Systemic and cerebral hemodynamic responses to the noncompetitive N-methyl-D-aspartate (NMDA) antagonist CNS 1102. *J. Cardiovasc. Pharmacol.* **25**, 705–709.

Grotta, J., W. Clark, B. Coull, C. Pettigrew, B. Mackay, L.B. Goldstein *et al.* (1995). Safety and tolerability of the glutamate antagonist CGS19755 (selfotel) in patients with ischemic stroke. *Stroke* **26**, 602–605.

Haldeman, S., R.D. Huffman, K.C. Marshall, and H. McLennan (1972). The antagonism of the glutamate-induced and synaptic excitations of thalamic neurones. *Brain Res.* **39**, 419–425.

Haldeman, S. and H. McLennan (1972). The antagonistic action of glutamic acid diethylester towards amino acid-induced and synaptic excitations of central neurones. *Brain Res.* **45**, 393–400.

Hayashi, T. (1952). A physiological study of epileptic seizures following cortical stimulation in animals and its application to human clinics. *Jpn. J. Physiol.* **3**, 46–64.

Helton, D.R., J.P. Tizzano, J.A. Monn, D.D. Schoepp, and M.J. Kallman (1998). Anxiolytic and side-effect profile of LY354740: A potent, highly selective, orally active agonist for group II metabotropic glutamate receptors. *J. Pharmacol. Exp. Ther.* **284**, 651–660.

Henrich-Noack, P., C.D. Hatton, and K.G. Reymann (1998). The mGlu receptor ligand (S)-4C3HPG protects neurons after global ischaemia in gerbils. *Neuroreport* **9**, 985–988.

Hesselink, M.B., H. Smolders, A.G. De Boer, D.D. Breimer, and W. Danysz (1999). Modifications of the behavioral profile of non-competitive NMDA receptor antagonists, memantine, amantadine and (+)MK-801 after chronic administration. *Behav. Pharmacol.* **10**, 85–98.

Hollmann, M., A. O'Shea-Greenfield, S.W. Rogers, and S. Heinemann (1989). Cloning by functional expression of a member of the glutamate receptor family. *Nature* **342**, 643–648.

Hurko, O. and F.S. Walsh (2000). Novel drug development for amyotrophic lateral sclerosis. *J. Neurol. Sci.* **180**, 21–28.

Iversen, L., E. Mulvihill, B. Haldeman, N.H. Diemer, F. Kaiser, M. Sheardown *et al.* (1994). Changes in metabotropic glutamate receptor mRNA levels following global ischemia: Increase of a putative presynaptic subtype (mGluR4) in highly vulnerable rat brain areas. *J. Neurochem.* **63**, 625–633

Johnston, G.A., D.R. Curtis, J. Davies, and R.M. McCulloch (1974). Spinal interneurone excitation by conformationally restricted analogues of L-glutamic acid. *Nature* **248**, 804–805.

Johnstone, M., V. Evans, and S. Baigel (1959). Sernyl (CI-395) in clinical anaesthesia. *Br. J. Anaesth.* **31**, 433–439.

Kawasaki-Yatsugi, S., S. Yatsugi, M. Takahashi, T. Toya, C. Ichiki, M. Shimizu-Sasamata *et al.* (1998). A novel AMPA receptor antagonist, YM872, reduces infarct size after middle cerebral artery occlusion in rats. *Brain Res.* **793**, 39–46.

Kawasaki-Yatsugi, S., C. Ichiki, S. Yatsugi, M. Takahashi, M. Shimizu-Sasamata, T. Yamaguchi *et al.* (2000). Neuroprotective effects of an AMPA receptor antagonist YM872 in a rat transient middle cerebral artery occlusion model. *Neuropharmacology* **39**, 211–217.

Kirk, C.J., N.L. Reddy, J.B. Fischer, T.C. Wolcott, A.G. Knapp, and R.N. McBurney (1994). In vitro neuroprotection by substituted guanidines with varying affinities for the N-methyl-D-aspartate receptor ionophore and for sigma sites. *J. Pharmacol. Exp. Ther.* **271**, 1080–1085.

Kohara, A., M. Okada, K. Ohno, S. Sakamoto, J. Shisikura, H. Inami *et al.* (1996). In vitro characterization of YM872: A selective, potent and highly water soluble AMPA receptor antagonist. *Soc. Neurosci. Abst.* **22**, 1528.

Konitsiotis, S., P.J. Blanchet, L. Verhagen, E. Lamers, and T.N. Chase (2000). AMPA receptor blockade improves levodopa-induced dyskinesia in MPTP monkeys. *Neurology* **54**, 1589–1595.

Kornhuber, J., J. Bormann, and S.A. Lipton (1993). Neuroprotective effects of memantine. *Neurology* **43**, 1054–1055.

Kretschmer, B.D., U. Kratzer, K. Breithecker, and M. Koch (1997). ACEA 1021, a glycine site antagonist with minor psychotomimetic and amnestic effects in rats. *Eur. J. Pharmacol.* **331**, 109–116.

Kristensen, J.D., B. Svensson, and T. Gordh Jr. (1992). The NMDA-receptor antagonist CPP abolishes neurogenic "wind-up pain" after intrathecal administration in humans. *Pain* **51**, 249–253.

Krogsgaard-Larsen, P., T. Honore, J.J. Hansen, D.R. Curtis, and D. Lodge (1980). New class of glutamate agonist structurally related to ibotenic acid. *Nature* **284**, 64–66.

Kuryatov, A., B. Laube, H. Betz, and J. Kuhse (1994). Mutational analysis of the glycine-binding site of the NMDA receptor: Structural similarity with bacterial amino acid-binding proteins. *Neuron* **12**, 1291–1300.

Lam, A.G., M.A. Soriano, J.A. Monn, D.D. Schoepp, D. Lodge, and J. McCulloch (1998). Effects of the selective metabotropic glutamate agonist LY354740 in a rat model of permanent ischaemia. *Neurosci. Lett.* **254**, 121–123.

Lea, T.J. and P.N. Usherwood (1973). The site of action of ibotenic acid and the identification of two populations of glutamate receptors on insect muscle-fibres. *Comp. Gen. Pharmacol.* **4**, 333–350.

Lees, K.R., A.G. Dyker, A. Sharma, G.A. Ford, M.E. Ardron, and D.G. Grosset (2001). Tolerability of the low-affinity, use-dependent NMDA antagonist AR-R15896AR in stroke patients: A dose-ranging study. *Stroke* **32**, 466–472.

Levay, D.E. (1999). Stroke Treatment with Ancrod Trial (STAT). *Stroke* **30**, 234.

Li, H. and A.M. Buchan (1993). Treatment with an AMPA antagonist 12 hours following severe normothermic forebrain ischemia prevents CA1 neuronal injury. *J. Cereb. Blood Flow Metab.* **13**, 933–939.

Lodge, D., P.M. Headley, and D.R. Curtis (1978). Selective antagonism by D-alpha-aminoadipate of amino acid and synaptic excitation of cat spinal neurons. *Brain Res.* **152**, 603–608.

McBurney, R.N. (1997). Development of the NMDA ion channel blocker, aptiganel hydrochloride, as a neuroprotective agent for acute CNS injury. In A.R. Green and A.J. Cross (eds), Neuroprotective Agents and Cerebral Ischemia. Academic Press, pp. 173–195.

McLennan, H., R.D. Huffman, and K.C. Marshall (1968). Patterns of excitation of thalamic neurones by amino-acids and by acetylcholine. *Nature* **219**, 387–388.

McLennan, H., and D. Lodge (1979). The antagonism of amino acid-induced excitation of spinal neurones in the cat. *Brain Res.* **169**, 83–90.

Merchant, R.E., M.R. Bullock, C.A. Carmack, A.K. Shah, K.D. Wilner, G. Ko *et al.* (1999). A double-blind, placebo-controlled study of the safety, tolerability and pharmacokinetics of CP-101,606 in patients with a mild or moderate traumatic brain injury. *Ann. NY Acad. Sci.* **890**, 42–50.

Miller, S., J.P. Kesslak, C. Romano, and C.W. Cotman (1992). Roles of metabotropic glutamate receptors in brain plasticity and pathology. *Ann. NY Acad. Sci.* **757**, 460–474.

Monaghan, D.T. and C.W. Cotman (1982). The distribution of [^{3}H]kainic acid binding sites in rat CNS as determined by autoradiography. *Brain Res.* **252**, 91–100.

Monyer, H., R. Sprengel, R. Schoepfer, A. Herb, M. Higuchi, H. Lomeli. *et al.* (1992). Heteromeric NMDA receptors: molecular and functional distinction of subtypes. *Science* **256**, 1217–1221.

Muir, K.W. and K.R. Lees (1995). Clinical experience with excitatory amino acid antagonist drugs. *Stroke* **26**, 503–513.

Muir, K.W. and K.R. Lees (1998). Dose optimization of intravenous magnesium sulfate after acute stroke. *Stroke* **29**, 918–923.

Nicoletti, F., V. Bruno, A. Copani, G. Casabona, and T. Knopfel (1996). Metabotropic glutamate receptors: A new target for the therapy of neurodegenerative disorders? *Trends Neurosci.* **19**, 267–271.

Nicoletti, F., V. Bruno, M. Catania, G. Battaglia, A. Copani, G. Barbagallo *et al.* (1999). Goup-I metabotropic glutamate receptors: Hypotheses to explain their dual role in neurotoxicity and neuroprotection. *Neuropharmacology* **38**, 1477–1484.

O'Hara, P.J., P.O. Sheppard, H. Thogersen, D. Venezia, B.A. Haldeman, V. McGrane *et al.* (1993). The ligand-binding domain in metabotropic glutamate receptors is related to bacterial periplasmic binding proteins. *Neuron* **11**, 41–52.

Olney, J.W., J. Labruyere, and M.T. Price (1989). Pathological changes induced in cerebrocortical neurons by phencyclidine and related drugs. *Science* **244**, 1360–1362.

Palmer, G.C., E.F. Cregan, P. Bialobok, S.G. Sydserff, T.J. Hudzik, and D.J. McCarthy (1999). The low-affinity, use-dependent NMDA receptor antagonist AR-R 15896AR. An update of progress in stroke. *Ann. NY Acad. Sci.* **890**, 406–420.

Palmer, G.C. (2001). Neuroprotection by NMDA receptor antagonists in a variety of neuropathologies. *Curr. Drug Targets* **2**, 241–271.

Park, C.K., D.G. Nehls, D.I. Graham, G.M. Teasdale, and J. McCulloch (1988). The glutamate antagonist MK-801 reduces focal ischemic brain damage in the rat. *Ann. Neurol.* **24**, 543–551.

Parsons, C.G., W. Danysz, and G. Quack (1999). Memantine is a clinically well tolerated N-methyl-D-aspartate (NMDA) receptor antagonist—a review of preclinical data. *Neuropharmacology* **38**, 735–767.

Popke, E.J., R.R. Allen, E.C. Pearson, T.G. Hammond, and M.G. Paule (2001a). Differential effects of two NMDA receptor antagonists on cognitive-behavioral development in nonhuman primates I. *Neurotox. Teratol.* **23**, 319–332.

Popke, E.J., R.R. Allen, E.C. Pearson, T.G. Hammond, and M.G. Paule (2001b). Differential effects of two NMDA receptor antagonists on cognitive-behavioral development in nonhuman primates II. *Neurotox. Teratol.* **23**, 333–347.

Popke, E.J., R. Patton, G.D. Newport, L.G. Rushing, C.M. Fogle, R.R. Allen *et al.* (2002). Assessing the potential toxicity of MK-801 and remacemide: Chronic exposure in juvenile rhesus monkeys. *Neurotox. Teratol.* **24**, 193–207.

Pratt, J., P. Jimonet, G.A. Bohme, A. Boireau, D. Damour, M.W. Debono *et al.* (2000). Synthesis and potent anticonvulsant activities of 4-oxo-imidazo[1,2-a]inden. *Bioorg. Med. Chem. Lett.* **10**, 2749–2754.

Rauca, C., P. Henrich-Noack, K. Schafer, V. Hollt, and K.G. Reymann (1998). (S)-4C3HPG reduces infarct size after focal cerebral ischemia. *Neuropharmacology* **37**, 1649–1652.

Schaumburg, H.H., R. Byck, R. Gerstl, and J.H. Mashman (1969). Monosodium L-glutamate: Its pharmacology and role in the Chinese restaurant syndrome. *Science* **163**, 826–828.

Schelkun, R.M., P.W. Yuen, K. Serpa, L.T. Meltzer, L.D. Wise, E.R. Whittemore *et al.* (2000). Subtype-selective N-methyl-D-aspartate receptor antagonists: Benzimidazalone and hydantoin as phenol replacements. *J. Med. Chem.* **43**, 1892–1897.

Sharp, F.R., M. Butman, K. Aardalen, J. Nickolenko, R. Nakki, S.M. Massa *et al.* (1994). Neuronal injury produced by NMDA antagonists can be detected using heat shock proteins and can be blocked with antipsychotics. *Psychopharmacol. Bull.* **30**, 555–560.

Sheng, M., J. Cummings, L.A. Roldan, Y.N. Jan, and L.Y. Jan (1994). Changing subunit composition of heteromeric NMDA receptors during development of rat cortex. *Nature* **368**, 144–147.

Shinozaki, H. and S. Konishi (1970). Actions of several anthelmintics and insecticides on rat cortical neurones. *Brain Res.* **24**, 368–371.

Shinozaki, H. and I. Shibuya (1974). A new potent excitant, quisqualic acid: Effects on crayfish neuromuscular junction. *Neuropharmacology* **13**, 665–672.

Sillevis Smitt, P., A. Kinoshita, B. De Leeuw, W. Moll, M. Coesmans, D. Jaarsma *et al.* (2000). Paraneoplastic cerebellar ataxia due to autoantibodies against a glutamate receptor. *N. Engl. J. Med.* **342**, 21–27.

Simmons, R.M., D.L. Li, K.H. Hoo, M. Deverill, P.L. Ornstein, and S. Iyengar (1998). Kainate GluR5 receptor subtype mediates the nociceptive response to formalin in the rat. *Neuropharmacology* **37**, 25–36.

Sirakawa, K., O. Aki, S. Tsushima, and K. Konishi (1966). Synthesis of ibotenic acid and 3-deoxyibotenic acid. *Chem. Pharm. Bull. (Tokyo)* **14**, 89–91.

Skerry, T.M. and P.G. Genever (2001). Glutamate signaling in non-neuronal tissues. *Trends Pharmacol. Sci.* **22**, 174–181.

Small, D.L., P. Morley, and A.M. Buchan (1999a). Glutamate receptor antagonists for the treatment of acute cerebral ischemia. In A. Shuaib and L.B. Goldstein (eds), *Management of Acute Stroke*. Marcel Dekker, New York, pp. 341–361.

Small, D.L., P. Morley, and A.M. Buchan (1999b). Preclinical studies of glutamate receptor antagonists for the treatment of cerebral ischemia. In L.P. Miller (ed.), *Stroke Therapy: Basic, Preclinical and Clinical Directions*. Wiley-Liss, New York, pp. 65–99.

Steinberg, G.K., M.A. Perez-Pinzon, C.M. Maier, G.H. Sun, E. Yoon, D.M. Kunis *et al.* (1994 July 20). CGS 19755: Correlation of in vitro neuroprotection, protection against experimental ischemia and CSF levels in cerebrovascular surgery patients. In *Proceedings of the 5th International Symposium on Pharmacology of Cerebral Ischemia*, Marburg, Germany. GLU 4. Abstract.

Synthelabo Reserche (1996, February 6) Press Release, Paris.

Takahashi, M., J.W. Ni, S. Kawasaki-Yatsugi, T. Toya, C. Ichiki, S.I. Yatsugi *et al.* (1998). Neuroprotective efficacy of YM872, an alpha-amino-3-hydroxy-5-methylisoxazole-4-propionic acid receptor antagonist, after permanent middle cerebral artery occlusion in rats. *J. Pharmacol. Exp. Ther.* **287**, 559–566.

Taniguchi, K., K. Shinjo, M. Mizutani, K. Shimada, T. Ishikawa, F.S. Menniti et al. (1997). Antinociceptive activity of CP-101,606, an NMDA receptor NR2B subunit antagonist. *Br. J. Pharmacol.* **122**, 809–812.

Taylor, C.P. (1994). Mechanism of action of new anti-epileptic drugs. In D. Chadwick (ed.), *New Trends in Epilepsy Management: The Role of Gabapentin*. Royal Society of Medicine, London, England, pp. 13–40.

Tomitaka, S., K. Hashimoto, N. Narita, A. Sakamoto, Y. Minabe, and A. Tamura (1996). Memantine induces heat shock protein HSP70 in the posterior cingulate cortex, retrosplenial cortex and dentate gyrus of rat brain. *Brain Res.* **740**, 1–5.

Tomitaka, S.I., K. Hashimoto, N. Narita, Y. Minabe, and A. Tamura (1997). Regionally different effects of scopolamine on NMDA antagonist-induced heat shock protein HSP70. *Brain Res.* **763**, 255–258.

Toms, N.J., P.J. Roberts, T.E. Salt, and P.C. Staton (1996). Latest eruptions in metabotropic glutamate receptors. *Trends Pharmacol. Sci.* **17**, 429–435.

Tortella, F.C., M. Pellicano, and N.G. Bowery (1989). Dextromethorphan and neuromodulation: old drug coughs up new activities. *Trends Pharmacol. Sci.* **10**, 501–507.

Umemura, K., K. Kondo, Y. Ikeda, Y. Teraya, H. Yoshida, M. Homma (1997). Pharmacokinetics and safety of the novel amino-3-hydroxy-5-methylisoxazole-4-propionate receptor antagonist YM90K in healthy men. *J. Clin. Pharmacol.* **37**, 719–727.

Wafford, K.A., C.J. Bain, B. Le Bourdelles, P.J. Whiting, and J.A. Kemp (1993). Preferential co-assembly of recombinant NMDA receptors composed of three different subunits. *Neuroreport* **4**, 1347–1349.

Warner, D.S., H. Martin, P. Ludwig, A. McAllister, J.F.W. Keana, and E. Weber (1995). In vivo models of cerebral ischemia: Effects of parenterally administered NMDA receptor glycine site antagonists. *J. Cereb. Blood Flow Metab.* **15**, 188–196.

Wesemann, W., K.H. Sontag, and J. Maj (1983). Pharmacodynamics and pharmacokinetics of memantine. *Arzneimittelforschung* **33**, 1122–1134.

Wright, J.L., T.F. Gregory, P.A. Boxer, L.T. Meltzer, K.A. Serpa, and L.D. Wise (1999). Discovery of subtype-selective NMDA receptor ligands: 4-benzyl-1-piperidinylalkynylpyrroles, pyrazoles and imidazoles as NR1A/2B antagonists. *Bioorg. Med. Chem. Lett.* **9**, 2815–2818.

Wright, J.L., T.F. Gregory, S.R. Kesten, P.A. Boxer, K.A. Serpa, L.T. Meltzer *et al.* (2000). Subtype-selective N-methyl-D-aspartate receptor antagonists: Synthesis and biological evaluation of 1-(heteroarylalkynyl)-4-benzylpiperidines. *J. Med. Chem.* **43**, 3408–3419.

Xue, D., Z.G. Huang, K. Barnes, H.J. Lesiuk, K.E. Smith, and A.M. Buchan (1994). Delayed treatment with AMPA, but not NMDA, antagonists reduces neocortical infarction. *J. Cereb. Blood Flow Metab.* **14**, 251–261.

Xue, D., B.K. Miyazaki, G. Daniell, X. Cheng, and R.M. Woodward (1999). Co-administration of probenecid with licostinel increases the steady state plasma level and reduces the minimum effective dose of neuroprotection in a rat model of transient focal ischemia. *Soc. Neurosci. Abstr.* **25**, 234.6.

Zhou, Z.L., S.X. Cai, E.R. Whittemore, C.S. Konkoy, S.A. Espitia, M. Tran *et al.* (1999). 4-Hydroxy-1-[2-(4-hydroxyphenoxy)ethyl]-4-(4-methylbenzyl)piperidine: A novel, potent, and selective NR1/2B NMDA receptor antagonist. *J. Med. Chem.* **42**, 2993–3000.

3

Expression of Non-Organelle Glutamate Transporters to Support Peripheral Tissue Function

James C. Matthews

1. Introduction

The goal of this chapter is to highlight some critical physiological relationships that exist among glutamate transport systems (biochemically defined activities), glutamate transporters (molecularly identified proteins capable of the biochemically defined activities), and the metabolism of glutamate by several peripheral (noncentral nervous system) tissues. Collectively, these processes account for much of the whole-body flux of nitrogen and carbon. Presented is (a) an overview of the importance of glutamate metabolism to the function of peripheral tissues, (b) molecular and functional characteristics, and expression patterns, of transport proteins capable of glutamate transport, (c) a detailed examination of how glutamate transport activities and proteins support the function of hepatic (mature and fetal), placental, white adipose, and muscle tissues, and (d) a listing of underexplored areas of research that this author thinks are important to more fully understand the integrated role of glutamate transport capacity and peripheral tissue function.

This chapter does not review the vastly greater body of literature that exists about how the expression and function of glutamate transporters support central nervous system or renal tissue metabolism, nor about research regarding the importance of glutamate transport capacity to pulmonary tissue metabolism. For convenience, the "L" form of amino acids is implied, unless indicated otherwise.

2. Overview: Glutamate Metabolism is a Critical Function of many Peripheral Tissues

L-glutamate (glutamate) is extensively metabolized to support whole-animal energy and nitrogen homeostasis (Heitmann and Bergman, 1981; Wu, 1998; Nissim, 1999). The predominant metabolic role/fate of glutamate is tissue specific. Glutamate can serve as a source

---

James C. Matthews • Department of Animal Sciences, University of Kentucky, Lexington.

Glutamate Receptors in Peripheral Tissue, edited by Santokh Gill and Olga Pulido.
Kluwer Academic / Plenum Publishers, New York, 2005.

of oxidizable fuel, precursor for amino acid synthesis, carbon receptor for alpha-amino- and ammonia-nitrogen derived from the catabolism of other amino acids, or substrate for *de novo* protein synthesis. For example, in the intestinal epithelia, dietary and enteric glutamate serve as important sources of carbons for oxidation-derived metabolic energy, and for endogenous synthesis of alanine and other amino acids (Wu, 1998). In the liver, glutamate serves as a precursor for glutamine (which is exported to the plasma), gluconeogenic carbons, and incorporation into *de novo* protein synthesis (Haussinger and Gerok, 1983). In kidney tissue, plasma and kidney-synthesized glutamate acts to regulate ammoniagenesis, serves as a precursor for incorporation into glutathione and extracellular matrix protein, and serves as a source of carbons for mitochondrial oxidation-derived energy (Welbourne and Matthews, 1999). In placental tissue, both fetal- (primarily) and maternal-derived glutamate serve as sources of oxidizable fuel. Consequently, more maternal glucose is available for fetal use, while neurotoxic levels of glutamate are removed from the fetal circulation (Moores *et al.*, 1994; Vaughn *et al.*, 1995). In skeletal muscle (Biolo *et al.*, 1995; Vesli, 2002) and subcutaneous fat (Kowalski and Watford, 1994), the production of glutamine from plasma glutamate is thought to constitute an important route for whole-body ammonia recovery and detoxification.

Due to the extensive utilization of dietary glutamate by the intestine (Tagari and Bergman, 1978; Reeds *et al.*, 1996; Wu, 1998), the principal source of circulating glutamate is the liver. More specifically, under normal physiological conditions, glutamate produced from the deamination of glutamine by liver-type glutaminase, and subsequently exported into sinusoidal blood by periportal vein hepatocytes, is thought to be the principal source of glutamate for peripheral tissue use. In contrast to this net export of glutamate by the liver, net users of glutamate include the lung, subcutaneous adipose, skeletal muscle, and intestine, whereas the brain and kidney generate high fluxes of glutamate but without any net utilization or production (Figure 3.1; Hediger and Welbourne, 1999; Patterson *et al.*, 2002). It is of importance to note that tissues which are net users of glutamate are also either net exporters

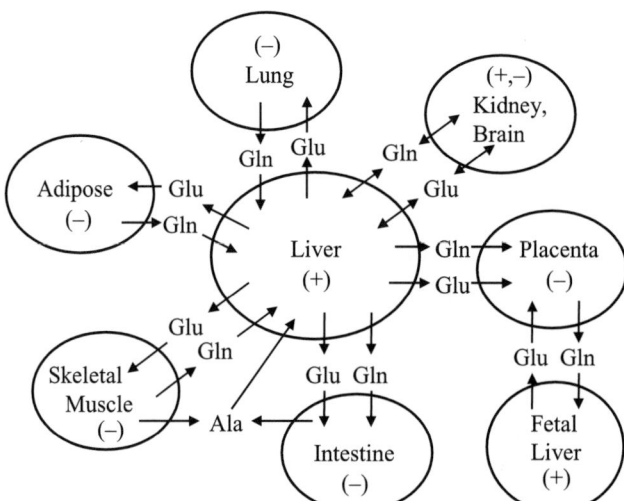

Figure 3.1. Diagrammatic representation of the net utilization ($-$, $+$) of plasma glutamate (Glu) by gut and peripheral tissues in relationship to glutamine (Gln) and alanine (Ala) use and/or production, as described principally by Hediger and Welbourne (1999), Patterson *et al.* (2002), and the text.

of glutamine (lung, adipose, skeletal muscle) and/or alanine (skeletal muscle, intestine). This understanding plus the observations that the liver, brain, and kidney possess well-defined glutamate/glutamine cycles (Nissim, 1999), and that a decreased glutamate exchange capacity appears responsible for elevated plasma glutamate levels of patients suffering from age- and disease-induced cachexia of skeletal muscle (Hack *et al.*, 1996), emphasize the importance of transport capacity to normal peripheral tissue metabolism of glutamate (Bronson, 2000; van Milgen, 2002).

3. Molecular and Functional Characterization of Proteins Capable of Glutamate Transport

Given the central role of glutamate to support tissue and whole-body growth and maintenance, plus the understanding that plasma transport capacity may well limit glutamate metabolism (Low *et al.*, 1994; Bode, 2001; Welbourne and Nissim, 2001) knowledge regarding the cellular distribution and functional properties of biochemically and molecularly characterized glutamate transport systems is critical. In mammals, mediated absorption of anionic amino acids occurs by Na^+-dependent and Na^+-independent biochemical activities. High-affinity and concentrative uptake of anionic amino acids principally occurs by Na^+-dependent X_{AG}^- and X_A^- transport systems, which recognize only anionic amino acids and some of their analogues. System X_{AG}^- activity is defined as the high affinity, Na^+-dependent, K^+ coupled, D-aspartate-inhibitable transport of L-glutamate or L-aspartate. System X_{AG}^- mediates the uniport uptake of Na^+ and L-glutamate, D-aspartate, or L-aspartate. In contrast, system X_A^- does not transport glutamate or its analogues (as reviewed by Matsuo *et al.*, 2002). In addition, systems ASC and B^O also transport L-glutamate and L-aspartate, but only in acidic (pH < 5.5) environments. Na^+-independent uptake of anionic amino acids is mediated by system x_C^- and x_G^-. System x_C^- transport is an obligate exchange activity that typically mediates the exchange of L-glutamate for L-cystine, but which also recognizes D-glutamate. In contrast, system x_G^- only transports L-glutamate. Within the last 10 years, the molecular identification and functional properties of five system X_{AG}^- proteins, one system ASC proteins, two system B^0, one system x_C^-, and one protein (AGT1) that possesses a previously undescribed Na^+-independent uptake of L-glutamate and L-aspartate activity, have been characterized in mammals (Table 3.1). These tremendous accomplishments have allowed the functional activities of glutamate transporters to be further delineated, and facilitated the ability to study their patterns of regulation and expression. Consequently, our understanding of how critical glutamate transport is to tissue-specific metabolism and whole-animal nutrient homeostasis continues to expand.

3.1. System X_{AG}^- Transporters

The nonhuman system X_{AG}^- transporter orthologs are known as GLAST1, GLT-1, EAAC1, EAAT4, and EAAT5, whereas the human orthologs are referred to as *excitatory amino acid transport* (EAAT1-5), respectively. The expression and function of these high-affinity transporters has been extensively studied in the brain to understand how their presence facilitates the use of glutamate as a neurotransmitter (thoroughly reviewed in Danbolt, 2001). Functionally, the process for transport by GLAST1, GLT-1, or EAAC1 is thought to involve extracellular binding and translocation of one amino acid and three Na^+, with

Table 3.1. Functional Properties of Cloned Proteins Capable of Glutamate Transport

Clone name	Alternate name	Transport activity	Substrate[a]	Affinity	Co-substrate coupling	References
Na[+]-dependent						
EAAT1	GluT, GLAST	X^-_{AG}	D-L-Asp, L-Glu	μM	Na^-_{in}, K^+_{out}, $OH^-/HCO^-_{3\,out}$	Storck et al. (1992)
EAAT2	GLT, GLAST2, GLTR	X^-_{AG}	D-L-Asp, L-Glu	μM	Na^+_{in}, K^+_{out}, $OH^-/HCO^-_{3\,out}$	Pines et al. (1992)
EAAT3	EAAC1	X^-_{AG}	D-L-Asp, L-Glu	μM	Na^+_{in}, K^+_{out}, $OH^-/HCO^-_{3\,out}$	Kanai and Hediger (1992)
EAAT4		X^-_{AG}	D-L-Asp, L-Glu	μM	Na^+_{in}, K^+_{out}, $OH^-/HCO^-_{3\,out}$	Fairman et al. (1995)
EAAT5		X^-_{AG}	D-L-Asp, L-Glu	μM	Na^+_{in}, K^+_{out}, $OH^-/HCO^-_{3\,out}$	Arriza et al. (1997)
ASCT1[b]	SATT	ASC	L-Asp, L-Glu, when pH < 5.5	mM	$Na^+_{in} AA_{in}$, AA_{out}	Arriza et al. (1993); Zerangue and Kavanaugh (1996a)
ASCT2[b]	AAAT	B^0	L-Asp, L-Glu, when pH < 5.5	mM	$Na^+_{in} AA_{in}$, AA_{out}	Liao and Lane (1995); Utsunomiya-Tate et al. (1996)
ATB[0 b,c]		B^0	L-Asp, L-Glu, when pH < 5.5		$Na^+_{in} AA_{in}$, AA_{out}	Kekuda et al. (1997); Avissar et al. (2001)
Na[+]-independent						
XCT[d]	4F2-lc4	x^-_C	L-CssC[e], L-Glu, L-Asp	μM	$AA1_{CssC}$-$AA2_{Glu}$	Sato et al. (1999)
AGT1[f]		Novel, unnamed	L-Asp, L-Glu; L-cysteine (?)			Bridges et al. (2001); Matsuo et al. (2002)

Notes
[a] ".," denotes physiologically significant differences in degree of substrate affinity.
[b] Above pH 6, this protein transports neutral amino acids, which is its predominant physiological function.
[c] This is the intestinal form of ATB[0], which recognizes anionic amino acids. The JAR (Kekuda et al., 1996) and NBL-1 (Pollard et al., 2002) cell ATB[0] orthologs do not.
[d] Member of the glycoprotein-associated amino acid transporter family; associates with the 4F2hc glycoprotein.
[e] "CssC" = L-cystine.
[f] Member of the glycoprotein-associated amino acid transporter family that associates with nonidentified glycoproteins (Chairoungdua et al., 2001).

reorientation of the transporter to the extracellular face of the membrane being driven by the intracellular-to-extracellular countertransport of one K^+. Also, either one OH^- is counter-transported or one H^+ cotransported. In addition to these ion fluxes associated with GLAST1, GLT-1, and EAAC1, EAAT4 and EAAT5 isoforms have a large inward chloride ion flux associated with their function, which may aid in the re-establishment of membrane potential by influencing cellular chloride permeability. Currently, the functional activities of EAAT4 and EAAT5 are thought to be most accurately described as that of glutamate-gated chloride channels (Arriza *et al.*, 1997). As is further discussed in Section 4.4., an important feature of the system X_{AG}^- transporters is that they are capable of reverse transport when the normally high intracellular-to-extracellular K^+ gradient is reduced.

Except for EAAT5, all the system X_{AG}^- transporters are highly expressed by brain tissue, but with distinct distribution patterns. GLAST1 and GLT-1 represent glial-specific glutamate transporters, whereas EAAC1 and EAAT4 represent neuron-specific activities. Accordingly, GLAST1 and GLT-1 are involved in the neurotransmission whereas EAAC1 and EAAT4 are thought to be responsible for more general metabolic functions of brain tissues. EAAC1 is the most widely distributed of the glutamate transporters outside of the brain. For example, in the rabbit, EAAC1 mRNA is expressed by the duodenum, jejunum, ileum, heart, liver, and placenta, but not the colon, lung, or spleen (Kanai and Hediger, 1992). In sheep and cattle, EAAC1 and GLT-1 protein is present in homogenates of forestomach (rumen, oma-sum) and small (duodenum, jejunum, ileum) and large (cecum, colon) intestinal epithelia, liver, kidney, and pancreatic tissues (Howell *et al.*, 2001). GLAST1 mRNA is also fairly widely expressed outside of the brain, having been reported in heart, lung, skeletal muscle, placenta, and retinal Müller and astrocyte cells. In cattle and sheep, outside of the brain, GLAST1 protein is not detectable by immunoblot analysis in tissues that express EAAC1 and GLT-1, except for the pancreas (Howell *et al.*, 2001). In contrast to the rather wide expression of GLAST1, GLT-1, and EAAC1, EAAT4 expression appears limited to brain and placenta. EAAT5 also has a restricted pattern of expression, being expressed primarily in retinal tissue, where it is thought to play a vital role in the retinal light response. However, expression of EAAT5 mRNA has been detected in liver and skeletal muscle tissue, but at levels about 20-fold less than for the retina (Arriza *et al.*, 1997).

Besides the brain, the one tissue that has been shown to express all the non-retinal mammalian system X_{AG}^- transporters is the placenta. In the placenta, system X_{AG}^- activity is thought to be responsible for the absorption of maternal- and fetal-derived glutamate (and aspartate), thereby providing trophoblasts with an important source of oxidizable fuel in the form of glutamate, generated by fetal nitrogen assimilation from glutamine (Matthews *et al.*, 1998a; Battaglia, 2000). Therefore, to the extent that the placenta can meet its demand for a readily oxidizable fuel by glutamate, the demand for maternal glucose for placental oxidation may be moderated. Transporters capable of system X_{AG}^- activity (EAAT1–3) have been identified on both the maternal- and fetal-facing membranes of the syncytial trophoblast of rat placenta (Matthews *et al.*, 1998b), whereas EAAT4 has been found in the visceral endothelium of late-term rat placental yolk sac (Novak and Beveridge, 2000). In the chorioalloantoic placenta, the pattern of transporter expression changes, depending on the day of gestation. Accordingly, it is hypothesized (Matthews *et al.*, 1998b) that the increase in GLAST1 and EAAC1 content on the apical membrane, and the increase in EAAC1 and decrease in GLT-1 content on the basal membrane provide an increased capacity to absorb glutamate (and aspartate) from the maternal and fetal circulations during the end of gestation. Additionally, the expression of GLAST1, GLT-1, and EAAC1 protein by maternal decidual, giant, and spongiotrophoblast cells also was altered with gestation. Overall, the expression of these glutamate transport proteins in placental cells was altered in a manner consistent with increasing the capacity for greater glucose passage to the fetus, during the period when fetal energy demands are greatest.

3.2. System ASC/B^0 Transporters

3.2.1. System ASC Transporters

As noted above, Na$^+$-dependent system ASC transports *a*lanine, *s*erine, *c*ysteine, and other neutral α-amino acids. ASCT1 cDNA, originally cloned from human brain tissue, encodes for system ASC-like activity and shares nearly 40% amino acid sequence identity with the anionic EAAT transporters. Similar to the EAATs, ASCT1 has a counter-chloride transport activity. A second cDNA (ASCT2) has been cloned from mouse and human tissues that encodes a system ASC-like activity. Human ASCT2 shares 61% identity with human ASCT1 and displays a broader pattern of substrate recognition but similar transport function with ASCT1. The ASCT1 and ASCT2 cDNAs encode open reading frames of 532 and 541 amino acids, respectively. As with the EAAT family, hydrophobicity modeling of the ASCT transporter sequences predicts six well-defined transmembrane sequences near the N-terminal of the protein and additional, less well-defined, hydrophobic stretches near the C-terminal. Although system ASC normally recognizes neutral amino acids, glutamate and aspartate are transported at pH levels of 5.5 or lower. This pH-dependent recognition and transport pattern is displayed by ASCT1, ASCT2, and ATB^0. Although the ability to transport anionic amino acids by ASCT1, ASCT2, and ATB^0 at low pH appears to be an example of similarities in structure/function that exist between the systems X$^-_{AG}$ and ASCT/B^0 transporter families (Palacin *et al.*, 1998), an important distinction between these member groups of the Glutamate : Na$^+$ symporter family (Saier, 1999) is that ASCT1, ASCT2, and ATB^0 are obligate amino acid exchangers.

System ASC activity has been identified in nearly every mammalian tissue tested and can account for the majority of uptake for several neutral amino acids in a number of cell types. For many years it was thought that Na$^+$-dependent system ASC functioned as a concentrative transporter. However, based on functional expression studies of ASCT1 in defolliculated *Xenopus laevis* oocytes, it is now thought that ASCT1-mediated system ASC activity is that of an obligatory exchanger (Zerangue and Kavanaugh, 1996a). As such, one extracellular amino acid would be exchanged for one intracellular transporter, resulting in no net accumulation of amino acids. Theoretically, however, because of potential differences in extracellular and intracellular binding affinities, the concentration of a given amino acid in the cytosol could be achieved at the expense of others being transported out of the cell. Whether glutamate can be transported in either direction, in physiologically significant amounts, and whether there is a "preferred" counter substrate, remains to be determined.

ASCT1 mRNA is expressed highly in the brain, skeletal muscle, and pancreas, moderately in the heart, and very weakly in liver, lung, placenta, and kidney human tissues (Arriza *et al.*, 1993). In mice, expression of ASCT2 mRNA has been identified by adipose, skeletal muscle, placenta, kidney, large intestine, lung, pancreas, and testes (Liao and Lane, 1995; Utsunomiya-Tate *et al.*, 1996).

3.2.2. System B^0 Transporters

That ASCT1 and ASCT2 have not been reported in the small intestine, but that system ASC-like activity is high in intestinal tissue, suggested that other members of this family of neutral amino acid transporters had yet to be identified. Recently, several cDNAs (ASCT2, ATB^0) have been isolated (Kekuda *et al.*, 1996, 1997; Pollard *et al.*, 2002) that encode for the system B^0 activity, an activity similar to system ASC. Specifically, system B^0 (originally named as system NBB, for *N*eutral *B*rush *B*order, Stevens *et al.*, 1982) mediates the Na^+-coupled uptake of a broad spectrum of zwitterionic amino acids. However, the intestinal iso-type of B^0 family members appears capable of anionic amino acid transport, if the pH is below 5.5 (Kekuda *et al.*, 1997). Recently, immunohistochemical analysis has demonstrated that ATB^0 is localized to the brush border membrane in proximal tubule cells and enterocytes (Avissar *et al.*, 2001), an observation that mirrors the previously characterized pattern of bio-chemically defined system B^0 activity (Stevens *et al.*, 1982; Lynch and McGivan, 1987).

3.3. Na^+-Independent Glutamate Transporters

System x_C^- is the most widely expressed Na^+-independent transport activity for anionic amino acids. Although system x_C^- activity is typically low, this activity mediates the highly specific exchange of L-glutamate or L-aspartate for the anionic form of L-cystine (Bannai and Kitamura, 1980, 1981). Although system x_C^- can import or export these amino acids, the pre-dominant function appears to be the counterexchange of extracellular L-cystine for intracel-lular L-glutamate, at a 1:1 molar ratio (Bannai and Kitamura, 1980). Because L-cystine is rapidly reduced in the cytosol, one cycle of system x_C^- activity yields two molecules of L-cysteine. Because cysteine is thought to be the limiting reagent for glutathione synthesis (Sato *et al.*, 1999; Bridges *et al.*, 2001), it follows that system x_C^- activity greatly influences the redox potential of a cell. Indeed, system x_C^- activity is upregulated in response to oxida-tive stress (Bridges *et al.*, 2001).

Two species orthologs of a protein (xCT) capable of system x_C^- activity have been identified. The mouse xCT transporter (Sato *et al.*, 1999) is predicted to possess 12-membrane-spanning domains, but is a member of the glycoprotein-associated amino acid transporter family (Verrey *et al.*, 2000), associating with the 4F2hc glycoprotein. The dual expression of xCT and 4F2hc in *Xenopus laevis* oocytes results in biochemically defined system x_C^- activity, displaying import K_m values of 81 and 160 µM for L-cystine and L-glutamate, respectively (Sato *et al.*, 1999). Similarly, human xCT (Bassi *et al.*, 2001) dis-plays K_m values of 43 and 93 µM for L-cystine and L-glutamate, respectively. Although K_m values were not reported for L-aspartate uptake, the velocity of L-aspartate uptake was about 25% that for L-glutamate, for both mouse and human xCT orthologs.

Northern blot analyses have identified the expression of three xCT mRNA transcripts (2.5, 3.5, and 12 kb) by activated mouse macrophages, but expression of only the 12 kb tran-script by mouse and human brain tissue (Sato *et al.*, 1999; Bassi *et al.*, 2001). With the more sensitive RT-PCR analysis, however, expression of xCT mRNA was first reported for human ARPE-19 (retinal pigment epithelial) cells after Southern detection (Bridges *et al.*, 2001). Subsequently, using ethidium bromide visualization techniques (Bassi *et al.*, 2001), expres-sion of xCT mRNA was detected in mouse kidney and intestine, and human intestinal epithelium, undifferentiated and differentiated Caco-2 cells, HEK293 (embryonic kidney) and HepG2 (hepatocarcinoma) cells, and pancreatic islets. Significantly, these analyses did not detect expression of xCT mRNA in adult mouse or human liver tissue. As indicated by Bassi *et al.* (2001), the overall expression pattern of xCT mRNA is consistent with reported system x_C^- in various cell types and lines. Therefore, the identification of xCT mRNA in pancreatic islet cells, the detection of EAAC1, GLT-1, and GLAST1 protein in pancreatic tissue (Howell *et al.*, 2001), and the observation that mitochondrial glutamate levels may "potentiate" the release of insulin granules (Macchler and Wolheim, 1999) collectively indicate that plasma glutamate concentrations act as a signal for acute whole-body nutritional status.

Recently, a protein (AGT1) capable of Na^+-independent uptake of L-aspartate and L-glutamate has been identified in the kidney (Matsuo *et al.*, 2002), which displays a functional activity that does not match any previous biochemical activity. In contrast to the strong recognition and transport of L-cystine, relatively weak transport of L-aspartate, and dependence on association with 4F2hc exhibited by xCT-mediated system x_C^- activity, AGT1 does not recognize L-cystine, displays a K_m value of about 22 μM for L-aspartate and transports L-aspartate with a velocity that is at least as great as for L-glutamate, when expressed as a chimera of either 4F2hc or rBAT. Importantly, when coexpressed with either 4F2hc or rBAT, AGT1 does not possess a functional activity. These functional activity assays, plus denaturing vs non-denaturing immunoblot assays, indicate that AGT1 likely is the second member of what appears to be an emerging family of transporters that associate with a glycoprotein(s) other than 4F2hc or rBAT (Chairoungdua *et al.*, 2001; Matsuo *et al.*, 2002). In terms of tissue distribution and putative physiological significance, AGT1 mRNA and protein have only been reported in the kidney, as compared to other evaluated epithelial tissues (Matsuo *et al.*, 2002). In the kidney, immunohistochemical analysis localized AGT1 to the basolateral membrane of the proximal straight tubules, in a distribution pattern similar to that for EAAC1. Thus, because EAAC1 localizes to the apical membrane, the obvious putative role of AGT1 is to mediate the transport of EAAC1-resorbed anionic amino acids across the basolateral membrane and into renal blood.

4. Glutamate Transport Supports the Function of many Peripheral Tissues

4.1. Differential (Zonal) Expression of Glutamate Metabolizing Enzymes and Transporters along the Hepatic Acinus

4.1.1. Mature Liver Model

A primary function of the mammalian liver is to coordinate whole-body energy and nitrogen metabolism. Hepatic transport and intermediary metabolism of glutamate are critical to these processes as glutamate is a central substrate for hepatic ureagenesis, gluconeogenesis, glutathione production, *de novo* protein synthesis, and nitrogen shuttling via glutamine (Meijer *et al.*, 1990; Watford, 2000). An important aspect of glutamate metabolism in the liver is its role in the hepatic "Intercellular Glutamine Cycle" (Figure 3.2), defined by Haussinger and colleagues (Haussinger *et al.*, 1985; Haussinger, 1990). Critical components of this cycle include periportal zone removal and disposal of portal ammonia through urea synthesis by periportal zone hepatocytes (hepatocytes that reside along the acinus from the terminal portal venules to deep within the acinus), periportal zone glutamine degradation, and perivenous zone synthesis of glutamine by hepatocytes that form rings of cells about one to three cells

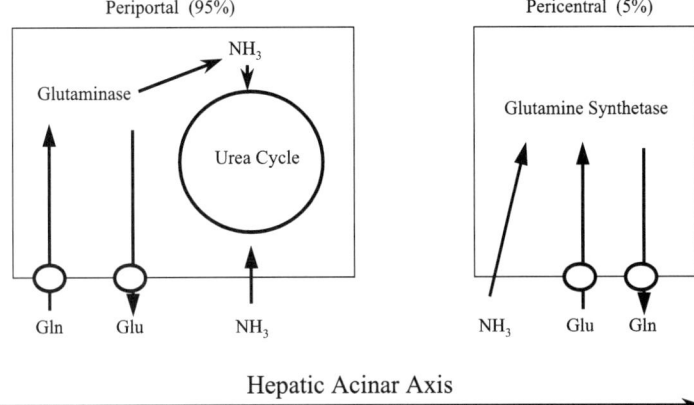

Hepatic Acinar Axis

Figure 3.2. Diagrammatic representation of the "Intercellular Glutamine Cycle" described by Haussinger *et al.*, (1985) and Haussinger (1990), but with emphasis on site of glutamine (Gln) and glutamate (Glu) transport events. As with the expression of glutaminase and glutamine synthetase, the net influx and efflux of glutamine and glutamate are thought to differ between periportal and pericentral hepatocyte populations. Both glutamine and glutamate influx and efflux events are thought to be mediated by specific transport proteins, whereas absorption of ammonia is thought to occur by transmembrane diffusion.

deep around terminal hepatic venules. Thus, this degradation and re-synthesis of glutamine results from the differentially expression of glutamine- and glutamate-metabolizing enzymes and transporters along the hepatic acinus.

Specifically, glutamine, arriving from the peripheral tissues via the portal vein, is efficiently absorbed by periportal hepatocytes by the hepatic system N activity (Kilberg *et al.*, 1980) and deaminated by glutaminase (liver-type) to release ammonia and glutamate. The released ammonia is incorporated into carbamoyl phosphate (by carbamoyl phosphate synthetase) for ureagenesis, whereas glutamate is available for export, conversion to alpha-ketoglutarate as an anapleurotic reaction to replenish the citric acid content, used for gluconeogenesis, used for protein synthesis, or transported back into the sinusoids, depending on the whole-body energy and/or ammonia status, and blood pH. Once released into the sinusoidal blood, glutamate is available for absorption by the "downstream" pericentral zone hepatocytes, and/or use by peripheral tissues, after exiting the liver in the hepatic vein. If absorbed by pericentral zone hepatocytes, glutamate and the "scavenged" sinusoidal ammonia (that which escapes incorporation into urea by periportal hepatocytes), can be incorporated into glutamine by glutamine synthetase.

Since periportal glutaminase and pericentral glutamine synthetase are active at the same time, an intercellular glutamine cycle results. The total flux of glutamine through this cycle is sensitive to portal vein pH and concentrations of ammonia and glutamine (Haussinger *et al.*, 1985). An important feature of the hepatic glutamine cycle is the capacity to modulate blood pH. That is, when the portal blood is relatively alkaline, the production of urea by periportal hepatocytes results in loss of blood carbonate and, hence, a drop in blood pH. In contrast, when blood is relatively acidic, pericentral synthesis of glutamine allows removal of blood ammonia without loss of carbonate (Haussinger *et al.*, 1985, 1989).

As noted above, an important feature of overall hepatic glutamate metabolism is that the liver is a net exporter of glutamate. A primary route for glutamate production is through the activity of glutaminase, which is restricted to the periportal zone (Watford, 1993).

Because the pericentral hepatocytes are thought to be responsible for more than half of the net portal export of glutamate from the liver (Haussinger *et al.*, 1989), it follows that periportal hepatocytes likely provide the remaining net exported glutamate. Accordingly, pericentral and periportal hepatocyte populations quantitatively contribute about equally to provisioning of peripheral tissues with glutamate. However, given that in mature (adult-like) livers the pericentral hepatocytes constitute about 5–10% of the total hepatocyte population, the proportional capacity of pericentral hepatocytes to export glutamate appears to be about five times more than for periportal hepatocytes. The relatively lower glutamate export capacity of periportal hepatocytes may well reflect their relatively higher gluconeogenic and ureogenic metabolic capacities (Meijer *et al.*, 1990; Kilberg and Haussinger, 1992). That is, the lesser proportional ability of periportal hepatocytes to export glutamate likely results from their greater fractional use of glutamate to support amino acid intermediary metabolism (Watford, 2000).

The characteristic expression of glutaminase and carbamoyl phosphate synthetase by periportal hepatocytes, and the concomitant expression of glutamine synthetase by pericentral hepatocytes, are not known to overlap, at least along the hepatic acinii of rats, pigs, and humans (Moorman *et al.*, 1989, 1990; Wagenaar *et al.*, 1994). Consequently, this zonal distribution of metabolic capacities is often used as a hallmark of "adult/mature" liver function. From this understanding, at least two predominant questions arise with regard to the need for, and presence of, glutamate transport capacities to support the different metabolic capacities of periportal and perivenous hepatocyte populations. First, how does the glutamate transporter capacity differ (if at all) along the hepatic acinus? Second, where are the transport capacities and proteins localized, in terms of the hepatic acinus and membrane subdomains? That is, do the basal (sinusoidal-facing) and/or canalicular (bile-facing) membranes of pericentral and/or periportal hepatocytes possess the glutamate transport capacity?

As just discussed, hepatocyte populations differ in their metabolism of glutamate. In parallel, periportal and perivenous hepatocytes likely also differ in their needs for glutamate transport capacity. The understanding that hepatocytes possess Na^+-dependent and/or Na^+-independent anionic amino acid transport came from studies using isolated, nonpolarized, rat hepatocytes (Gaseously *et al.*, 1981; Gebhardt and Meckel, 1983). The understanding that perivenous hepatocytes are responsible for the majority of the net uptake of glutamate in the liver came from studies (Taylor and Rennie, 1987; Haussinger *et al.*, 1989) with rats that used antegrade vs retrograde perfusion of digitonin to selectively kill periportal or pericentral hepatocytes, respectively. Subsequently, a histoautoradiographic study (Stoll *et al.*, 1991) of perfused glutamate, aspartate, and alpha-ketoglutarate confirmed that the hepatic cell-type that synthesizes glutamine also possesses the greatest, and almost exclusive, ability to accumulate glutamate. Although these studies clearly demonstrated the dominant ability of pericentral hepatocytes to accumulate glutamate, this research also indicated that periportal hepatocytes can absorb glutamate, most probably by an exchange transport activity, and therefore, probably system x_C^- (Burger *et al.*, 1989; Haussinger *et al.*, 1989).

As hepatocytes are polarized cells, if the importance of glutamate transport capacity to hepatic function is to be understood, then it is important to know if glutamate transport capacities are localized in sinusoidal (basolateral) and/or canalicular (apical) membranes of hepatocytes. Localization of transport activities in the basolateral membrane begets the ability to extract and excrete glutamate into the sinusoidal blood, whereas apically localized activities allow flux across the canalicular membrane and into bile. Using isolated membrane vesicles, system X_{AG}^- activity has been demonstrated in canalicular-enriched fractions and an Na^+-independent activity in the sinusoidal membrane domain (Ballatori *et al.*, 1986b; Cariappa and Kilberg, 1992). In contrast, others have measured both system X_{AG}^- activity and an Na^+-independent (exchanger) glutamate transport activity, in sinusoidal membranes of rat hepatocytes (Low *et al.*, 1992). Furthermore, the system X_{AG}^- activity was upregulated in response to catabolic hormones (corticosteroid, glucagon), whereas the Na^+-independent, glutamate exchanger activity (system x_C^--like) was insensitive to alteration of whole-body catabolic status. These results suggest that system X_{AG}^- activity on the sinusoidal membrane facilitates adaptation of hepatocytes to handle the increase in sinusoidal ammonia loads that accompany catabolic stress. That is, possession of an increased concentrative glutamate transport capacity would supply more glutamate to serve as an acceptor of ammonia and/ or a source of gluconeogenic carbons, through the activities of glutamine synthetase and glutamate dehydrogenase, respectively. Unfortunately, response of canalicular membrane glutamate transport capacity to induction of whole-body catabolic states was not evaluated.

The primary purpose of concentrative canalicular glutamate transport activity (presumably, system X_{AG}^- activity) is enigmatic to this author. The importance of possessing canalicular glutamate transport capacity seems apparent, given that glutamate is the most abundant amino acid in bile, being present at about 1 mM (Folsch and Wormsley, 1977; Ballatori *et al.*, 1986a). However, given the high glutamate concentration in bile, and the understanding that all heretofore molecularly identified system X_{AG}^- transport proteins demonstrate K_m values for glutamate from about 5 to 50 μM (*Z. laevis* vs mammalian cell expression systems), it is not immediately clear as to why a high-affinity, but low capacity, glutamate transport activity (system X_{AG}^-) is expressed, instead of a low-affinity, high capacity activity (e.g., system x_C^-).

One potential explanation (Ballatori *et al.*, 1986b) is that the primary role of system X_{AG}^--mediated transport in canalicular membranes of hepatocytes is to recover from bile the glutamate generated by the activity of γ-glutamyltransferase. However, the activity of γ-glutamyltransferase is reported (Lindros *et al.*, 1989) to be three to four times higher in the periportal hepatocytes than in pericentral hepatocytes. Given this separation of γ-glutamyltransferase and system X_{AG}^- activities, and that bile flow is opposite to sinusoidal blood flow, it seems that possession of a high-affinity glutamate transport activity, through system X_{AG}^- activity in the canalicular membrane of pericentral hepatocytes, serves a purpose other than the scavenging of glutamate from recycling of bile glutathione. For example (as noted in Section 3.3), because the synthesis of glutathione is limited by intracellular availability of cysteine, and system x_C^- is the supplier of extracellular cysteine molecules (as cystine) through the counterexchange of intracellular glutamate, the role of system X_{AG}^- activity is to ensure that intracellular glutamate concentrations are high enough to optimize system x_C^- activity.

In contrast to the rather enigmatic presence of system X_{AG}^- activity in canalicular membrane, and in disagreement with the conclusion of others (Watford, 2000), expression of this high-affinity, concentrative transport activity on the sinusoidal membrane of pericentral hepatocytes does seem appropriate to ensure adequate supplies of substrate to support both glutamine and glutathione synthesis. Although it initially may seem counterintuitive that a high-affinity system is capable of a large transport capacity, especially given that the system X_{AG}^- transport proteins individually possess relatively low V_{max} values, when expressed in high abundance system X_{AG}^- transporters mediate the extraction of large amounts of glutamate (Danbolt, 2001; Gegelashvili *et al.*, 2001). Moreover, system X_{AG}^- transport activity is known to be extremely high in cells that also express high activities of glutamine synthetase (Hertz, 1979), as is the case for pericentral hepatocytes. This argument is strengthened by the observation that sinusoidal system X_{AG}^- activity is upregulatable (Low *et al.*, 1992). In addition, the presence of the apparently constitutive expression of an Na^+-independent transport activity (Low *et al.*, 1992) would function to both facilitate the import of glutamate when sinusoidal concentrations are relatively high, and serve to mediate the export glutamate when cytosolic levels exceeded that of the sinusoids. As noted above, this ability to mediate the export of glutamate from both periportal and pericentral hepatocytes is of physiological importance.

In conclusion, much still remains to be determined about the zonal distribution and submembrane localization of glutamate transport activities in hepatocytes, and how this distribution of transport activities supports whole-body intermediary metabolism of glutamate. A diagrammatic representation of this author's working hypothesis for the distribution of glutamate transporter activities along the hepatic acinus of mature liver is presented in Figure 3.3. The assignment of EAAC1 as the protein responsible for system X_{AG}^- activity is based on the immunoblot identification of EAAC1 in adult rat (McGivan, 1998), sheep, and cattle liver homogenates (Howell *et al.*, 2001, 2003; preliminary data, Gissendanner *et al.*, 2003a), and basal membranes isolated from primary cultures of cattle hepatocytes (J.C. Matthews, unpublished data).

4.1.2. Fetal Liver Model

As in adult liver, the deamination of glutamine to glutamate and ammonia by glutaminase-expressing hepatocytes is an essential function of the fetal liver. However, unlike hepatocytes from adult liver which express liver-type glutaminase, hepatocytes in the developing fetal liver are not organized into hepatic acinii, do not express carbomoyl synthetase, and express the kidney-type of glutaminase (Horowitz and Knox, 1968; Katunuma *et al.*, 1968), rather than

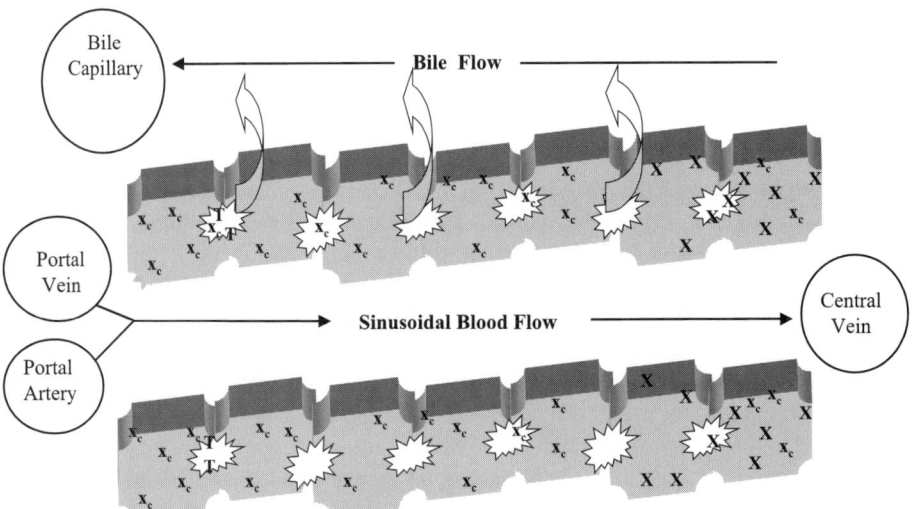

Figure 3.3. Diagrammatic representation of a working hypothesis for the distribution of γ-glutamyltransferase (T) and systems X_{AG}^- (X) and x_C^-, (x_C) transport activities along the hepatic acinus. See text for justification for the assignment of activities in periportal and/or pericentral hepatocytes, and within a hepatocyte population, assignment to canalicular (bile-facing) or basal (sinusoidal blood-facing) membrane subdomains.

liver-type. Consequently, the deamination of placental glutamine, which is predominantly of maternal origin, produces ammonia that is retained by the liver, whereas the resulting glutamate is released into the fetal circulation. Glutamate is then avidly taken up across the basal membrane of the placenta and the majority subsequently catabolized to CO_2 and H_2O by the placenta for energy production (Broeder *et al.*, 1994; Moores *et al.*, 1994). Overall, this glutamine/glutamate "cycle" (see Section 4.2), which was originally described using a sheep model by Battaglia and Meechia and colleagues (Moores *et al.*, 1994; Vaughn *et al.*, 1995), results in the net flux of nitrogen to the fetal liver and glutamate to the placenta. Two beneficial energetic consequences of the simultaneous generation of glutamate by the liver and oxidation by the placenta are the provisioning of the placenta with a source of metabolic fuel, thus sparing maternal glucose for use by the fetal liver (which has a low gluconeogenic capacity), and the removal from fetal circulation of neurotoxic levels of glutamate.

For most mammals, the shift from kidney-type to liver-type glutaminase expression occurs as the hepatocytes organize into acinar plates. For example, in the rat (20–21 days gestational period), kidney-type activity is expressed until the last 2 days of gestation, when it is replaced by the liver-specific form (Linder-Horowitz, 1969). The characteristically "adult" periportal distribution of liver-type glutaminase is then completed by the end of the first week of life (Cooper *et al.*, 1988).

Overall, this ontogeny-controlled glutaminase isoform expression is considered to be an important indicator that the fetal liver is developing adult characteristics, as liver-type glutaminase essentially couples ammonia production to urea synthesis, whereas the activity of kidney-type glutaminase produces free ammonia (Curthoys and Watford, 1995). As with the ammonia-generating glutaminase, the expression patterns of carbamoylphosphate synthetase, glutamine synthetase, and glutamate dehydrogenase do not start to develop their zonal metabolic heterogeneity until about the last 5–10% of gestation. However, within 2 weeks of birth for rats (Moorman *et al.*, 1990), and 4 weeks for pigs (Wagenaar *et al.*, 1994) and humans (Moorman *et al.*, 1989), the adult-like distribution of these ammonia utilizing enzymes essentially has occurred.

As for the adult liver, the ability of the developing fetal liver to export glutamate is critical to both the developing fetal liver and growing placenta (see Section 4.2). Unlike the adult liver, because hepatocytes and their metabolic capacities of the developing liver are not well organized, the correlation of enzymatic function with glutamate transport capacity is harder to characterize. In rats and humans, it is now thought that the onset of glutamine synthetase expression by pericentral hepatocytes lags behind accumulation of carbamoylphosphate synthetase by periportal hepatocytes, and that the relative abundance of c-met (the tyrosine kinase receptor of hepatocyte growth factor) controls onset of hepatocyte differentiation (D'Errico *et al.*, 1996; Spijkers *et al.*, 2001). Unfortunately, because the influence of ontogeny on the acinar distribution of glutamate transporters expressed in fetal liver has not been characterized, definitive statements can not be made regarding the role or type of glutamate transporters expressed by developing liver. However, ontogenetically, as the feto-placental unit develops, it would seem advantageous to have a concomitant increase in glutamate export capacity, so that the flux of glutamine through the placental glutamine/glutamate cycle could be increased to support both fetus and placenta. In contrast, as the fetal liver organizes and develops its ammonia-utilizing capacity, it would seem advantageous to have a parallel increase in concentrative uptake capacity (e.g., system X_{AG}^-), as the liver can now use glutamate as a metabolic fuel and/or to recover ammonia nitrogen as glutamine. In addition, it is convenient to think that a putative delay in the expression of concentrative glutamate transport capacity occurs until after onset of liver-type glutaminase expression, thus avoiding the "negative feedback" mechanism of glutamate concentration on kidney-type (but not the liver-type) glutaminase activity (Horowitz and Knox, 1968; Snodgrass and Lund, 1984; Smith and Watford, 1988, 1990).

In summary, data defining mature or fetal hepatic glutamate transport activities and mechanisms are relatively sparse. Although hepatic glutamate metabolizing enzymes are known to be regulated during liver development, nothing is known about how ontogeny regulates hepatic glutamate transport activities and proteins that supply and/or remove enzymatic substrates and products. We do not know which of the four non-retinal glutamate transporter genes are expressed during development, whether their acinar distribution in mature liver differs with regard to hepatocyte heterogeneity, or whether their localization within the plasma membrane subdomains differs. However, the expression of GLT-1 and EAAC1, but not GLAST1 or EAAT4, by the liver of mature sheep and cattle has been identified by Northern and immunoblot analysis of whole-liver homogenates and/or crude membrane vesicles (Howell *et al.*, 2001). Subsequently, we have determined that this same pattern of differential system X_{AG}^{-} transporter isoform expression also occurs in young but fully-ruminating sheep (Howell *et al.*, 2003), and throughout the development of growing cattle (preliminary data, Gissendanner *et al.*, 2003a).

4.2. Placental Epithelia

Three primary functions of the placenta are to mediate the passage of nutrients from the maternal to the fetal circulation, to facilitate endocrine communication between maternal and fetal units, and to absorb end products of fetal metabolism, thereby establishing metabolic cycles between the fetus and placenta (Hay, 1995; Battaglia and Renault, 2001). In contrast to most amino acids, the fetus does not require a continuous supply of glutamate from the maternal circulation to support normal growth. Instead, the fetus depends on the placenta to absorb maternal glutamine and remove fetally-derived glutamate to establish the placental–fetal glutamine/glutamate cycle, which is essential for the proper growth and development of both the placenta and fetus (Vaughn *et al.*, 1995; Malandro *et al.*, 1996; Matthews *et al.*, 1998b). Along with glutamine absorbed by the placenta from the maternal circulation, glutamine synthesized and released by the placenta is absorbed by the fetal liver and deaminated to glutamate as a source of nitrogen (Moores *et al.*, 1994). Glutamate is then released back into the fetal circulation and actively extracted by the placenta. Much of the returned glutamate is oxidized for metabolic fuel and is also used to generate NADPH for placental fatty acid and steroid synthesis, thus sparing glucose for use by the fetus (Moores *et al.*, 1994; Vaughn *et al.*, 1995). Additionally, active placental absorption of circulating fetal glutamate may serve to protect the fetus from potentially neurotoxic levels of glutamate (Moe and Smith, 1989).

Consequently, knowledge concerning the identity and distribution of the proteins responsible for glutamate uptake in the placenta is essential to understand how the fetoplacental glutamine/glutamate cycle is maintained and regulated. The concentration of glutamate measured in placental tissue (mM) exceeds that measured in maternal or fetal blood (μM) (Dierks-Ventling *et al.*, 1971). In humans, system X_{AG}^{-} activity has been identified in both the apical (Hoeltzli *et al.*, 1990) and basal (Moe and Smith, 1989) membranes of labyrinth zone syncytiotrophoblasts and is responsible for concentrating glutamate in the placenta. In the rat labyrinth tissue, trophoblast cells are arranged in three distinct layers with respect to blood supply. Layer I is composed of fenestrated syncytiotrophoblasts that face maternal blood. Because of the fenestration, these cells are not thought to represent a barrier to nutrient transfer. Layer II syncytiotrophoblasts form a contiguous epithelium, with their apical membranes facing the maternal blood. This apical domain is thought to be responsible for active extraction of nutrients from the maternal bloodstream. Layer III syncytiotrophoblasts also form a contiguous syncytia and the basal membrane surface faces the fetal mesenchyme and vasculature. The layer III basal membrane is responsible for both the release of extracted maternal nutrients into the fetal circulation and the active extraction of nutrients from the fetal circulation into the placental tissue. The labyrinth zone trilaminar syncytia increases in tissue mass until day 20 of gestation, and represents the structural and functional barrier to substrate passage between maternal and fetal circulations (Davies and Glasser, 1968).

As noted in Section 3.1, the last trimester development of membrane-specific system X_{AG}^- activities and the distribution of transporters capable of this activity has been described (Matthews *et al.*, 1998b). An important understanding gained from this study was that the relative increase in system X_{AG}^- activity at the end of gestation (day 20) vs the beginning of the last trimester (day 14) resulted from the differential expression of three different system X_{AG}^- transporters. The greater expression of system X_{AG}^- activity and proteins in the fetal-facing (basal) than in the maternal-facing (apical) plasma membrane domains at day 14 may represent an adaptation substrate supply. That is, the average concentration of glutamate in fetal rat blood during the last trimester is twice that of maternal blood. In apical membranes, the increase in system X_{AG}^- activity was paralleled by an increase in GLAST1 and EAAC1, but not GLT-1 protein content. In the basal membranes, the relatively small increase in system X_{AG}^- activity was paralleled by an increase in EAAC1 and GLAST1 content, but a large decrease in GLT-1 protein. Overall, this pattern of differential regulation of glutamate transporter expression in placental membranes probably reflects an increased reliance on the maternal source of glutamate for metabolic fuel as gestation proceeds, concomitant with the development of glutamate oxidative capacity by the fetal liver and, thus, reduced fetal supply of glutamate to the placenta (Matthews, 2000). Subsequent research with a placental cell line also has documented the sensitivity of these three system X_{AG}^- transport proteins to amino acid supply (Novak *et al.*, 2001).

4.3. Adipose Tissues

Compared to epithelial tissue, relatively little is known about the *in vivo* amino acid metabolism of adipose tissues. One reason for this relatively limited knowledge may be the inherent difficulties of catheterizing the small arterioles and venules of fat tissues, leading to inconsistent measurement of arterial-venous nutrient fluxes across fat tissues. Consequently, the recent development and use of microdialysis sampling techniques has allowed the determination of differences between concentrations of extracellular and arterial nutrients to be determined. However, the readers are cautioned to understand that potential limitations to the use of these new techniques may not be fully realized. Consequently, their accurate use may be restricted to specific nutrients (Summers *et al.*, 1998; Rolinski *et al.*, 2001; Lange *et al.*, 2002).

In terms of basal (postabsorptive) adipocyte tissue metabolism, subcutaneous fat is a net user of glutamate and net producer of glutamine, as determined by arteriovenous analysis (Frayn *et al.*, 1991). The inguinal fat pad/adipocyte tissue of fed rats also appears to be a net user of glutamate and ammonia, and a producer of glutamine but not of alanine (Kowalski and

Watford, 1994). Accordingly, these two studies gave rise to the concept that the fat tissues of normally nourished mammals absorb glutamate to synthesize glutamine and, therefore, contribute to whole-body nitrogen metabolism through the production of glutamine and removal of ammonia from the blood. In contrast, it has been observed that there is no net release of glutamine or uptake of ammonia by the inguinal pad, if rats are starved and refed (Kowalski *et al.*, 1997). Therefore, this group concluded that the substrates for glutamine synthesis come from intracellular proteolysis, not plasma. If this interpretation is proven to be correct, then mitochondrial transport of glutamate (Begum *et al.*, 2002; del Arco *et al.*, 2002) may be the critical transport process to support glutamate metabolism by adipocytes. Alternatively, as seen for kidney proximal tubule epithelia, it may be that the extent of mitochondrial metabolism of glutamate by adipose tissue is dictated by the interaction/competition between plasma and mitochondrial glutamate transport activities (Welbourne and Nissim, 2001).

Clearly, much remains to be determined about the metabolic pathways and physiological consequences of glutamate metabolism by fat tissue, especially when the normal postabsorptive state is perturbed. To this end, the influence of a protein-catabolic state (fasted for 22 hr) on protein metabolism in lean vs obese women was measured (Patterson *et al.*, 2002). Although the subcutaneous abdominal adipose tissue of both treatment groups demonstrated net uptake of glutamate and production of glutamine, a net production of alanine also was measured. Whether this difference in the pattern of alanine production observed for postabsorptive vs fasted subjects results from differences in the types of fat tissue measured, or metabolic state, remains to be determined.

Collectively, the rather limited literature on glutamate metabolism by fat tissue indicates that at least various white-fat tissues are net users of glutamate and producers of glutamine and, perhaps, alanine. Accordingly, it seems reasonable to suggest that the relative contribution of fat-synthesized glutamine to whole-body nitrogen metabolism is strongly influenced by the relative activity of glutamate and glutamine transporters (Ritchie *et al.*, 2001), glutamine synthetase, and alanine transaminase. However, a limitation of all arteriovenous measurements is that the metabolic activities and relative capacities of all the cell types that constitute a given tissue cannot be delineated. Accordingly, the extent to which the above studies actually represent adipocyte metabolic activities that are separate from glutamate metabolism by vasculature endothelial cells, stromal, or white cells (Patterson *et al.*, 2002), also remains to be determined.

With regard to specific glutamate transport processes that contribute to glutamine and alanine syntheses in fat tissues, an interesting and seemingly paradoxical observation was that ASCT2 mRNA expression is dramatically increased in differentiated vs undifferentiated 3T3-L1 adipocytes (Liao and Lane, 1995). Given that ASCT2 is thought to only transport anionic amino acids when the pH is less than 5.5, the physiological relevance of ASCT2 mRNA expression (and, presumably, protein) to glutamate transport capacity is unclear to this author. However, another important observation of this study was that insulin stimulated the uptake of serine, alanine, and glutamate by 96%, 38%, and 60% in differentiated 3T3-L1 cells that were transiently transfected with ASCT2 cDNA. Thus, either ASCT2 is capable of glutamate transport at pH greater than pH 5.5 under whole-cell conditions, expression of ASCT2 mRNA expression was relevant to transport processes of another membrane across which a strong pH gradient exists (e.g., lysomes), or the increased glutamate transport capacity was a secondary consequence of ASCT2-enhanced serine and alanine uptake (e.g., see Section 3.3).

As for system X_{AG}^- transporters, this author is ignorant of any published reports of plasma membrane glutamate transport activity or transporters by fat tissues. However, in pursuit of knowledge regarding the importance of glutamate transport to adipose tissue production of glutamine, our lab has recently evaluated the potential expression of system X_{AG}^- transporters by fat tissues in growing cattle and detected the expression of EAAC1 mRNA and protein, and GLT-1 mRNA, by subcutaneous, perirenal, omental, mesenchymal, and/or interstitial adipose tissues of growing cattle (preliminary data, Gissendanner *et al.*, 2003b).

4.4. Striated Muscle Tissues

The uptake of plasma glutamate, intracellular conversion of glutamate to glutamine, and subsequent export to the plasma of synthesized glutamine by skeletal muscle is critical to the proper functioning of intestinal epithelia and the immune system (Newsholme *et al.*, 1990; Hack *et al.*, 1996). Besides diminishing the synthesis and export of glutamine, a reduction in the capacity for plasma glutamate uptake by skeletal muscle is thought to cause intracellular deficiencies of glutamate, glutamine, glutathione, and citric acid cycle intermediates, but increased carbon dioxide levels (Hack *et al.*, 1996; Holm *et al.*, 1997; Ushmorov *et al.*, 1999).

Glutamate absorption by mammalian skeletal muscle initially was thought to occur by an Na^+-independent process, based on conclusions drawn from a rat hindlimb perfusion study (Hundal *et al.*, 1989). Subsequently, and in accordance with that observed in the muscle of barnacles (Revest and Baker, 1988), the expression of both Na^+-dependent and -independent glutamate uptake activities was identified in skeletal muscle using primary cultures of rat myotubules (Low *et al.*, 1994). The Na^+-dependent uptake of L-glutamate activity was inhibited by both D- and L-aspartate and displayed an affinity constant for L-glutamate of 0.7 mM. Therefore, although the profile of substrates that inhibit Na^+-dependent uptake of L-glutamate by myotubules is consistent with system X_{AG}^- transporters (Table 3.1), the affinity is too low. As for cultured rat myotubules (Low *et al.*, 1994), cultured myogenic C2C12 cells also possess an Na^+-coupled glutamate uptake capacity with an affinity constant of about 0.6 mM for L-glutamate (Frank *et al.*, 2002). Importantly, the velocity of this activity could be stimulated one-fold in the presence of 2-chloro adenosine, by a process that was not mediated by glutamate receptors (Frank *et al.*, 2000), and which resulted in a physiologically relevant increase of intracellular Ca^{2+} concentrations (Frank *et al.*, 2002). This latter observation suggests that the uptake of glutamate by skeletal muscle may serve as a more immediate form of inter-tissue signaling, as compared to just an increased capacity for intermediary metabolism that results from elevated intracellular levels of glutamate. Furthermore, if it is shown that the rise of intracellular Ca^{2+} concentrations results from the act of glutamate transport per se, and not a generic alteration of intracellular Na^+, K^+, and/or Cl^- concentrations, and/or membrane potential, which would occur with the induction of any ion-coupled transport process, then an important understanding about the differences in supplying anaplerotic glutamate carbons as glutamine vs glutamate (Rennie *et al.*, 2001) may have been gained. Unfortunately, the biochemical characterization of potential system X_{AG}^- transporter isoforms was insufficient to allow delineation of specific family members, whereas molecular analyses for the expression of putative glutamate transporters was not performed. Therefore, both the glutamate transporter type expressed by differentiated C2C12 cells, and the pathway by which increased glutamate transport activity affects elevated intracellular Ca^{2+} concentrations remain to be determined.

Specific information about the molecular identity of potential glutamate transporters expressed by muscle tissue is very limited. EAAC1 mRNA has been identified in skeletal muscle (Kanai and Hediger, 1992), thus confirming the possibility that EAAC1 could account for the system X_{AG}^- activity found in skeletal (Revest and Barker, 1989; Low *et al.*, 1994; Frank *et al.*, 2002) myocytes. Recently, we have identified the presence of both mRNA and protein for two system X_{AG}^- transporters (EAAC1 and GLT-1) in the longissimus dorsi muscle of cattle (preliminary data, Gissendanner *et al.*, 2003b). From a developmental and/or aging perspective, an important observation was that the amount of both EAAC1 and GLT-1 expressed was inversely proportional to developmental age. That is, the content of both EAAC1 and GLT-1 in this skeletal muscle of suckling animals (30 days old) was greater than in weanling age (184 days old), which was greater than that by postweaning (284 days) or fattened (423 days old) cattle. Whether this reduction in apparent glutamate transport capacity in skeletal muscle reflects a loss of peripheral tissue sensitivity to insulin as cattle physiologically adjust from a monogastric-like (suckling animal) to a fully ruminant (postweaning and fattened) metabolism, or is an ontogenetic (aging) effect, remains to be determined. In humans, it has been well demonstrated that the capacity for skeletal muscle uptake of glutamate is decreased in both elderly and non-insulin-dependent diabetic mellitus individuals (Hack *et al.*, 1996; Holm *et al.*, 1997). However, evaluation of system X_{AG}^- transporter protein expression by human skeletal muscle is unknown to this author.

As just described, a discrepancy apparently exists between the biochemically and molecularly defined identities of Na^+-dependent transporters expressed by skeletal muscle tissue and myocytes. In contrast, for cardiac muscle, the high-affinity and substrate competition profiles observed for rat cardiac myocytes and sarcolemma vesicle uptake of L-aspartate well matches the identification of EAAC1 in both cell and vesicle protein preparations (King *et al.*, 2001). Specifically, the observed affinities 6.5 and 9.8 μM for L-aspartate uptake are in reasonable agreement with those achieved by overexpression of rat EAAC1 cDNA in oocytes and mammalian cells. These results provide strong evidence that at least EAAC1 is expressed by cardiac myocytes.

Overall, as reviewed above, skeletal muscle plays an important role in supplying glutamine and alanine for use by several other organs and cell-types, while the uptake of plasma glutamate by skeletal muscle seems essential to these processes. However, critical differences may exist in how the uptake of glutamate is maintained during postprandial vs postabsorptive states. In the postprandial state, insulin typically stimulates the absorption of amino acids through a variety of direct and indirect events, including stimulation of specific amino acid (Cheeseman, 1991) and Na^+ transporters (Hundal *et al.*, 1992; Moore, 1993). In contrast, during the postabsorptive state, the muscle exports most amino acids, except for glutamate (Holm *et al.*, 1997).

Importantly, a working hypothesis has been proposed by E. Holm and W. Droge which theorizes that normal postabsorptive state (when insulin levels are low) catabolism of muscle protein is responsible for Na^+-dependent glutamate uptake by skeletal muscle (Holm *et al.*, 1997). Essentially, the hypothesis argues that during postabsorptive states, cellular energy levels are not high enough to support the requisite elevated functioning of Na^+/K^+ ATPase. Consequently, the normally high extracellular-to-intracellular Na^+ gradient is not maintained during functioning of the Na^+-dependent concentrative transport of glutamate by system X_{AG}^- transporters. Thus, instead of Na^+/K^+ ATPase activity, it is proposed that the Na^+-coupled export (not uptake) of amino acids (especially glutamine, alanine, and glycine) from skeletal muscle serves to maintain the transmembrane Na^+ gradient.

Inherent to this interesting proposal is the understanding that the typical direction of several Na^+-dependent amino acid transport systems would be reversed. In addition to a reduced ability to re-establish the normal transmembrane Na^+ gradient, a lowered Na^+/K^+ ATPase activity also would reduce the normally high intracellular-to-extracellular K^+ gradient. Consequently, glutamate uptake capacity would be reduced because system X^-_{AG} transporters reverse their activity when extracellular K^+ levels become atypically high (Rossi *et al.*, 2000; Jabaudon *et al.*, 2000; Trotti *et al.*, 2001). This latter concept may have significant whole-body physiological consequences when individuals are metabolically stressed. For example, as noted above, patients undergoing normal aging processes or afflicted with non-insulin-dependent diabetes mellitus (Hack *et al.*, 1996), cancer (Droge *et al.*, 1988), HIV (Eck *et al.*, 1991), and amyotrophic lateral sclerosis (Plaitakis and Caroscio, 1987; Trotti *et al.*, 2001) have elevated plasma glutamate concentrations, presumably resulting from an impaired ability to absorb glutamate (Holm *et al.*, 1997). However, because of the especially low energy levels of the cachexic muscle tissues of these patients, the elevation of plasma glutamate levels may result from the export of glutamate from muscle tissue, rather than a reduced uptake of glutamate from the plasma.

Based on the above concepts by Holm and colleagues (1997), and a variety of recently identified biochemical properties and expression profiles (see Bode, 2001, and Matthews and Anderson, 2002, for recent reviews) of molecularly identified amino acid transporters, a composite model is presented in Figure 3.4 that identifies transporters present in skeletal muscle that could function to support the efflux of glutamine, alanine, and glycine from skeletal muscle, and import glutamate during normal postabsorptive states. In this model, all three system A transporter isoforms (ATA1–3) are indicated, as their mRNA has been identified in muscle tissues or myotubules. All of these transporters function to mediate the cotransport of Na^+ and glutamine, alanine, or glycine, although differences exist in their relative affinities for substrates and regulation of expression. ATA1 (Wang *et al.*, 2000) may in fact be the major glutamine exporter as it prefers glutamine to other substrates. In contrast, ATA2

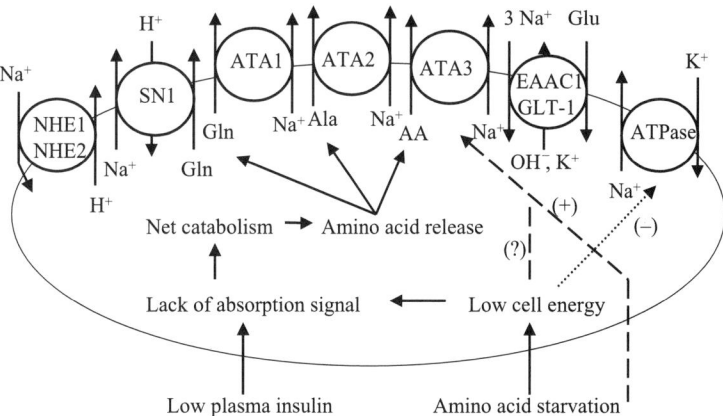

Figure 3.4. Diagrammatic representation of a working model for the putative complement of amino acid and ion transporters present in skeletal muscle (see text for justification of transporter assignments), whose combined function could mediate the support of coupled efflux of Na^+ and glutamine (Gln), alanine (Ala), and glycine (AA) from skeletal muscle to drive the Na^+-coupled import of glutamate during the postabsorptive state, as postulated by Holm *et al.* (1997). Amino acid transport systems A (ATA1, ATA2, ATA3), N (SN1), and X^-_{AG} (EAAC1, GLT-1) isotypes are indicated, as are the presence of the Na^+/H^+ exchanger (NHE1, NHE2) and Na^+/K^+ ATPase ion transporters.

(Hatanaka *et al.*, 2000; Sugawara *et al.*, 2000a; Yao *et al.*, 2000) may be the predominant alanine transporter as it expresses the more classic system A transport activity, by preferring alanine over glutamine as a substrate. Similar to ATA2, ATA3 also prefers alanine and glycine rather than glutamine as substrates. Although ATA3 has a greater preference for alanine than does ATA2 (Sugawara *et al.*, 2000b), expression of ATA2 mRNA and activity is upregulated during amino acid starvation, an effect that is reversible upon refeeding of system A substrates (Gazzola *et al.*, 2001). Therefore, collectively, the functional properties of the system A isotypes appear to well endow muscle tissue with the ability to export and/or import glutamine, alanine, and glycine during postprandial and postabsorptive states.

In addition to glutamine transport by the system A transporter, ATA1, the mRNA for one (SN1) of the two known system N transporter isotypes is expressed by muscle tissue (Gu *et al.*, 2000). SN1 mediates the cotransport of glutamine and two Na^+ in exchange for H^+ (Fei *et al.*, 2000), and prefers glutamine over alanine as a substrate. Thus, although SN1 would seemingly act to better re-establish the extracellular-to-intracellular two Na^+ gradient than the system A transporters, by exchanging two Na^+ for a single amino acid, the need to remove the counterexchanged H^+ (presumably by an Na^+/H^+ exchanger, NHE) would negate this benefit. Both NHE1 and NHE2 are indicated as potential Na^+/H^+ exchangers, although NHE2 is likely the predominant skeletal muscle isoform (Malakooti *et al.*, 1999). EAAC1 and GLT-1 are included in the model as the system X_{AG}^- transporters because both their mRNA and protein are expressed by skeletal muscle (preliminary data, Gissendanner *et al.*, 2003b) and they are known to mediate the export, rather than import, of glutamate when intracellular K^+ levels are low (Zerangue and Kavanaugh, 1996b; Levy *et al.*, 1998).

5. *In Vivo* Regulation of Glutamate Transport Capacity

5.1. Whole-Body and Tissue Regulation

Information regarding substrate regulation of system X_{AG}^- transporter expression and activity is limited, especially with regard to whole-animal studies. With respect to substrate regulation of EAAC1 and GLT-1 expression, evidence exists that the content of GLT-1 and EAAC1 in brain tissue is differentially regulated by L-glutamate in brain tissue (Duan *et al.*, 1999; Munir *et al.*, 2000). In cell culture models, EAAC1 and/or GLT-1 expression by epithelial cells are sensitive to extracellular glutamate concentrations (Nicholson and McGivan, 1996; Novak *et al.*, 2001). Even less is known about how the *in vivo* function of system X_{AG}^- transporters is regulated to support whole-body growth and maintenance. In *eaac-1$^{-/-}$* mice, intestinal and kidney phenotypic development is normal, as are plasma amino acid levels (Peghini *et al.*, 1997). However, the concentration of glutamate and aspartate in the urine of these animals is elevated 1,400- and 10-fold, respectively, as compared to wild-type mice. In contrast, in *glt-1$^{-/-}$* mice, urine glutamate and aspartate concentrations are unchanged but kidney glutamate and glutathione concentrations are markedly reduced (Tanaka and Welbourne, 2001).

Together, these gene knockout studies clearly indicate that at least two of the system X_{AG}^- transporters support different functions within the same tissue. We recently conducted a growth trial using sheep to determine (a) whether the ileal epithelium, hepatic, and/or renal tissue expression of GLT-1 and EAAC1 differed in wethers fed to gain vs maintain body weight and (b) whether a relationship existed between tissue content of D- and L-glutamate and EAAC1 or GLT-1 (Howell *et al.*, 2003). A pertinent observation from this study was the increased expression of EAAC1 by ileal epithelial and GLT-1 by liver tissue membranes of growing vs nongrowing lambs. Coincident with increased glutamate transporter content was an increased concentration of L-glutamate in ileal, and D-glutamate in hepatic, cellular membranes. That EAAC1 content was elevated in ileal epithelia that contained elevated L-glutamate, but GLT-1 was not, suggests that *in vivo* EAAC1 expression may have been stimulated by the presence of L-glutamate, whereas GLT-1 was not. Alternatively, the elevated glutamate levels could have resulted from elevated EAAC1 function, which in turn, was increased in response to some other cellular signal. By whatever physiological stimulus, this whole-animal study supports the understanding that the dual expressions of EAAC1 and GLT-1 by small intestinal epithelia and the liver are independently regulated.

With regard to intracellular signals, the influence of intracellular L-glutamate on EAAC1 and GLT-1 content has been investigated using primary and immortalized cell cultures. In contrast to elevated EAAC1 amounts that were coincident with increased L-glutamate concentrations (29.3 vs 43.7 nmol/mg) in ileal epithelium of growing sheep (Howell *et al.*, 2003), expression of EAAC1 protein by immortalized bovine renal tubule epithelial cells is increased when intracellular L-glutamate is decreased from about 30 to 15 nmol/mg protein (Nicholson and McGivan, 1996). In another apparent difference between whole-animal and *in vitro* cell culture models, expression of EAAC1 by primary "astrocyte-poor" neuronal cultures was decreased in the presence of L-glutamate (Munir *et al.*, 2000). However, further research led these investigators to conclude that intracellular glutamate level is not a sufficient stimulus for alteration of plasma membrane expression of EAAC1. Instead, only when the PKC pathway is activated, would increased intracellular L-glutamate result in decreased EAAC1 expression (Robinson, 2002).

An obvious candidate for this stimuli and difference between experimental models, is the influence of hormones. However, little is known regarding hormonal control of system X_{AG}^- transporter expression. Specifically, culture of C6 glioma cells with platelet-derived growth factor (PDGF) induces increased system X_{AG}^- activity and EAAC1 content in appearance in plasma membranes, consistent with glutamate supporting PDGF-stimulated proliferation responses (Sims *et al.*, 2000). Research with transgenic mice (Matthews *et al.*, 1999) indicates that placental expression of GLT-1 is stimulated by atypically high growth hormone levels and that IGF-II is required for GLT-1 and EAAC1 expression. Preliminary data from our lab (J.C. Matthews, unpublished data) indicates that expression of system X_{AG}^- activity by primary cultures of steer hepatocytes is stimulated by insulin. Common to both PDGF- and insulin-induced secondary messenger pathways is the dependence on tyrosine kinase and phosphatidylinositol-3 kinase activation. Therefore, given the elevated nutritional status of the growing vs nongrowing lambs of the present study, it is likely that both insulin/PDGF and L-glutamate levels were elevated in the ileal epithelium in growing lambs, and that a combination of these regulatory protein factors resulted in the increased expression of EAAC1.

5.2. Transcriptional vs Posttranscriptional Regulation of Glutamate Transporter Content and Capacity

An important question of gene expression is at what level the regulation occurs. Our research with the expression of system X_{AG}^- transporters by placenta, small intestine epithelia, liver, and kidney (Matthews *et al.*, 1998b, 1999; Howell *et al.*, 2003; preliminary data, Gissendanner *et al.*, 2003a) indicates that posttranscriptional regulation is an important form of regulating EAAC1, GLT-1, and EAAT4, as the presence or relative content of mRNA is often not mirrored by the presence or relative content of transporter protein. Research from other groups also indicate that steady-state levels of glutamate transporters do not always correlate with mRNA levels. For example, increased amounts of EAAC1 protein expression by AA-deprived NBL-1 cells is accompanied by a decrease in EAAC1 mRNA levels (Plakidou-Dymock and McGivan, 1993). In contrast to these findings, expression of GLT-1 mRNA by primary astrocytes was paralleled by increased GLT-1 protein content when the medium was supplemented with dibutyryl-cAMP (dbcAMP) (Schlag *et al.*, 1998).

In terms of the potential for regulation of the activity of transport proteins already localized in cell-wall membranes, membrane-binding proteins that separately regulate the specific activities of EAAC1 (GTRAP3-18; Lin *et al.*, 2001; Butchbach *et al.*, 2002) and EAAT4 (GTRAP41, GTRAP48; Jackson *et al.*, 2001) have been identified. In addition, a cytosolic protein (Ajuba) known to activate MAP kinase and colocalizes with GLT-1 in plasma membranes, but which does not influence the K_m or V_{max} values of glutamate uptake has been characterized (Marie *et al.*, 2002). Although speculative, Ajuba is thought to function by migrating to the nuclear membrane to activate GLT-1 transcription when GLT-1 plasma membrane content is too low to quantitatively bind Ajuba (Marie *et al.*, 2002). Collectively, all these data clearly indicate that regulation of system X_{AG}^- proteins is complex and likely involves transcriptional, posttranscriptional, and posttranslational regulation, depending on the particular effects.

6. Future Questions about Glutamate Transporter Physiology in Support of Peripheral Tissue Function

This brief review has identified either explicitly or implicitly, a number of gaps and/or chasms in our knowledge regarding which glutamate transport activities and transporters are resident in specific membranes, cell-types, and tissues. Overall, we know very little about the physiology of glutamate transporters in peripheral tissues and, critically, how and which transporters interact with glutamate metabolizing enzymes to support peripheral tissue function. Several specific research topic areas and concepts which this author thinks are critical to next address if we are to fill the voids of our knowledge include:

1. What proportion of total glutamate transport is achieved by low-affinity ASC/B^0 transporter function? Is this a subcellular phenomenon? For example, is this low pH-dependent event localized to specific plasma membrane functional environments such as caveolae or to cell organelles such as lysosomes?

2. Are there proteins analogous to the GTRAP family members that control the functional activity of all glutamate transporters?

3. In peripheral tissues, does the multiple expression of system X_{AG}^- transporter isoforms indicate that specific glutamate transport events support specific metabolic pathways or biochemical activities, such as EAAC1 "sensing" of cellular energy/nutritional status and GLT-1-dependent tight junction function (Welbourne and Matthews, 1999)?

4. Do cells express *de facto* functional units of plasma membrane transporters and cytosolic metabolizing enzymes, and/or mitochondrial transporters and metabolizing enzymes exist to handle the cellular flux of glutamate (Welbourne *et al.*, 2001)?

5. Is/how is the capacity of glutamate transport and metabolism coordinated with the capacity to transport and metabolize precursors and products of glutamate metabolism?

6. Assuming that regulatory proteins do exist that coordinate the expression and capacity of the functional units of transporters and enzymes, how does their activity differ among cell- and tissue-types?

References

Arriza, J.L., M.P. Kavanaugh, W.A. Fairman, Y.N. Wu, G.H. Murdoch, R.A. North *et al.* (1993). Cloning and expression of a human neutral amino acid transporter with structural similarity to the glutamate transporter gene family. *J. Biol. Chem.* **268**, 15329–15332.

Arriza, J.L., S. Eliasof, M.P. Kavanaugh, and S.C. Amara (1997). Excitatory amino acid transporter 5, a retinal glutamate transporter coupled to a chloride conductance. *Proc. Natl. Acad. Sci. USA* **94**, 4155–4160.

Avissar, N.E., C.K. Ryan, V. Ganapathy, and H. Sax (2001). Na^+-dependent neutral amino acid transporter ATB^0 is a rabbit epithelial cell brush-border protein. *Am. J. Physiol. Cell Physiol.* **281**, C963–C971.

Ballatori, N., R. Jacob, and J.L. Boyer (1986a). Intrabiliary glutathione hydrolysis. A source of glutamate in bile. *J. Biol. Chem.* **261**, 7860–7865.

Ballatori, N., R.H. Moseley, and J.L. Boyer (1986b). Sodium gradient-dependent L-glutamate transport is localized to the canalicular domain of liver plasma membranes. Studies in rat liver sinusoidal and canalicular membrane vesicles. *J. Biol. Chem.* **261**, 6216–6221.

Bannai, S. and E. Kitamura (1980). Transport interaction of L-cystine and L-glutamate in human diploid fibroblasts in culture. *J. Biol. Chem.* **255**, 2372–2376.

Bannai, S. and E. Kitamura (1981). Role of proton dissociation in the transport of cystine and glutamate in human diploid fibroblasts in culture. *J. Biol. Chem.* **256**, 5770–5772.

Bassi, M.T., E. Gasol, M. Manzoni, M. Pineda, M. Riboni, R. Martin *et al.* (2001). Identification and characterization of human xCT that co-expresses, with 4F2 heavy chain, the amino acid transport activity system x_C^-. *Pflugers Arch.* **442**, 286–296.

Battaglia, F.C. (2000). Glutamine and glutamate exchange between the fetal liver and the placenta. *J. Nutr.* **130**, 974S–977S.

Battaglia, F.C. and T.R.H. Renault (2001). Placental transport and metabolism of amino acids. *Placenta* **22**, 145–161.

Begum, L., M.A. Jalil, K. Kobayashi, M. Iijima, M.X. Li, T. Yasuda *et al.* (2002). Expression of three mitochondrial solute carriers, citrin, aralar1 and ornithine transporter, in relation to urea cycle in mice. *Biochim. Biophys. Acta* **1574**, 283–292.

Biolo, G., X.-J. Zhang, and R.R. Wolfe (1995). Role of membrane interorgan amino acid flow between muscle and small intestine. *Metabolism* **44**, 719–724.

Bode, B. (2001). Recent molecular advances in mammalian glutamine transport. *J. Nutr.* **131**, 2475S–2485S.

Bridges, C.C., R. Kekuda, H. Wang, P.D. Prasad, P. Mehta, W. Huang *et al.* (2001). Structure, function, and regulation of human cystine/glutamate transporter in retinal pigment epithelial cells. *Invest. Opthalmol. Vis. Sci.* **42**, 47–54.

Broeder, J.A., C.H. Smith, and A.J. Moe (1994). Glutamate oxidation by trophoblasts *in vitro. Am. J. Physiol.* **267**, C189–C194.

Bronson, J.T. (2000). Glutamate, at the interface between amino acid and carbohydrate metabolism. *J. Nutr.* **130**, 988S–990S.

Burger, H.-J., R. Gebhardt, C. Mayer, and D. Meckel (1989). Different capacities for amino acid transport in periportal and perivenous hepatocytes isolated by digitonin/collagenase perfusion. *Hepatology* **9**, 22–28.

Butchbach, M.E., L. Lai, and C.L. Lin (2002). Molecular cloning, gene structure, expression profile and functional characterization of the mouse glutamate transporter (EAAT3) interacting protein GTRAP3–18. *Gene* **292**, 81–90.

Cariappa, R. and M.S. Kilberg (1992). Plasma membrane domain localization, and transcytosis of the glucagon-induced hepatic system A carrier. *Am. J. Physiol.* **263**, E1021–E1028.

Chairoungdua, A., Y. Kanai, H. Matsuo, J. Inatomi, D.K. Kim, and H. Endou (2001). Identification and characterization of a novel member of the heterodimeric amino acid transporter family presumed to be associated with an unknown heavy chain. *J. Biol. Chem.* **276**, 49390–49399.

Cheeseman, C.I. (1991). Molecular mechanisms involved in the regulation of amino acid transport. *Prog. Biophysiol. Molec. Biol.* **55**, 71–84.

Cooper, A.J., E. Nieves, K.C. Rosenspire, S. Filc-DeRicco, A.S. Gelbard, and S.W. Brasilia (1988). Short-term metabolic fate of ^{13}N-labeled glutamate, alanine, and glutamine(amide) in rat liver. *J. Biol. Chem.* **263**, 12268–12273.

Curthoys, N.P. and M. Watford (1995). Regulation of glutaminase activity and glutamine metabolism. *Annu. Rev. Nutr.* **15**, 133–159.

Danbolt, N.C. (2001). Glutamate uptake. *Prog. Neurobiol.* **65**, 1–105.

Dantzig, A.H., M.C. Finkelstein, E.A. Adelberg, and C.W. Slayman (1978). The uptake of L-glutamic acid in a mormal mouse lymphocyte line and in a transport mutant. *J. Biol. Chem.* **253**, 5813–5819.

Davies, J. and S.R. Glasser (1968). Histological and fine structural observations on the placenta of the rat. *Acta Anat.* **69**, 542–608.

del Arco, A., J. Morcillo, J.R. Martinez-Morales, C. Galian, V. Martos, P. Bovolenta *et al.* (2002). Expression of the aspartate/glutamate mitochondrial carriers aralar1 and citrin during development and in adult rat tissues. *Eur. J. Biochem.* **269**, 3313–3320.

D'Errico, A., M. Fiorentino, Y. Daikuhara, H. Tsubouchi, C. Brechot, J.-Y. Scoaazec *et al.* (1996). Liver hepatocyte growth factor does not always correlate with hepatocellular proliferation in human liver legions; its specific receptor c-met does. *Hepatology* **24**, 60–64.

Dierks-Ventling, C., A.L. Cone, and R.A. Wapnir (1971). Placental transfer of amino acids in the rat. 1. L-Glutamic acid and L-glutamine. *Biol. Neonate* **17**, 361–372.

Droge, W., H-P. Eck, M. Betzler, P. Schlag, P. Drings, and W. Ebert (1988). Plasma glutamate concentration and lymphocyte activity. *J. Cancer Clin. Oncol.* **114**, 124–128.

Duan, S., C.M. Anderson, B.A. Stein, and R.A. Swanson (1999). Glutamate induces rapid upregulation of astrocyte glutamate transport and cell-surface expression of GLAST. *J. Neurosci.* **19**, 10193–10200.

Eck, H.-P., C. Stahl-Hennig, G. Hunsmann, and W. Droge (1991). Metabolic disorder as an early consequence of simian immunodeficiency virus infection in rhesus macques. *Lancet* **338**, 346–357.

Fairman, W.A., R.J. Vandenberg, J.L. Arriza, M.P. Kavanaugh, and S.G. Amara (1995). An excitatory amino-acid transporter with properties of a ligand-gated chloride channel. *Nature* **375**, 599–603.

Fei, Y.-J., M. Sugawara, T. Nakanishi, W. Huang, H. Wang, P.D. Prasad *et al.* (2000). Primary structure, genomic organization, and functional and electrogenic characteristics of human system N1, a Na$^+$-and H$^+$-coupled glutamine transporter. *J. Biol. Chem.* **275**, 23707–23717.

Folsch, U.R. and K.G. Wormsley (1977). The amino acid composition of rat bile. *Experientia* **33**, 1055–1056.

Frank, C., A.M. Giammarioli, L. Falzano, S. Rufini, A. Camurri *et al.* (2000). 2-Chloro-adenosine induces a glutamate-dependent calcium response in C2C12 myotubes. *Biochem. Biophys. Res. Comm.* **227**, 546–551.

Frank, C., A.M. Giammarioli, L. Falzano, C. Fiorentini, and S. Rufini (2002). Glutamate-induced calcium increase in myotubes depends on up-regulation of a sodium-dependent transporter. *FEBS Lett.* **527**, 269–273.

Frayn, K.N., K. Khan, S.W. Coppack, and M. Elia (1991). Amino acid metabolism in human subcutaneous adipose tissue in vivo. *Clin. Sci.* **80**, 471–474.

Gaseously, G.C., V. Dall'Asta, O. Bussolati, M. Makowske, and H.N. Christensen (1981). A stereoselective anomaly in dicarboxylic amino acid transport. *J. Biol. Chem.* **256**, 6054–6059.

Gazzola, R.F., R. Sala, O. Bussolati, Visigalli, V. Dall'Asta, V. Ganapathy *et al.* (2001). The adaptive regulation of amino acid transporter system A is associated to changes in ATA2 expression. *FEBS Lett.* **490**, 11–14.

Gebhardt, R. and D. Meckel (1983). Glutamate uptake by cultured rat hepatocytes is mediated by hormonally inducible, sodium-dependent transport systems. *FEBS Lett.* **161**, 275–278.

Gegelashvili, G., M.B. Robinson, D. Trotti, and T. Rauen (2001). Regulation of glutamate transporters in health and disease. *Prog. Brain Res.* **132**, 267–286.

Gissendanner, S.J., N.M.P. Etienne, and J.C. Matthews (2003a). Differential expression of EAAC1 and GLT-1 glutamate transporters by bovine epithelial tissues is not altered by physiological development. *FASEB J.* **17**, A305.

Gissendanner, S.J., N.M.P. Etienne, K.R. McLeod, and J.C. Matthews (2003b). The pattern of EAAC1 and GLT-1 glutamate transporter expression by skeletal muscle and adipose tissues of fattening cattle differs from that of glutamine synthetase. *FASEB J.* **17**, A738.

Gu, S., H.L. Roderick, P. Camacho, and J.X. Jiang, (2000). Identification and characterization of an amino acid transporter expressed differentially in liver. *Proc. Natl. Acad. Sci. USA* **97**, 3230–3235.

Hack, V., O. Stutz, R. Kinscherf, M. Schykowski, M. Kellerer, E. Holm *et al.* (1996). Elevated venous glutamate levels in (pre)catabolic conditions result at least partly from a decreased glutamate transport activity. *J. Mol. Med.* **74**, 337–343.

Hatanaka, T., W. Huang, H. Wang, M. Sugawara, P.D. Prasad, F.H. Leibach *et al.* (2000). Primary structure, functional characteristics and tissue expression pattern of a human ATA2, a new subtype of amino acid transport system A. *Biochim. Biophys. Acta* **1467**, 1–6.

Haussinger, D. and W. Gerok (1983). Hepatocyte heterogeneity in glutamate uptake by isolated perfused rat liver. *Eur. J. Biochem.* **136**, 421–425.

Haussinger, D., H. Sies, and W. Gerok (1985). Functional hepatocyte heterogeneity in ammonia metabolism. The intercellular glutamine cycle. *J. Hepatol.* **1**, 3–14.

Haussinger, D., B. Stoll, T. Stehle, and W. Gerok (1989). Hepatocyte heterogeneity in glutamate transport in perfused rat liver. *Eur. J. Biochem.* **185**, 189–195.

Haussinger, D. (1990). Liver glutamine metabolism. *J. Parent. Enteral. Nut.* **14**, 56S–62S.

Hay, W.W.J. (1995). Current topic: Metabolic interelationships of placenta and fetus. *Placenta* **16**, 19–30.

Hediger M.A. and T.C. Welbourne (1999). Introduction: Glutamate transport, metabolism, and physiological responses. *Am. J. Physiol.* **277**, F477–F480.

Heitmann, R.N. and E.N. Bergman. (1981). Glutamate interconversions and glucogenicity in the sheep. *Am. J. Physiol.* **241**, 465–472.

Hertz, L. (1979). Functional interactions between neurons and astrocytes. I. Turnover and metabolism of putative amino acid transmitters. *Prog. Neurobiol.* **13**, 277–323.

Hoeltzli, S.D., L.K. Kelley, A.J. Moe, and C.H. Smith (1990). Anionic amino acid transport systems in isolated basal plasma membrane vesicles of human placenta. *Am. J. Physiol.* **259**, C47–C55.

Holm, E., V. Hack, M. Tokus, R. Breitkreutz, A. Babylon, and W. Droge (1997). Linkage between postabsorptive amino acid release and glutamate uptake in skeletal muscle tissue of healthy young subjects, cancer patients, and the elderly. *J. Mol. Med.* **75**, 454–461.

Horowitz, M.L. and W.E. Knox (1968). A phosphate activated glutaminase in rat liver different from that in kidney and other tissues. *Enzymol. Biol. Clin.* **9**, 241–255.

Howell, J.A., A.D. Matthews, K.C. Swanson, D.L. Harmon, and J.C. Matthews (2001). Molecular identification of high-affinity glutamate transporters in sheep and cattle forestomach, intestine, liver, kidney, and pancreas. *J. Anim. Sci.* **79**, 1329–1336.

Howell, J.A., A.D. Matthews, T.C. Welbourne, and J.C. Matthews (2003). Content of ileal EAAC1 and hepatic GLT-1 high-affinity glutamate transporters is increased in growing versus non-growing lambs, paralleling increased tissue concentrations of D- and L-glutamate and plasma glutamine and alanine. *J. Anim. Sci.* **81**, 1030–1039.

Hundal, H.S., M.J. Rennie, and P.W. Watt (1989). Characteristics of acidic, basic and neutral amino acid transport in the perfused rat hindlimb. *J. Physiol. (London)* **408**, 93–114.

Hundal, H.S., A. Marette, Y. Mitsumoto, T. Ramlal, R. Blostein, and A. Klip (1992). Insulin induces translocation of the α2 and β1 subunits of the Na$^+$/K$^+$-ATPase from intracellular compartments to the plasma membrane in mammalian skeletal muscle. *J. Biol. Chem.* **267**, 5040–5043.

Jabaudon, D., M. Scanziana, B.H. Gahwiler, and U. Gerber (2000). Acute decrease in net glutamate uptake during energy deprivation. *PNAS* **97**, 5610–5615.

Jackson, M., W. Song, M.-Y. Yu, L. Jin, M. Dykes-Hoberg, C.G. Lin *et al.* (2001). Modulation of the neuronal glutamate transporter EAAT4 by two interacting proteins. *Nature* **410**, 89–93.

Kanai, Y., and M.A. Hediger (1992). Primary structure and functional characterization of a high-affinity glutamate transporter. *Nature* **360**, 467–471.

Katunuma, N., I. Tomino, and Y. Sanada (1968). Differentiation of organ specific glutaminase isozyme during development. *Biochem. Biophys. Res. Commun.* **32**, 426–432.

Kekuda, R., P.D. Prasad, Y.-J. Fei, V. Torres-Zamorano, S. Sinha, T.L. Yang-Feng *et al.* (1996). Cloning of the sodium-dependent, broad-scope, neutral amino acid transporter B^0 from a human placental choriocarcinoma cell line. *J. Biol. Chem.* **271**, 18661–18675.

Kekuda, R., V. Torres-Zamorano, Y.-J. Fei, P.D. Prasad, H.W. Li, D. Mader *et al.* (1997). Molecular and functional characterization of intestinal Na$^+$-dependent neutral amino acid transporter B^0. *Am. J. Physiol.* **272**, G1463–G1472.

Kilberg, M.S., M.E. Handlogten, and H.N. Christensen (1980). Characteristics of an amino acid transport system in rat liver for glutamine, asparagine, histidine, and closely related analogs. *J. Biol. Chem.* **255**, 4011–4019.

Kilberg, M.S. and D. Haussinger (1992). Amino acid transport in liver. In M.S. Kilberg and D. Haussinger (eds), *Mammalian Amino Acid Transport: Mechanisms and Control*. Academic Press, San Diego, CA, pp. 133–148.

King, N., H. Williams, J.D. McGivan, and M.S. Suleiman (2001). Characteristics of L-aspartate transport and expression of EAAC-1 in sarcolemma vesicles and isolated cells from rat heart. *Cardiovasc. Res.* **52**, 84–94.

Kowalski, T.J. and M. Watford (1994). Production of glutamine and utilization of glutamate by rat subcutaneous adipose tissue in vivo. *Am. J. Physiol.* **266**, E151–E154.

Kowalski, T.J., G. Wu, and M. Watford (1997). Rat adipose tissue amino acid metabolism in vivo as assessed by microdialysis and arteriovenous techniques. *Am. J. Physiol.* **272**, E613–E622.

Lange, K.H., J. Loresten, F. Isaksson, L. Simonsen, J. Bulow, and M. Kaer (2002). Lipolysis in human adipose tissue during exercise: comaparison of microdialysis and a-v measurements. *J. Appl. Physiol.* **92**, 1310–1316.

Levy, L.M., O. Warr, and D. Attwell (1998). Stoichiometry of the glial glutamate transporter GLT-1 expressed inducibly in a chinese hamster ovary cell line selected for low endogenous Na$^+$-dependent glutamate uptake. *J. Neurosci.* **18**, 962–9628.

Liao, K. and D. Lane (1995). Expression of a novel insulin-activated amino acid transporter gene during differentiation of 3t3-L1 preadipocytes into adipocytes. *Biochem. Biophys. Res. Comm.* **208**, 1008–1015.

Lin, C.I., I. Orlov, A.M. Ruggiero, M. Dykes-Hoberg, A. Lee, M. Jackson *et al.* (2001). Modulation of the neuronal glutamate transporter EAAC1 by the interacting protein GTRAP3–18. *Nature* **410**, 84–88.

Linder-Horowitz, M. (1969). Changes in glutaminase activities of rat liver and kidney during pre- and post-natal development. *Biochem. J.* **114**, 65–70.

Lindros, K.O., K.E. Penttila, J.J.W. Gaasbeek, A.F. Moorman, H. Speisky, and Y. Isreal (1989). The gamma-glutamyltransferase/glutamine synthetase activity ratio. A powerful marker for the acinar origin of hepatocytes. *J. Hepatol.* **8**, 338–343.

Low, S.Y., P.M. Taylor, H.S. Hundal, C.I. Pogson, and M.J. Rennie (1992). Transport of L-glutamine and L-glutamate across sinusoidal membranes of rat liver. Effects of starvation, diabetes and corticosteroid treatment. *Biochem. J.* **284**, 333–340.

Low, S.Y., M.J. Rennie, and P.M. Taylor (1994). Sodium-dependent glutamate transport in cultured rat myotubes increases after glutamine deprivation. *FASEB J.* **8**, 127–131.

Lynch, A.M. and J.D. McGivan (1987). Evidence for a single common Na$^+$-dependent transport system for alanine, glutamine, leucine and phenylalanine in brush-border membrane vesicles from bovine kidney. *Biochim. Biophys. Acta* **899**, 176–184.

Maechler, P. and C. Wolheim (1999). Mitochondrial glutamate acts as a messenger in glucose-induced insulin exocytosis. *Nature* **402**, 685–689.

Malakooti, J., R.Y. Dahdal, L. Schmidt, T.J. Layden, P.K. Dudeja, and K. Ramaswamy (1999). Molecular cloning, tissue distribution, and functional expression of the human Na$^+$/H$^+$exchanger NHE2. *Am. J. Physiol.* **277**, G383–G390.

Malandro, M.S., M.J. Beveridge, D.A. Novak, and M.S. Kilberg (1996). Effect of low-protein diet-induced intrauterine growth retardation on ra placental amino acid transport. *Am. J. Physiol.* **271**, C295–C303.

Marie, H., D. Billups, F.K. Bedford, A. Dumoulin, R.K. Goyal, G.D. Longmore *et al.* (2002). The amino terminus of the glial glutamate transporter GLT-1 interacts with the LIM protein Ajuba. *Mol. Cell. Neurosci.* **19**, 152–164.

Matsuo, H., Y. Kanai, J.Y. Kim, A. Chairoungdua, D. K. Kim, J. Inatomi *et al.* (2002). Identification of a novel Na$^+$-independent acidic amino acid transporter with structural similarity to the member of a heterodimeric amino acid transporter family associated with unknown heavy chains. *J. Biol. Chem.* **277**, 21017–21026.

Matthews, J.C., M.J. Beveridge, M.S. Malandro, D.A. Novak, and M.S. Kilberg (1998a). Response of placental amino acid transport to gestational age and intrauterine growth retardation. *Proc. Nutr. Soc.* **57**, 257–263.

Matthews, J.C., M.J. Beveridge, M.S. Malandro, J.D. Rothstein, M. Campbell-Thompson, J.W. Verlander *et al.* (1998b). Activity and protein localization of multiple glutamate transporters in gestation day 14 vs. day 20 rat placenta. *Am. J. Physiol.* **274**, C603–C614.

Matthews, J.C., M.J. Beveridge, E. Dialynas, A. Bartke, M.S. Kilberg, and D.A. Novak (1999). Placental anionic and cationic amino acid transporter expression in growth hormone overexpressing and null IGF-II or null IGF-I receptor mice. *Placenta* **20**, 639–650.

Matthews, J.C. (2000). Amino acid and peptide transport systems. In J.P.F.D'Mello (ed.), *Farm Animal Metabolism and Nutrition*. CABI Publishing, New York, pp. 3–23.

Matthews, J.C. and K.A. Anderson (2002). Recent advances in amino acids transporters and excitatory amino acid receptors. *Curr. Opin. Clin. Nutr. Metab. Care* **5**, 77–84.

McGivan, J.D. (1998). Rat hepatoma cells express novel transport systems for glutamine and glutamate in addition to those present in normal hepatocytes. *Biochem J.* **330**, 255–260.

Meijer, A.J., W.H. Lamers, and R.A.F.M. Chamuleau (1990). Nitrogen metabolism and ornithine cycle function. *Physiol. Rev.* **70**, 701–748.

Moe, A.J. and C.H. Smith (1989). Anionic amino acid uptake by microvillus membrane vesicles from human placenta. *Am. J. Physiol.* **257**, C1005–C1011.

Moore, R.D. (1993). Effect of insulin upon the sodium pump in frog skeletal muscle. *J. Physiol.* **232**, 23–45.

Moores, R.R., P.R. Vaughn, F.C. Battaglia, P.V. Fennessey, R.B. Wilkening, and G. Meschia (1994). Glutamate metabolism in fetus and placenta of late-gestation sheep *Am. J. Physiol.* **267**,:R89–R96.

Moorman, A.F., J.L. Vermeulen, R. Charles, and W.H. Lamers (1989). Localization of ammonia-metabolizing enzymes in human liver: ontogenesis of heterogeneity. *Hepatology* **9**, 367–372.

Moorman, A.F., P.A. de Boer, A.T. Das, W.T. Labruyere, R. Charles, and W.H. Lamers (1990). Expression patterns of mRNAs for ammonia-metabolizing enzymes in the developing rat: The ontogenesis of hepatocyte heterogeneity. *Histochem. J.* **22**, 457–468.

Munir, M., D.M. Correale, and M.B. Robinson (2000). Substrate-induced up-regulation of Na^+-dependent glutamate transport activity. *Neurochem. Int.* **37**, 147–162.

Newsholme, E.A. and M. Parry-Billings (1990). Properties of glutamate release from muscle and its importance for the immune system. *J. Parent. Ent. Nutr.* **14**, 63S–67S.

Nicholson, B. and J.D. McGivan (1996). Induction of high affinity glutamate transport activity by amino acid deprivation in renal epithelial cells does not involve an increase in the amount of transporter protein. *J. Biol. Chem.* **271**, 12159–12164.

Nissim, I. (1999). Newer aspects of glutamine/glutamate metabolism: The role of acute pH changes. *Am. J. Physiol.* **277**, F493–F497.

Novak, D.A. and M.J. Beveridge (2000). Anionic amino acid transporter expression in late gestation rodent yolk sac. *Placenta* **21**, 834–839.

Novak, D., F. Quiggle, C. Artime, and M. Beveridge (2001). Regulation of glutamate transport and transport proteins in a placental cell line. *Am. J. Physiol. Cell Physiol.* **281**, C1014–C1022.

Palacin, M., R. Estevez, J. Bertran, and A. Zorzano (1998). Molecular biology of mammalian plasma membrane amino acid transporters. *Physiol. Rev.* **78**, 969–1054.

Patterson, B.W., J.F. Horowitz, G. Wu, M. Watford, S.W. Coppack, and S. Klein (2002). Regional muscle and adipose tissue amino acid metabolism in lean and obese women. *Am. J. Endocrinol. Metab.* **282**, E931–E936.

Peghini, P., J. Janzen, and W. Stoffel (1997). Glutamate transporter EAAC-1-deficient mice develop dicarboxylic aminoaciduria and behavioral abnormalities but no neurodegeneration. *EMBO J.* **16**, 3822–3832.

Pines, G., N.C. Danbolt, M. Borjas, Y. Zhang, A. Bendahan, L. Eide *et al.* (1992). Cloning and expression of a rat brain L-glutamate transporter. *Nature* **360**, 464–467.

Plaitakis, A. and J.T. Caroscio (1987). Abnormal glutamate metabolism in amyotrophic lateral sclerosis. *Annu. Neurol.* **22**, 575–579.

Plakidou-Dymock, S. and J.D. McGivan (1993). Regulation of the glutamate transporter by amino acid deprivation and associated effects on the level of EAAC1 mRNA in the renal epithelial cell line NBL-I. *Biochem J.* **295**, 749–755.

Pollard, M., D. Meridith, and J.D. McGivan (2002). Characterization and cloning of a Na^+-dependent broad-specificity neutral amino acid transporter from NBL-1 cells: A novel member of the ASC/B^0 transporter family. *Biochim. Biophys. Acta* **1561**, 202–208.

Reeds, P.J., D.G. Burrin, F. Jahoor, L. Wykes, J. Henry, and E.M. Frazer (1996). Enteral glutamate is almost completely metabolized in first pass by the gastrointestinal tract of infant pigs. *Am. J. Physiol.* **270**, E413–E418.

Rennie, M.J., J.L. Bowtell, M. Bruce, and S.E.O. Khogali (2001). Interaction between glutamate availability and metabolism of glycogen, tricarboxylic acid cycle intermediates and glutathione. *J. Nutr.* **131**, 2488S–2490S.

Revest, P.A. and P.F. Baker (1988). Glutamate transport in large muscle fibres of Balanus nubilus. *J. Neurochem.* **50**, 94–102.

Ritchie, J.W., F.E. Baird, G.R. Christie, A. Stewart, S.Y. Low, H.S. Hundal *et al.* (2001). Mechanisms of glutamine transport in rat adipocytes and acute regulation by cell swelling. *Cell Physiol. Biochem.* **11**, 259–70.

Robinson, M.B. (2002). Regulated trafficking of neurotransmitter transporters: Common notes but different melodies. *J. Neurochem.* **80**, 1–11.

Rolinski, B., F.A. Baumeister, and A.A. Roscheer (2001). Determination of amino acid tissue concentrations by microdialysis: Method evaluation and relation to plasma values. *Amino Acids* **21**, 129–138.

Rossi, D.J., T. Oshima, and D. Attwell (2000). Glutamate release in severe brain ischemia is mainly by reversed uptake. *Nature* **403**, 316–321.

Saier, Jr., M.H. (1999). A proposed system for classification of transmembrane transport proteins in living organisms. In L.J. Van Winkle (ed.), *Biomembrane Transport.* Academic Press, San Diego, CA, pp. 265–276.

Sato, H., M. Tamba, T. Ishii, and S. Bannai (1999). Cloning and expression of a plasma membrane cystine/glutamate exchange transporter composed of two distinct proteins. *J. Biol. Chem.* **276**, 11455–11458.

Schlag, B.D., J.R. Vondrasek, M. Munir, A. Kalandadze, O.A. Zelenaia, J.D. Rothstein *et al.* (1998). Regulation of the glial Na$^+$-dependent glutamate transporters by cyclic AMP analogs and neurons. *Mol. Pharmacol.* **53**, 355–369.

Sims, K.D., D.J. Straff, and M.B. Robinson (2000). Platelet-derived growth factor rapidly increases activity and cell surface expression of the EAAC1 subtype of glutamate transporter through activation of phosphatidylinositol 3-kinase. *J. Biol. Chem.* **274**, 5228–5237.

Smith, E.M. and M. Watford (1988). Rat hepatic glutaminase: Purification and immunochemical characterization. *Arch. Biochem. Biophys.* **260**, 740–751.

Smith, E.M. and M. Watford (1990). Molecular cloning of a cDNA for rat glutaminase. Sequence similarity to kidney-type glutaminase. *J. Biol. Chem.* **265**, 10631–10636.

Snodgrass, P.J. and P. Lund (1984). Allosteric properties of phosphate-activated glutaminase of human liver mitochondria. *Biochim. Biophys. Acta* **798**, 21–27.

Spijkers, J.A.A., M.J.B. van den Hoff, T.B.M. Hakvoort, J.L.M. Vermeulen, S. Tesink-Taekema, and W.H. Lamers (2001) Foetal rise in hepatic enzymes follows decline in c-met and hepatic growth factor expression. *J. Hepatology* **34**, 699–710.

Stevens, B.R., H.J. Ross, and E.M. Wright (1982). Multiple transport pathways for neutral amino acids in rabbit jejunal brush border vesicles. *J. Membr. Biol.* **66**, 213–225.

Stoll, B., S. McNelly, H.P. Butcher, and D. Haussinger (1991). Functional hepatocyte heterogeneity in glutamate, aspartate and α-ketoglutarate uptake: A histoautoradiographical study. *Hepatology* **13**, 247–253.

Storck, T., S. Schulte, K. Hofmann, and W. Stoffel (1992). Structure, expression, and functional analysis of a Na$^+$-dependent glutamate/aspartate transporter from rat brain. *Proc. Natl. Acad. Sci. USA.* **89**, 10955–10959.

Sugawara, M., T. Nakanishi, Y.-J. Fei, W. Huang, M.E. Ganapathy, F.H. Leibach *et al.* (2000a). Cloning of an amino acid transporter with functional characteristics and tissue expression pattern identical to that of system A. *J. Biol. Chem.* **275**, 16473–16477.

Sugawara, M., T. Nakanishi, Y.-J. Fei, R.G. Martindale, M.E. Ganapathy, F.H. Leibach *et al.* (2000b). Structure and function of ATA3, a new subtype of amino acid transport system A, primarily expressed in the liver and skeletal muscle. *Biochim. Biophys. Acta* **1509**, 7–13.

Summers, L.K., P. Arner, V. Ilic, M.L. Clark, S.M. Humphreys, and K.N. Frayn (1998). Adipose tissue metabolism in the postprandial period: microdialysis and arteriovenous techniques compared. *Am. J. Physiol.* **274**, E651–655.

Tagari, H. and E.N. Bergman (1978). Intestinal disappearance and portal blood appearance of amino acids in sheep. *J. Nutr.* **108**, 790–803.

Tanaka, K. and I. Welbourne (2001). Enhanced ammonium excretion in mice lacking the glutamate transporter GLT-1. *JASN* **108**, A0287.

Taylor, P.M. and M.J. Rennie (1987). Perivenous localisation of Na-dependent glutamate transport in perfused rat liver. *FEBS Lett.* **221**, 370–374.

Trotti, D., M. Aoki, P. Pasinella, U.V. Berger, N.C. Danbolt, R.H. Brown, Jr. *et al.* (2001). Amyotrophic lateral sclerosis-linked glutamate transporter mutant has impaired glutamate clearance capacity. *J. Biol. Chem.* **276**, 576–582.

Ushmorov, A., V. Hack, and W. Druge (1999). Differential reconstitution of mitochondrial respiratory chain activity and plasma redox state by cysteine and ornithine in a model of cancer cachexia. *Cancer Res.* **59**, 3527–3534.

Utsunomiya-Tate, N., H. Endou, and Y. Kanai (1996). Cloning and functional characterization of a System ASC-like Na$^+$-dependent neutral amino acid transporter. *J. Biol. Chem.* **271**, 14883–14890.

van Milgen, J. (2002). Modeling biochemical aspects of energy metabolism in mammals. *J. Nutr.* **132**, 3195–2002.

Vaughn, P.R., C. Lobo, F.C. Battaglia, P.V. Fennessey, R.B. Wilkening, and G. Meschia (1995). Glutamine-glutamate exchange between placenta and fetal liver. *Am. J. Physiol.* **268**, E705–E711.

Verrey, F., C. Meir, G. Rossier, and L.C. Kuhn (2000). Glycoprotein-associated amino acid exchanges: Broadening the range of transport specificity. *Eur. J. Physiol.* **440**, 503–512.

Vesli, R.F., M. Klaude, O.E. Rooyackers, I. Tjader, H. Barle, and J. Wernerman (2002). Longitudinal pattern of glutamine/glutamate balance across the leg in long-stay intensive care patients. *Clin. Nutr.* **21**, 505–514.

Wagenaar, G.T., W.J. Geerts, R.A. Chamuleau, N.E. Deutz, and W.H. Lamars (1994). Lobular patterns of expression and enzyme activities of glutamine synthetase, carbamoylphosphate synthase and glutamate dehydrogenase during postnatal development of the porcine liver. *Biochim. Biophys. Acta* **1200**, 265–70.

Wang, H., W. Huang, M. Sugawara, L.D. Devoe, F.H. Leibach, P.D. Prasad *et al.* (2000). Cloning and functional expression of ATA1, a subtype of amino acid transporter A, from human placenta. *Biochem. Biophys. Res. Commun.* **14**, 1175–1179.

Watford, M. (1993). Hepatic glutaminase expression: relationship to kidney-type glutaminase and to the urea cycle. *FASEB J.* **7**, 1468–1474.

Watford, M. (2000). Glutamine and glutamate metabolism across the liver sinusoid. *J. Nutr.* **130**, 983S–987S.

Welbourne, T.C., and J.C. Matthews (1999). Glutamate transport and renal function. *Am. J. Physiol.* **277**, F501–F505.

Welbourne, T. and I. Nissim, (2001). Regulation of mitochondrial glutamine/glutamate metabolism by glutamate transport: Studies with ^{15}N. *Am. J. Physiol. Cell Physiol.* **280**, C1151–C1159.

Welbourne, T., R. Routh, M. Yudkoff, and I. Nissim (2001). The glutamine/glutamate couplet and cellular function. *News Physiol. Sci.* **16**, 157–160.

Wu, G. (1998). Intestinal mucosal amino acid catabolism. *J. Nutr.* **128**, 1249–1252.

Yao, D., B. Mackenzie, H. Ming, H. Varoqui, H. Zhu, M.A. Hediger *et al.* (2000). A novel system A isoform mediating Na$^+$/neutral amino acid transport. *J. Biol. Chem.* **275**, 22790–22797.

Zerangue, N. and M.P. Kavanaugh (1996a). ASCT-1 is a neutral amino acid exchanger with chloride channel activity. *J. Biol. Chem.* **271**, 27991–27994.

Zerangue, N. and M.P. Kavanaugh (1996b). Flux coupling in a neuronal transporter. *Nature* **383**, 634–637.

4

Anticancer Effects of Glutamate Antagonists

Wojciech Rzeski, Lechoslaw Turski, and Chrysanthy Ikonomidou

Despite significant progress achieved in chemotherapy, radiation technologies, surgical measures, bone marrow transplantation, and immunotherapy, the management of malignances in humans still constitutes a major challenge for contemporary medicine (Sporn, 1996; Scott and Cebon, 1997; Vijayakumar and Hellman, 1997; Vokes, 1997; Workman and Kaye, 2002). Gene therapy, after initial encouraging laboratory results, needs to overcome many practical obstacles to be successfully applied to humans (McCormic, 2001). Therefore, the development of novel and effective anticancer strategies continues to be eagerly pursued.

Glutamate is an essential amino acid and the major excitatory neurotransmitter in the mammalian nervous system (Watkins and Evans, 1981; Cavalheiro et al., 1988). N-methyl-D-aspartate (NMDA), α-amino-3-hydroxy-5-methyl-4-isoxasole-propionate (AMPA), kainate, and metabotropic receptors are activated by glutamate (Watkins and Evans, 1981; Cavalheiro et al., 1988). Glutamate is involved in interneuronal signaling, synaptic plasticity, sensory perception, learning and memory, and control of high cognitive functions (Turski et al., 2001). On the other hand, abnormally high extracellular glutamate levels and/or abnormally increased susceptibility of neurons to glutamate have been associated with neurological and psychiatric disorders including stroke, traumatic brain injury, epilepsy, and chronic neurodegenerative diseases, such as Parkinson's, Huntington's, and Alzheimer's disease (Lee et al., 1999; McNamara, 1999; Price, 1999; Zipfel et al., 1999). Glutamate antagonists were demonstrated to have anxiolytic, anticonvulsant, muscle relaxant, sedative/anesthetic, and neuroprotective properties, and their development for clinical applications in neurology has been eagerly pursued by the pharmaceutical industry (Turski et al., 2001).

A growing body of evidence indicates that glutamate and glutamate receptors are not restricted to the nervous system but also exist in non-neuronal tissues, including cancers. Glutamate receptors were identified in bone osteoblasts and osteoclasts, keratinocytes, megakaryocytes, pancreatic islets, the liver, the heart, in kidney cells, in adrenal tissue, and

Wojciech Rzeski • Department of Virology and Immunology, Institute of Microbiology and Biotechnology, Maria Curie-Sklodowska University, Akademicka 19, 20-033 Lublin, Poland.
Lechoslaw Turski • Solvay Pharmaceuticals, C. J. van Houtenlaan 36, 1382 CP Weesp, The Netherlands.
Chrysanthy Ikonomidou • Department of Pediatric Neurology, Charité, Virchow Clinics, Humboldt University, Augustenburger Platz 1, 13353 Berlin, Germany.

Glutamate Receptors in Peripheral Tissue, edited by Santokh Gill and Olga Pulido.
Kluwer Academic / Plenum Publishers, New York, 2005.

taste buds (Skerry and Genever, 2001). However, it is still elusive what role glutamate signaling plays in peripheral tissues.

During CNS development, glutamate regulates proliferation, migration, and survival of neuronal progenitors and immature neurons (Kleinschmidt et al., 1987; Komuro and Rakic, 1993; Behar et al., 1999; Ikonomidou et al., 1999). Certain characteristics of neuronal embryonic cells, including propensity to proliferate, migrate, and die, are shared by tumor cells, as is regulation of their invasive behavior by trophic factors (Welch et al., 1990). The above evidence, along with the knowledge that glutamate is present in non-neuronal tissues, including cancers, prompted investigations on whether glutamate and glutamate antagonists may influence growth and metastatic behavior of human cancer cells.

1. Glutamate Receptor Antagonists Inhibit Tumor Cell Proliferation

To evaluate whether glutamate antagonists may influence tumor cell proliferation, eight different tumor cell lines were exposed to different concentrations of the NMDA antagonists (+)dizocilpine (MK801), memantine, and ketamine or the AMPA antagonists GYKI52466, 2,3-dihydroxy-6-nitro-7-sulfamoyl-benzo(F)quinoxaline (NBQX), and 1-(4'-aminophenyl)-3,5-dihydro-7,8-dimethoxy-4H-2,3-benzodiazepin-4-one (CFM-2). Tumor cells derived from cancers of the nervous system (neuroblastoma, rhabdomyosarcoma/medulloblastoma, and astrocytoma) and a range of peripheral cancers including lung, breast, colon, and thyroid carcinoma were tested. Proliferation of all tumor cell lines was decreased in cultures exposed to antagonists in a concentration-dependent manner as measured by means of MTT assay (Figure 4.1). Time course studies revealed that antiproliferative effect of glutamate antagonists was already established after 24 hr. The antiproliferative effect of dizocilpine was reproduced in a three-dimensional neuroblastoma cell culture model, where cells were immobilized and allowed to grow on gelatin sponge (Spongostan®) (Rzeski et al., 2001). Compared to traditional cell cultures, such a three-dimensional culture model reflects much closer the in vivo conditions (Figure 4.1B).

The effect of glutamate antagonists on tumor cell proliferation was attributed both to decreased cell division, as determined by measurements of incorporation of BrdU, and increased cell death, as determined by measurements of LDH activity and trypan blue exclusion test (Rzeski et al., 2001). Colon adenocarcinoma (HT29), astrocytoma (MOGGCCM), breast carcinoma (T47D), and lung carcinoma (A549) were most sensitive to the cytostatic effect of the NMDA antagonist dizocilpine, whereas breast carcinoma (T47D), lung carcinoma (A549), colon adenocarcinoma (HT29), and neuroblastoma (SKNAS) were most sensitive to the AMPA antagonist GYKI52466. The threshold concentrations of dizocilpine or GYKI52466 required to elicit antiproliferative effects ranged from 1 to 50 μM (Rzeski et al., 2001). Significant antiproliferative effect of dizocilpine was detected at concentrations as low as 1 μM in colon adenocarcinoma (HT29) cells and as low as 10 μM in astrocytoma (MOG-GCCM) cells. In lung carcinoma (A549), neuroblastoma (SKNAS), breast carcinoma (T47D), Caucasian colon adenocarcinoma (LS180), rhabdomyosarcoma/medulloblastoma (TE671), and thyroid carcinoma (FTC238) cells, antiproliferative effects of dizocilpine were seen at 50 μM (Figure 4.1; Rzeski et al., 2001). Similarly, GYKI52466 significantly inhibited proliferation of colon adenocarcinoma cells (HT29) at concentrations as low as 1 μM and breast carcinoma (T47D) and Caucasian adenocarcinoma (LS180) cells at concentrations as

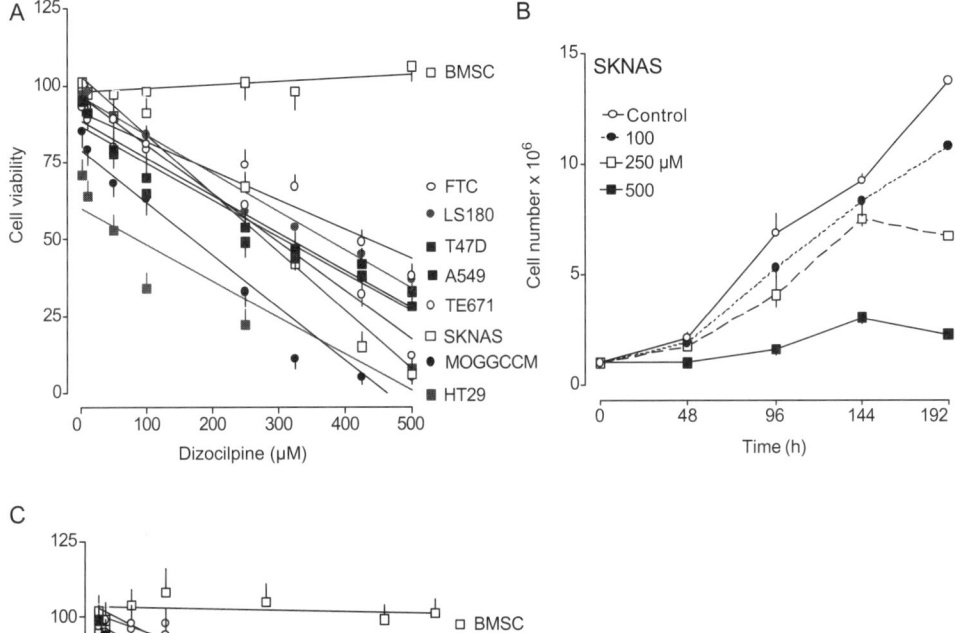

Figure 4.1.   The NMDA antagonist dizocilpine (A) and the AMPA antagonist GYKI52466 (C) exert concentration dependent antiproliferative effect in human tumor cell lines. Cells were exposed to either culture medium alone (control), dizocilpine (10–250 μM) or GYKI52466 (10–250 μM) over 96 hrs and viability was measured photometrically by means of the MTT assay. Data represent mean normalized optical densities ± SEM of 6–8 measurements and were analyzed by means of linear regression. In (B) growth inhibition of neuroblastoma cells in three-dimensional cultures by dizocilpine (100–500 μM) is shown in relation to the time in culture. Numbers of cells were assessed by counting following trypsinization.

low as 10 μM (Figure 4.1; Rzeski *et al.*, 2001). Such concentrations of glutamate antagonists are required to modulate NMDA- or AMPA-mediated currents in non-neuronal tissues such as osteoblasts and osteoclasts (Laketic-Ljubojevic *et al.*, 1999; Peet *et al.*, 1999) or to inhibit migration of embryonic cortical neurons (Behar *et al.*, 1999).

The antiproliferative effect of dizocilpine and GYKI52466 was reproduced by other NMDA (memantine and ketamine) and AMPA (NBQX and CFM-2) antagonists (Rzeski *et al.*, 2001). (−)Dizocilpine, the less active enantiomer of dizocilpine, was significantly less effective than

(+)dizocilpine in lung carcinoma (A549) and rhabdomyosarcoma/medulloblastoma (TE671) cells, suggesting stereoselectivity for the NMDA channel blockade (Rzeski *et al.*, 2001).

It is of crucial importance that anticancer drugs display antiproliferative activity in tumor cells without affecting healthy proliferating cells. Chemotherapy very often causes severe side effects, which are in part a consequence of destruction of normal cells (Komarov *et al.*, 1999). We tested the effect of glutamate antagonists on human skin fibroblasts (HSF) and bone marrow stromal cells (BMSC). Proliferation of these cell lines was unaffected by concentrations of dizocilpine and GYKI52466 as high as 250 μM (Rzeski *et al.*, 2001). The fact that glutamate antagonists were not antiproliferative in normal cells suggests a beneficial side-effect profile.

2. Glutamate Elicits Trophic Effects on Tumor Cells

The discovery of an antiproliferative action of glutamate antagonists on tumor cells led us to investigate whether glutamate and other glutamate receptor agonists influence tumor cell growth. Glutamate stimulated proliferation of lung carcinoma cells in serum-deprived medium or culture medium in which serum supplement was added. The trophic effect of glutamate was concentration-dependent within the range of 0.5–10 mM with maximal stimulation of cellular proliferation of 21% at 10 mM (Rzeski *et al.*, 2001). Similarly, NMDA, serine, and AMPA stimulated proliferation of A549 cells in serum-free media (unpublished data). These observations indicate that glutamate and other glutamate receptor agonists themselves can elicit trophic effects on tumor cells. It seems that these results are consistent with recently presented data showing that glutamate stimulates growth of gliomas *in vivo* (Takano *et al.*, 2001).

From the above evidence, the question arose whether tumor cells express subunits that form NMDA or AMPA receptor/ion channel complexes. To verify this hypothesis, immunocytochemistry was applied using antibodies against either the NR1 subunit of the NMDA receptor/ion channels or the Glu2/3 subunit of the AMPA receptor/ion channel. Positive immunostaining for both subunits was detected in all tumor cell lines, whereas no immunoreactivity was present in HSFs (Rzeski *et al.*, 2001). These data strongly support the idea that anticancer effect of glutamate antagonists can be directly linked to the receptor/ion channel blockade. Further characterization of glutamate receptors expressed on tumor cells and comparison to those expressed in nervous system is needed.

3. Glutamate Antagonists Limit Tumor Cell Invasiveness and Motility

Limiting tumor metastasis has highest priority in cancer therapy, since metastatic disease and not local tumor growth determines mortality in most peripheral cancers. The situation is different in the treatment of CNS cancers, where antiproliferative action is of crucial importance in order to preserve brain tissue and function. To evaluate the effect of glutamate antagonists on tumor cell morphology, lung carcinoma (A549), rhabdomyosarcoma/ medulloblastoma (TE671), and thyroid carcinoma (FTC238) cells were exposed to the NMDA antagonist dizocilpine (100 and 250 μM) or the AMPA antagonist GYKI52466 (100 and 250 μM) and their morphology was examined by light and scanning electron microscopy. Light microscopy revealed that dizocilpine induced rounded cell appearance with prominent vacuoles in the cytoplasm, whereas GYKI52466 produced less prominent vacuoles and shrinkage of the cells (Figure 4.2). Scanning electron microscopy revealed that tumor cells displayed an invasive phenotype with membrane ruffling and numerous pseudopodia (Figure 4.3). In contrast, tumor cells exposed to glutamate antagonists displayed a non-invasive phenotype with reduced membrane ruffling and absence of pseudopodia (Figure 4.3).

The ability of tumor cells to migrate is an important marker of tumor metastatic potential. To test the hypothesis that glutamate antagonists may decrease tumor cell locomotion, lung carcinoma (A549), rhabdomyosarcoma/medulloblastoma (TE671), and thyroid carcinoma (FTC238) cells were plated on 3 μM pore size polycarbonate membrane filters in the

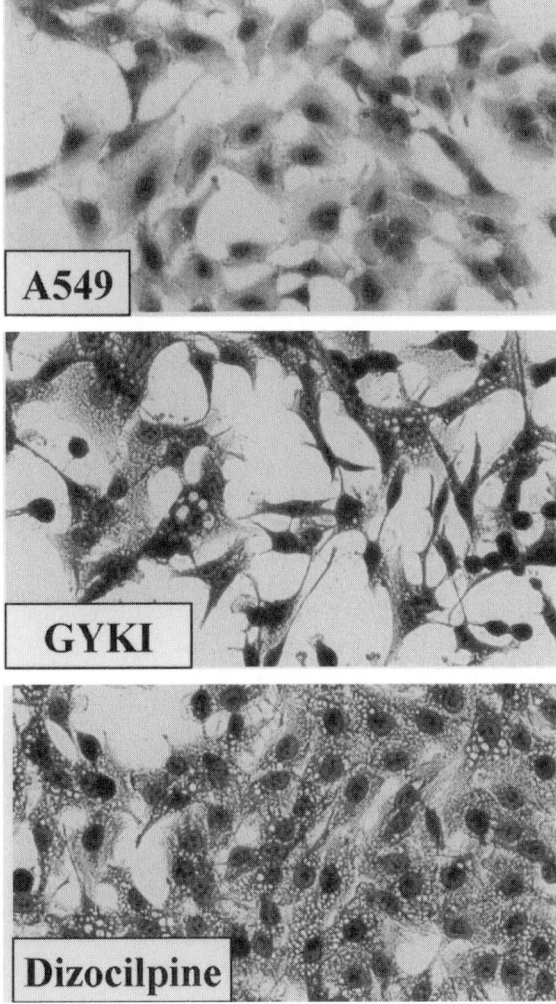

Figure 4.2. Glutamate antagonists alter tumor cell morphology. Light micrographs of lung carcinoma cells (A549) under control conditions and following exposure to dizocilpine (250 μM) and GYKI52466 (250 μM). Tumor cells exposed to dizocilpine display multiple cytoplasmic vacuoles. GYKI treated cells have an elongated appearance and less prominent cytoplasmic vacuolization.

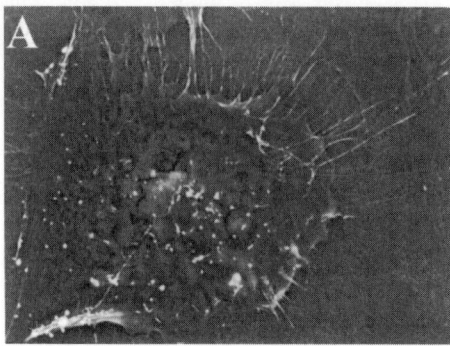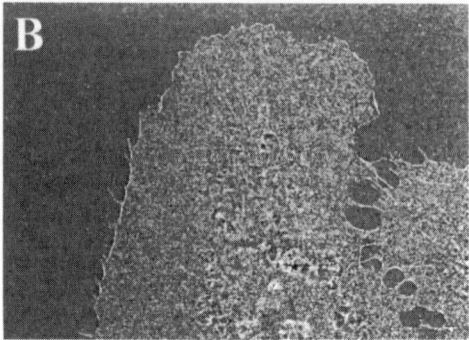

Figure 4.3. Scanning electron micrographs of (A) thyroid carcinoma cells (FTC238) under control conditions and (B) following exposure to dizocilpine (100 μM). The tumor cell in (A) displays numerous pseudopodia which are far less prominent following exposure to the glutamate antagonist (B).

presence and absence of dizocilpine and GYKI52466. In cultures exposed to glutamate antagonists, significantly fewer cells migrated through the filters than in control cultures. Of note is the fact that concentrations of glutamate antagonists influencing tumor cell locomotion were lower than those required to elicit an antiproliferative effect (Rzeski *et al.*, 2001).

4. Synergistic Effect of Glutamate Antagonists with Cytostatic Agents

Combination of different chemotherapeutic strategies often achieves superior cytostatic effects. Thus, we tested potential synergistic effect of glutamate antagonists with commonly used anticancer drugs. Lung carcinoma, astrocytoma, neuroblastoma, and rhabdomyosarcoma/ medulloblastoma cells were subjected to treatment with cyclophosphamide, cisplatin, thiotepa, or vinblastin and either dizocilpine or GYKI52466. As expected, the cytostatic drugs decreased tumor cell viability in a concentration-dependent manner. This effect was enhanced in all cell lines tested by glutamate antagonists (Rzeski *et al.*, 2001). Enhancement of antiproliferative effects of cytostatic agents by glutamate antagonists was due to enhanced tumor cell death and decreased cell division. Both, dizocilpine and GYKI52466, were found to enhance toxicity of cytostatic agents at concentrations as low as 10 μM (Rzeski *et al.*, 2001). These observations suggest that glutamate antagonists possess anticancer potential, which could add to existing chemotherapeutic regimens.

5. Possible Mechanism of Anticancer Activity of Glutamate Antagonists

There is no doubt that much work remains to be done to decipher the mechanisms involved in anticancer effect of glutamate antagonists. Calcium appears to play a crucial role, since in the absence of calcium in the culture medium, antiproliferative effect of glutamate antagonists was markedly diminished (Rzeski *et al.*, 2001). It is known that calcium can stimulate tumor growth (Meloni *et al.*, 1998; Celli *et al.*, 1999), that calcium is necessary for cell division and survival (Clapham, 1995; Gleason and Spitzer, 1998; Horner and Gage, 2000), and that it regulates protein trafficking through the nuclear membrane (Stehno-Bittel and Perez-Terzic, 1995). Calcium also controls axon extension and pathfinding and influences cell migration (Marks and Maxfield, 1990; Nakato *et al.*, 1992; Lawson and Maxfield 1995; Gomez and Spitzer, 1999). It is known from developmental work that glutamate receptor/ion channels on embryonic neurons are permeable to calcium (Gallo *et al.*, 1995; Bardoul *et al.*, 1998; Scherer and Gallo, 1998). Thus, the interesting hypothesis arises whether calcium trafficking through the cytoplasmic membrane of tumor cells via glutamate receptor/ion channel complexes may be occurring to a much higher extent than it does in neurons, since the resting membrane potential in tumor cells is in the range of -30 to -50 mV (Iwata *et al.*, 1999; Sonnier *et al.*, 2000). In neurons, at such depolarized membrane potentials, the Mg^{2+} block of the NMDA receptor ion channel is relieved and ion-permeability of both NMDA and AMPA glutamate receptor/ion channels increases compared to resting membrane potentials (Burnashev *et al.*, 1995; Jonas and Burnashev, 1995). Blockade by glutamate antagonists of such an active ion trafficking across the cytoplasmic membrane may indeed be one mechanism to explain their antiproliferative activity.

6. Conclusions

The discovery of anticancer activity of glutamate antagonists provides new challenges for cancer biologists and the pharmaceutical industry. One crucial issue to resolve is determining whether glutamate antagonists exert similar anticancer activity *in vivo*. It will be important to decipher the molecular pathways that glutamate antagonists utilize to limit tumor growth, invasiveness, and migration. The electrophysiological and binding properties of glutamate receptor/ion channels present on tumor cells will need to be investigated as well as their subunits better characterized and sequenced. Having achieved this, hopefully it will be possible to support existing chemotherapy armamentarium with a new class of drugs that have primarily been developed for neurological disorders.

Abbreviations

SKNAS, human neuroblastoma; TE671, human medulloblastoma; MOGGCCM, human brain astrocytoma; FTC238, human thyroid carcinoma; A549, human Caucasian lung carcinoma; LS180, human Caucasian colon adenocarcinoma; T47D, human breast carcinoma.

References

Bardoul, M., C. Levallois, and N. Konig (1998). Functional AMPA/kainate receptors in human embryonic and foetal central nervous system. *J. Chem. Neuroanat.* **14**, 79–85.

Behar, T.N., C.A. Scott, C.J. Greene, X. Wen, S.V. Smith, D. Maric *et al.* (1999). Glutamate acting at NMDA receptors stimulates embryonic cortical neuronal migration. *J. Neurosci.* **19**, 4449–4461.

Burnashev, N., Z. Zhou, E. Neher, and B. Sakmann (1995). Fractional calcium currents through recombinant GluR channels of the NMDA, AMPA and kainate receptor subtypes. *J. Physiol.* **485**, 403–418.

Cavalheiro, E.A., J. Lehmann, and L. Turski (1988). *Frontiers in Excitatory Amino Acid Research*. Alan R Liss, New York.

Celli, A., C. Treves, P. Nassi, and M. Stio (1999). Role of 1,25-dihydroxyvitamin D3 and extracellular calcium in the regulation of proliferation in cultured SH-SY5Y human neuroblastoma cells. *Neurochem. Res.* **24**, 691–698.

Clapham, D.E. (1995). Calcium signalling. *Cell* **80**, 259–268.

Gallo, V., M. Pende, S. Cherer, M. Molne, and P. Wright (1995). Expression and regulation of kainate and AMPA receptors in uncommitted and committed neural progenitors. *Neurochem. Res.* **20**, 549–560.

Gleason, E.L. and N.C. Spitzer (1998). AMPA and NMDA receptors expressed by differentiating Xenopus spinal neurons. *J. Neurophysiol.* **79**, 2986–2998.

Gomez, T.M. and N.C. Spitzer (1999). *In vivo* regulation of axon extension and pathfinding by growth-cone calcium transients. *Nature* **397**, 350–355.

Horner, P.J. and F.H. Gage (2000). Regenerating the damaged central nervous system. *Nature* **407**, 963–970.

Ikonomidou, C., F. Bosch, M. Miksa, P. Bittigau, J. Vöckler, and K. Dirkanian *et al.* (1999). Blockade of NMDA receptors and apoptotic neurodegeneration in the developing brain. *Science* **283**, 70–74.

Iwata, M., S. Komori, T. Unno, N. Minamoto, and H. Ohashi (1999). Modification of membrane currents in mouse neuroblastoma cells following infection with rabies virus. *Br. J. Pharmacol.* **126**, 1691–1698.

Jonas, P. and N. Burnashev (1995). Molecular mechanisms controlling calcium entry through AMPA-type glutamate receptor channels. *Neuron* **5**, 987–990.

Kleinschmidt, A., M.F. Bear, and W. Singer (1987). Blockade of "NMDA" receptors disrupts experience-dependent plasticity of kitten striate cortex. *Science* **238**, 355–358.

Komarov, P.G., E.A. Komarova, R.V. Kondratov, K. Christov-Tselkov, J.S. Coon, M.V. Chernov *et al.* (1999). A chemical inhibitor of p53 that protects mice from the side effects of cancer therapy. *Science* **285**, 1733–1737.

Komuro, H. and P. Rakic (1993). Modulation of neuronal migration by NMDA receptors. *Science* **260**, 95–97.

Laketic-Ljubojevic, I., L.J. Suva, F.J.M. Maathuis, D. Sanders, and T.M. Skerry (1999). Functional characterization of N-methyl-D-aspartic acid-gated channels in bone cells. *Bone* **25**, 631–637.

Lawson, M.A. and F.R. Maxfield (1995). Ca^{2+}- and calcineurin-dependent recycling of an integrin to the front of migrating neutrophils. *Nature* **377**, 75–79.

Lee, J.-M., G.J. Zipfel, and D.W. Choi (1999). The changing landscape of ischaemic brain injury mechanisms. *Nature* **399**, A7–A14.

Marks, P.W. and F.R. Maxfield (1990). Transient increases in cytosolic free calcium appear to be required for the migration of adherent human neutrophils. *J. Cell. Biol.* **110**, 43–52.

McCormic, F. (2001). Cancer gene therapy: Fringe or cutting edge? *Nature Rev. Cancer* **1**, 130–141.

McNamara, J.O. (1999). Emerging insights into the genesis of epilepsy. *Nature* **399**, A15–A22.

Meloni, F., A. Brochieri, P.C. Ballabio, A. Tua, G. Grignani, and G.G. Grassi (1998). Bombesin, calcium homeostasis and tumour growth. *Monaldi Arch. Chest Dis.* **53**, 405–409.

Nakato, K., T. Furuno, K. Inagaki, R. Teshima, R. Terao, and M. Nakanishi (1992). Cytosolic and intranuclear calcium signals in rat basophilic leukemia cells as revealed by a confocal fluorescence microscope. *Eur. J. Biochem.* **209**, 745–749.

Peet, N.M., P.S. Grabowski, I. Laketic-Ljubojevic, and T.M. Skerry (1999). The glutamate receptor antagonist MK801 modulates bone resorption *in vitro* by a mechanism predominantly involving osteoclast differentiation. *FASEB J.* **13**, 2179–2185.

Price, D.L. (1999). New order from neurological disorders. *Nature* **399**(Suppl.), A3–A5.

Rzeski, W., L. Turski, and C. Ikonomidou (2001). Glutamate antagonists limit tumor growth. *Proc. Natl. Acad. Sci USA* **98**, 6372–6377.

Scherer, S.E. and V. Gallo (1998). Expression and regulation of kainate and AMPA receptors in the rat neural tube. *J. Neurosci. Res.* **52**, 356–368.

Scott, A.M. and J. Cebon (1997). Clinical promise of tumour immunology. *Lancet* **349**, S19–S22.

Skerry, T.M. and P.G. Genever (2001). Glutamate signalling in non-neuronal tissues. *Trends in Pharmacol. Sci.* **22**, 174–181.

Sonnier, H., O.V. Kolomytkin, and A. Marino (2000). Resting potential of excitable neuroblastoma cells in weak magnetic fields. *Cell. Mol. Life Sci.* **57**, 514–520.

Sporn, M.B. (1996). The war on cancer. *Lancet* **347**, 1377–1381.

Stehno-Bittel, L. and C. Perez-Terzic (1995). Diffusion across the nuclear envelope inhibited by depletion of the nuclear Ca^{2+} store. *Science* **270**, 1835–1838.

Takano, T., J.H.-C. Lin, G. Arcuino, Q. Gao, J. Yang, and M. Nedergaard (2001). Glutamate release promotes growth of malignant gliomas. *Nat. Med.* **7**, 1010–1015.

Turski, L., E.A. Cavalheiro, and D.D. Schoepp (2001). *Excitatory Amino Acids: Ten Years Later*. IOS Press, Amsterdam.

Vijayakumar, S. and S. Hellman (1997). Advances in radiation oncology. *Lancet* **349**, S1–S3.

Vokes, E.V. (1997). Combined modality therapy of solid tumours. *Lancet* **349**, S4–S6.

Watkins, J.C. and R.H. Evans (1981). Excitatory amino acid transmitters. *Ann. Rev. Pharmacol. Toxicol.* **21**, 165–204.

Welch, D.-R., A. Fabra, and M. Nakajima (1990). Transforming growth factor beta stimulates mammary adenocarcinoma cell invasion and metastatic potential. *Proc. Natl. Acad. Sci. USA* **93**, 7688–7692.

Workman, P. and S. Kaye (2002). Translating basic cancer research into new cancer therapeutics. *Trends Mol. Med.* **8**, 1–9.

Zipfel, G.L., J.M. Lee, and D.W. Choi (1999). Reducing calcium overload in the ischemic brain. *N. Engl. J. Med.* **341**, 543–544.

5

Glutamate Receptors and their Role in Acute and Inflammatory Pain

Susan M. Carlton

1. Introduction

During the last decade, there has been a resurgence of interest in the contribution of peripheral receptors to sensory transduction and transmission. In particular, discovery of both ionotropic and metabotropic glutamate receptors on peripheral nociceptors suggests that these families of receptors may provide novel targets for modulation of sensory input.

2. Localization of Glutamate Receptors on Primary Sensory Neurons

2.1. Ionotropic

Ionotropic glutamate (iGlu) receptors are associated with ion channels and are responsible for fast synaptic transmission. The first evidence that iGlu receptors are present on primary sensory neurons comes from *in vitro* dorsal root recordings from both young and adult rats where it is demonstrated that kainate depolarizes unmyelinated C fibers (Davies *et al.*, 1979; Evans, 1985; Agrawal and Evans, 1986; Evans *et al.*, 1987). Whole cell patch clamp technique demonstrates that rat sensory neurons are depolarized by glutamate, N-methyl-D-aspartate (NMDA), kainate, α-amino-3-hydroxy-5-methyl-4-isoxazole propionic acid (AMPA), and quisqualate (Lovinger and Weight, 1988; Huettner, 1990). The presence of iGlu receptors in primary sensory neurons is confirmed in anatomical studies using *in situ* hybridization and localization of mRNA, Western blots, and real-time RT-PCR, or antibodies and immunohistochemistry, allowing localization of NMDA, AMPA, and kainate receptor subunits in dorsal root ganglion (DRG; Shigemoto *et al.*, 1992; Sato *et al.*, 1993; Liu *et al.*, 1994; Marvizon *et al.*, 2002) and trigeminal ganglion cells (Sahara *et al.*, 1997). Using double labeling techniques, subunit 1 of the NMDA receptor (NR1) is localized in DRG cells innervating the colon (McRoberts *et al.*, 2001), and the 2B subunit (NR2B) localized in DRG cells giving rise to small caliber sciatic nerve fibers (Ma and Hargreaves, 2000). Western blots

Susan M. Carlton • University of Texas Medical Branch, Galveston, TX 77555-1069.

Glutamate Receptors in Peripheral Tissue, edited by Santokh Gill and Olga Pulido.
Kluwer Academic / Plenum Publishers, New York, 2005.

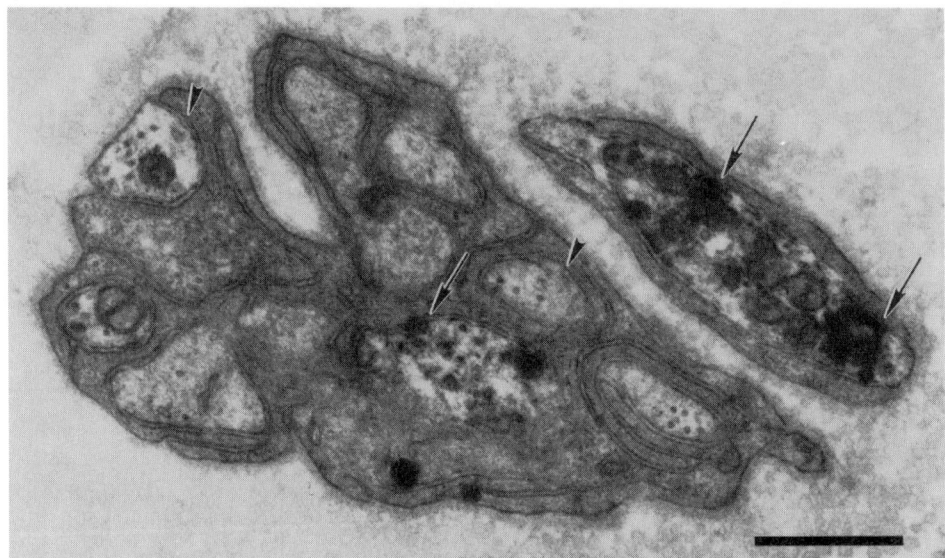

Figure 5.1. Electron micrograph demonstrating a bundle of unmyelinated peripheral axons; two are immunohistochemically labeled for the NMDAR1 subunit identifying NMDA receptors (arrows). Unlabeled axons are also present (arrowheads). Bar = 0.5 μm. (From Carlton *et al.*, 1995.)

and RT-PCR indicate that DRG cells have two different NMDA receptors, one containing the NR1, NR2D and possibly NR2C subunits, present only in C fibers and the other containing the NR1/NR2B subunits present in both A and C fibers (Marvizon *et al.*, 2002). That glutamate receptors and in particular NMDA receptors are transported both centrally and peripherally is demonstrated with autoradiography and receptor binding (Lewis *et al.*, 1987; Cincotta *et al.*, 1989). Immunohistochemical analysis at the electron microscopic level demonstrates the presence of NMDA receptors on the central terminals of primary afferents (Liu *et al.*, 1994). The authors hypothesize that these NMDA receptors could function as autoreceptors, increasing neurotransmitter release since many of the NMDA receptor-labeled terminals contain glutamate.

The presence of glutamate receptors on peripheral terminals is predicted from earlier studies showing bidirectional transport. Accordingly, it is shown that *L*-glutamate and kainate applied to rat tail skin stimulates nociceptive reflexes in an isolated spinal cord-tail preparation (Ault and Hildebrand, 1993a, b). Anatomical studies in rat confirm that about 20% of unmyelinated cutaneous axons at the dermal–epidermal junction label for the NMDA, AMPA, or kainate receptors (Figure 5.1; Carlton *et al.*, 1995). Both myelinated and unmyelinated axons in the sural and plantar nerves (Coggeshall and Carlton, 1998) and in human skin (Kinkelin *et al.*, 2000) label for iGlu receptors.

2.2. Metabotropic

Glutamate activates a family of G protein-coupled receptors referred to as metabotropic glutamate (mGlu) receptors (Pin and Duvoisin, 1995). There are currently eight receptors in this family and they are believed to play a role in the fine-tuning of synapses. The mGlu receptor family is divided into three groups on the basis of their sequence homology, transduction mechanisms, and pharmacology (Conn and Pin, 1997). Group I (mGlu1 and mGlu5) receptors are linked to phosphoinositide hydrolysis and intracellular calcium mobilization via phospholipase C (Pin and Duvoisin, 1995; Conn and Pin, 1997), whereas Group II (mGlu2 and mGlu3) and Group III (mGlu4, mGlu6–8) receptors are negatively coupled to adenylyl cyclase, and activation results in a decrease in cyclic AMP production (Pin and Duvoisin, 1995; Conn and Pin, 1997). Thus, based on CNS studies, activation of Group I mGlu receptors can result in increased neuronal excitability whereas activation of Group II or III mGlu receptors can result in decreased neuronal excitability. Anatomical studies demonstrate that Group I mGlu receptors are localized on peripheral unmyelinated primary afferent axons in the rat and mouse (Bhave *et al.*, 2001; Walker *et al.*, 2001; Zhou *et al.*, 2001). Antibodies recognizing Group II mGlu receptors demonstrate labeling of 50% of rat small diameter DRG cells and 32% of peripheral unmyelinated axons in the digital nerve (Carlton *et al.*, 2001). Furthermore, 28% of myelinated axons are labeled for mGluR2/3. In the spinal cord, the Group III mGlu7 receptor is localized presynaptically on presumed nociceptive primary afferents fibers (Li *et al.*, 1997), but no data are currently available concerning the distribution of Group III mGlu receptors on peripheral axons (Ohishi *et al.*, 1995).

2.3. Conclusions

Both iGlu and mGlu receptors are localized on the cell bodies and the peripheral processes of primary sensory neurons. If functional, this positioning would allow iGlu receptors to participate in the fast transmission of sensory impulses from the periphery. In contrast, the mGlu receptors might modulate/modify sensory transduction before transmission of impulses to the central nervous system.

3. Peripheral Glutamate Receptors in Acute Pain

3.1. Behavioral Studies

Several lines of evidence indicate that activation of peripheral iGlu receptors results in nociception. The earliest report states that a single intraplantar injection of glutamate results in heat hyperalgesia that lasts for 8 days and is blocked by an NMDA antagonist (Follenfant and Nakamura-Craig, 1992). Subsequently, it was demonstrated that intraplantar glutamate results in mechanical (Carlton *et al.*, 1995) and thermal hyperalgesia (Jackson *et al.*, 1995). Intraplantar injection of specific ligands such as NMDA, AMPA, or kainate results in mechanical hyperalgesia and allodynia and these behaviors are blocked by coinjection of the appropriate antagonists in the hindpaw (Zhou *et al.*, 1996). Furthermore, injection of different combinations of excitatory amino acids (EAA) (i.e., glutamate/aspartate) into the rat knee joint cavity results in reduced paw withdrawal latencies and paw withdrawal threshold indicating the generation of thermal hyperalgesia and mechanical allodynia and these are blocked by intra-articular injection of either NMDA or non-NMDA antagonists (Lawand *et al.*, 1997). Subcutaneous injection of MK-801, an NMDA antagonist, produces a reversible reduction in the responses of lumbar wide dynamic range neurons to noxious and innocuous stimuli (Ushida *et al.*, 1999). This occurs through a peripheral action since injection of MK-801 in the contralateral hindpaw has no effect.

Peripheral NMDA receptors play a role in visceral nociception. Activation of NMDA receptors in colonic tissue causes Ca^{2+}-dependent release of proinflammatory peptides, and behavioral responses to noxious colonic mechanical stimulation are inhibited in a dose-dependent fashion by an NMDA antagonist (McRoberts *et al.*, 2001). Furthermore, this study shows that single fiber recordings from afferents in cut pelvic nerves demonstrate activity in response to colorectal distension that is inhibited by an NMDA antagonist. The authors conclude that peripheral NMDA receptors are important in normal visceral pain transmission and may provide a mechanism for development of peripheral sensitization and visceral hyperalgesia.

Administration of kainic acid through a variety of parenteral routes (intraperitoneal, subcutaneous), but not intrathecal, results in a persistent thermal and mechanical hyperalgesia in rats and mice that can be blocked by 6-cyano-7-nitroquinoxaline-2,3-dione (CNQX), a non-NMDA antagonist (Giovengo *et al.*, 1999). An intrathecal antagonist fails to attenuate the hyperalgesic effect of kainic acid administered intraperitoneally. Thus, the authors conclude that the spinal cord is not the primary site of action of the kainic acid. Activation of kainate receptors on peripheral projections of primary afferents is proposed.

Recent data also indicates a role for peripheral mGlu receptors in acute nociception. Intraplantar injection of a Group I agonist, activating mGlu1/5 receptors results in a significant and prolonged increase in mechanical but not thermal sensitivity in rats that is blocked by coinjection of (RS)-1-aminoindan-1,5-dicarboxylic acid (AIDA), an mGlu1α antagonist (Zhou *et al.*, 2001) or by coinjection of 2-methyl-6-(phenylethynyl) pyridine hydrochloride (MPEP), an mGlu5 receptor antagonist (Walker *et al.*, 2001). In contrast, intraplantar injection of a Group I agonist causes thermal hyperalgesia in mice (Bhave *et al.*, 2001). This difference in efficacy may represent a species difference or a dose effect.

3.2. Physiological Studies

Recordings from nociceptors using an *in vitro* skin-nerve preparation demonstrate that 43% Aδ and 68% C but no Aβ fibers are excited by an ascending series of glutamate concentrations (10–1,000 μM) (Du *et al.*, 2000). Furthermore, exposure of the receptive fields of nociceptors to glutamate results in sensitization of the units to heat evidenced by an increase in the discharge rate and a decrease in the threshold to heat activation (Figure 5.2; Du *et al.*, 2000). There is, however, no change in threshold to mechanical stimulation. Glutamate-induced excitation and sensitization of nociceptors in the skin provide a physiological basis for the nociceptive behaviors observed following intraplantar injection of glutamate or glutamate agonists in the hindpaw.

Evidence that EAA receptors within the temporomandibular joint (TMJ) are involved in the reflex activation of the jaw muscles is demonstrated by the application of EAA agonists glutamate, NMDA, kainate, or AMPA to the TMJ region. Jaw muscle responses are similar to those evoked by mustard oil application to the TMJ region and this activity is blocked by co-application NMDA or non-NMDA antagonists (Cairns *et al.*, 1998; Fiorentino *et al.*, 1999).

Wide dynamic range neurons in the spinal cord demonstrate increased activity following injection of a Group I agonist into their peripheral receptive fields (Walker *et al.*, 2001). Coinjection with a Group I antagonist inhibits this activity but injection of the antagonist alone has no effect.

3.3. Conclusions

These behavioral and physiological studies confirm that glutamate receptors localized on peripheral nociceptors are functionally relevant in the transmission of acute, physiologic pain

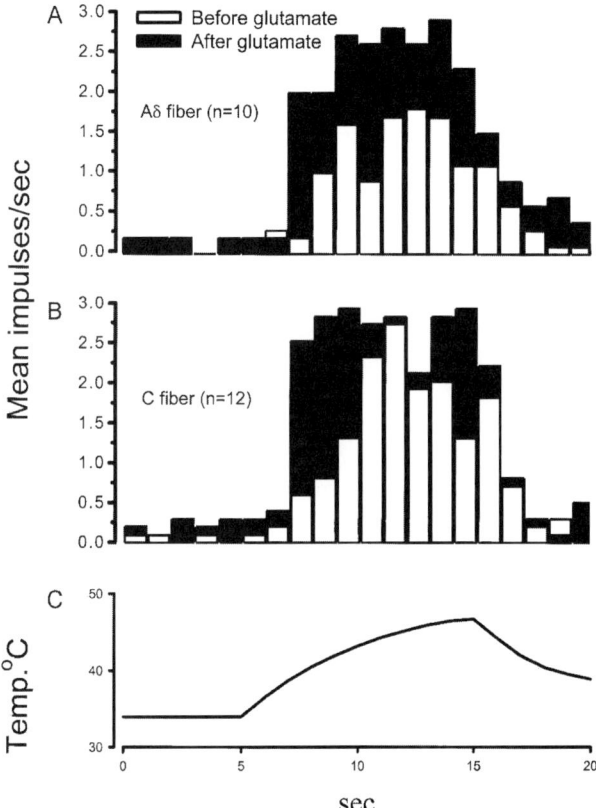

Figure 5.2. Exposure of the receptive fields of (A) Aδ and (B) C fibers to 0.3 mM glutamate for 2 min results in a shift to the left in the heat response such that higher discharge frequencies occur during the early part of (C) a 10-s heat trial. The decrease in heat thresholds and increase in discharge rates indicate glutamate-induced sensitization to heat. (From Du *et al.*, 2000.)

impulses in animal models of nociception. Furthermore, the data suggest that manipulation of peripheral glutamatergic systems could lead to therapies for maladaptive pain in humans.

4. Peripheral Glutamate Receptors in Inflammatory Pain

4.1. Anatomical Studies

Studies using persistent/chronic inflammatory animal models indicate that peripheral glutamate receptors may contribute in a significant way to inflammatory pain. Following intraplantar injection of compete Freund's adjuvant (CFA), there is a significant increase in the proportion of peripheral unmyelinated axons labeled for either NMDA, AMPA, or kainate receptors (Carlton and Coggeshall, 1999). In contrast to these findings, one study reports that the number of NMDAR1-immunostained cells in the DRG of CFA-inflamed animals is decreased for at least 20 days following the initial inflammation (Wang *et al.*, 1999). Based on the anatomical results in the periphery, it is highly likely that the decrease in DRG label is due to enhancement of anterograde transport of NMDAR1 from the cell bodies to peripheral terminals. Analysis of mRNA levels for NMDAR1 in the DRG would help confirm this hypothesis. Intraplantar injection of NMDA results in an increased expression of c-fos in the superficial dorsal horn, similar to that produced by formalin injection (Wang *et al.*, 1997). Coinjection of an NMDA receptor antagonist with formalin suppresses the formalin-induced c-fos expression, confirming the presence of NMDA receptors on peripheral nociceptors and a role in nociception.

4.2. Behavioral Studies

Investigations using various inflammatory models provide several lines of evidence that peripheral iGlu receptor activation contributes to inflammatory pain. The formalin test has an inflammatory component and it is demonstrated that intraplantar injection of NMDA or non-NMDA receptor antagonists attenuates formalin-induced nociceptive behaviors (Figure 5.3; Davidson *et al.*, 1997; Davidson and Carlton, 1998). In more persistent pain models induced by injection of carrageenan, carrageenan/kaolin, or CFA, the thermal and mechanical hyperalgesia induced by the inflammatory agent is attenuated following peripheral administration of NMDA and non-NMDA antagonists (Jackson *et al.*, 1995; Lawand *et al.*, 1997).

Manipulation of peripheral mGlu receptors demonstrates that intraplantar injection of Group I antagonists can reduce formalin pain behaviors (Bhave *et al.*, 2001; Zhou *et al.*, 2001) and CFA- and carrageenan-induced hyperalgesia (Walker *et al.*, 2001).

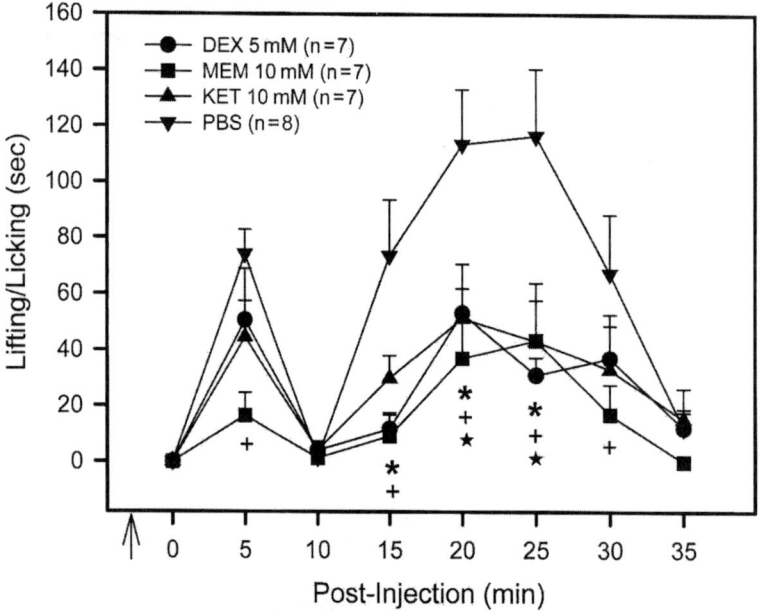

Figure 5.3.   Intraplantar injection of the NMDA antagonists, dextrorphan (DEX), memantine (MEM), or ketamine (KET), results in attenuation of lifting/licking, a formalin-induced nociceptive behavior. ↑ = Formalin injection. (From Davidson *et al.*, 1998.)

4.3. Physiological Studies

The C fiber excitant and inflammatory irritant mustard oil applied to the TMJ causes increased electromyographic (EMG) activity in the masseter and digastric muscles. Application of an NMDA antagonist to the TMJ region reduces the magnitude of the EMG activity evoked by mustard oil, indicating that peripheral NMDA receptors are involved in the generation of the activity (Yu *et al.*, 1996).

4.4. Conclusions

Inflammatory pain models in animals are accompanied by increases in peripheral iGlu receptors on nociceptors, behavioral signs of hyperalgesia and allodynia that can be attenuated by peripheral administration of iGlu and mGlu antagonists, and increased EMG activity from peripheral nerves that is EAA-dependent. Thus, peripheral glutamatergic systems contribute to inflammatory pain.

5. Clinical Relevance: Human Studies

It is reported that injection of glutamate into the masseter muscle of male and female volunteers evokes pain that is significantly greater than isotonic saline. Furthermore, the glutamate-evoked peak pain and the overall muscle pain are significantly greater in females compared to males (Cairns *et al.*, 2001). Studies repeated in rats of both sexes while recording from masseter muscle afferents show that glutamate evokes a significantly greater response in afferent activity in female rats compared to males. The results indicate that glutamate injection into the masseter muscle evokes pain responses in humans that are greater in women. These sex-related differences may be relevant to the higher prevalence of chronic muscle pain conditions in women (Cairns *et al.*, 2001).

Local peripheral treatment with the NMDA receptor antagonist ketamine reduces the primary hyperalgesia and prevents the development of secondary hyperalgesia following an experimental burn injury in volunteers (Warncke *et al.*, 1997). These data indicate that cutaneous NMDA receptors contribute to the development of primary and secondary hyperalgesia in humans.

Samples of synovial fluid taken from patients with active arthritis show increased levels of glutamate and aspartate suggesting that EAAs may contribute to the pathogenesis of the human arthritic condition (McNearney *et al.*, 2000).

Glutamate is present in human disc material and is significantly increased in herniated compared to non-herniated discs (Harrington *et al.*, 2000). Following epidural infusion of low concentrations of [^{3}H]-glutamate in rats, DRG cells demonstrate significant glutamate labeling indicating that release of glutamate from herniated discs could contribute to activation of glutamate receptors present on nociceptive DRG cells. The authors conclude from these studies that glutamate, released from herniated discs could contribute to back pain (Harrington *et al.*, 2000).

6. Sources of Peripheral Glutamate

There are many sources of ligand for activation of peripheral glutamate receptors. In the normal, glutamate is localized in the majority of DRG cells (Battaglia and Rustioni, 1988;

Keast and Stephensen, 2000) and glutamate and aspartate are present in myelinated and unmyelinated dorsal roots (Westlund *et al.*, 1989a,b). Several studies document release of glutamate and aspartate from DRG cells or their central processes (Skilling *et al.*, 1988; Jeftinija *et al.*, 1991; Kangrga and Randic, 1991; Sorkin and Moore, 1996). Glutamate content in the hindpaw increases following intraplantar injection of formalin (Omote *et al.*, 1998), after electrical stimulation of the sciatic nerve at A or C fiber strength, or after application of capsaicin to the sciatic nerve (deGroot *et al.*, 2000). Glutamate is also present in keratinocytes (Genever *et al.*, 1999) and serum (Erdo, 1991). During inflammation, plasma extravasation would increase the glutamate content in inflamed regions. Furthermore, during inflammation, glutamate content increases not only in nerves innervating the inflamed region (Westlund *et al.*, 1992), but also in dialysate from inflamed knee joints (Lawand *et al.*, 2000), and in dermal and epidermal cells in inflamed skin (Nordlind *et al.*, 1993).

7. Summary

Data from both animals and humans indicate that peripheral glutamate receptors are involved in nociceptive transmission in the normal and inflamed state. Studies in animal models indicate that peripheral glutamate receptors may provide novel targets for treatment of inflammatory pain. The variety of peripheral iGlu and mGlu receptors provides many targets for drug development. Manipulating the peripheral glutamatergic system could lead to advances in treating maladaptive pain while avoiding CNS side effects.

Acknowledgment

This work was supported by NS11255, NS27910, and NS40700 to SMC.

References

Agrawal, S.G. and R.H. Evans (1986). The primary afferent depolarizing action of kainate in the rat. *Br. J. Pharmacol.* **87**, 345–355.

Ault, B. and L.M. Hildebrand (1993a). Activation of nociceptive reflexes by peripheral kainate receptors. *J. Pharm. Exp. Ther.* **265**, 927–932.

Ault, B. and L.M. Hildebrand (1993b). L-glutamate activates peripheral nociceptors. *Agents Actions* **39**, 142–144.

Battaglia, G. and A. Rustioni (1988). Coexistence of glutamate and substance P in dorsal root ganglion neurons of the rat and monkey. *J. Comp. Neurol.* **277**, 302–312.

Bhave, G., F. Karim, S.M. Carlton, and R.W. Gereau (2001). Peripheral group I metabotropic glutamate receptors modulate nociception in mice. *Nature Neurosci.* **4**, 417–423.

Cairns, B.E., J.W. Hu, L. Arendt-Nielsen, B.J. Sessle, and P. Svensson (2001). Sex-related differences in human pain and rat afferent discharge evoked by injection of glutamate into the masseter muscle. *J. Neurophys.* **86**, 782–791.

Cairns, B.E., B.J. Sessle, and J.W. Hu (1998). Evidence that excitatory amino acid receptors within the temporomandibular joint region are involved in the reflex activation of the jaw muscles. *J. Neurosci.* **18**, 8056–8064.

Carlton, S.M. and R.E. Coggeshall (1999). Inflammation-induced changes in peripheral glutamate receptor populations. *Brain Res.* **820**, 63–70.

Carlton, S.M., G. Hargett, and R.E. Coggeshall (2001). Localization of metabotropic glutamate receptors 2/3 on primary afferent axons in the rat. *Neuroscience* **105**, 957–969.

Carlton, S.M., G.L. Hargett, and R.E. Coggeshall (1995). Localization and activation of glutamate receptors in unmyelinated axons of rat glabrous skin. *Neurosci. Lett.* **197**, 25–28.

Cincotta, M., P.M. Beart, R.J. Summers, and D. Lodge (1989). Bidirectional transport of NMDA receptor and ionophore in the vagus nerve. *Eur. J. Pharm.* **160**, 167–171.

Coggeshall, R.E. and S.M. Carlton (1998). Ultrastructural analysis of NMDA, AMPA and kainate receptors on unmyelinated and myelinated axons in the periphery. *J. Comp. Neurol.* **391**, 78–86.

Conn, P.J. and J.-P. Pin (1997). Pharmacology and functions of metabotropic glutamate receptors. *Ann. Rev. Pharmacol. Toxicol.* **37**, 205–237.

Davidson, E.M. and S.M. Carlton (1998). Intraplantar injection of dextrorphan, ketamine or memantine attenuates formalin-induced behaviors. *Brain Res.* **785**, 136–142.

Davidson, E.M., R.E. Coggeshall, and S.M. Carlton (1997). Peripheral NMDA and non-NMDA glutamate receptors contribute to nociceptive behaviors in the rat formalin test. *NeuroReport* **8**, 941–946.

Davies, J., R.H. Evans, A.A. Francis, and J.C. Watkins (1979). Excitatory amino acid receptors and synaptic excitation in the mammalian central nervous system. *J. Physiol. (Paris)* **75**, 641–654.

deGroot, J.F., S. Zhou, and S.M. Carlton (2000). Peripheral glutamate release in the hindpaw following low and high intensity sciatic stimulation. *NeuroReport* **11**, 497–502.

Du, J., M. Koltzenburg, and S.M. Carlton (2000). Glutamate-induced excitation and sensitization of nociceptors in rat glabrous skin. *Pain* **89**, 187–198.

Erdo, S.L. (1991). Excitatory amino acid receptors in the mammalian periphery. *TIPS* **12**, 426–429.

Evans, R.H. (1985). Kainate sensitivity of C-fibres in rat dorsal roots. *J. Physiol.* **358**, 42P.

Evans, R.H., S.J. Evans, P.C. Pook, and D.C. Sunter (1987). A comparison of excitatory amino acid antagonists acting as primary afferent C fibres and motoneurones of the isolated spinal cord of the rat. *Br. J. Pharmacol.* **91**, 531–537.

Fiorentino, P.M., B.E. Cairns, and J.W. Hu (1999). Development of inflammation after application of mustard oil or glutamate to the rat temporomandibular joint. *Arch. Oral Biol.* **44**, 27–32.

Follenfant, R.L. and M. Nakamura-Craig (1992). Glutamate induces hyperalgesia in the rat paw. *Br. J. Pharmacol.* **106**, 49P.

Genever, P.G., S.J. Maxfield, G.D. Kennovin, J. Maltman, C.J. Bowgen, and M.J.S.T.M. Raxworthy (1999). Evidence for a novel glutamate-mediated signaling pathway in keratinocytes. *J. Invest. Dermatol.* **112**, 337–342.

Giovengo, S.L., K.F. Kitto, H.J. Kurtz, R.A. Velázquez, and A.A. Larson (1999). Parenterally administered kainic acid induces a persistent hyperalgesia in the mouse and rat. *Pain* **83**, 347–358.

Harrington, J.F., A.A. Messier, D. Bereiter, B. Barnes, and M.H. Epstein (2000). Herniated lumbar disc material as a source of free glutamate available to affect pain signals through the dorsal root ganglion. *Spine* **25**, 929–936.

Huettner, J.E. (1990). Glutamate receptor channels in rat DRG neurons: Activation by kainate and quisqualate and blockade of desensitization by con A. *Neuron* **5**, 255–266.

Jackson, D.L., C.B. Graff, J.D. Richardson, and K.M. Hargreaves (1995). Glutamate participates in the peripheral modulation of thermal hyperalgesia in rats. *Eur. J. Pharm.* **284**, 321–325.

Jeftinija, S., K. Jeftinija, F. Liu, S.R. Skilling, D.H. Smullin, and A.A. Larson (1991). Excitatory amino acids are released from rat primary afferent neurons *in vitro*. *Neurosci. Lett.* **125**, 191–194.

Kangrga, I. and M. Randic (1991). Outflow of endogenous aspartate and glutamate from the rat spinal dorsal horn in vitro by activation of low-and high-threshold primary afferent fibers: Modulation by mu opioids. *Brain Res.* **553**, 347–352.

Keast, J.R. and T.M. Stephensen (2000). Glutamate and aspartate immunoreactivity in dorsal root ganglion cells supplying visceral and somatic targets and evidence for peripheral axonal transport. *J. Comp. Neurol.* **424**, 577–587.

Kinkelin, I., E.-B. Brocker, M. Koltzenburg, and S.M. Carlton (2000). Localization of ionotropic glutamate receptors in peripheral axons of human skin. *Neurosci. Lett.* **283**, 149–152.

Lawand, N.B., T. McNearney, and K.N. Westlund (2000). Amino acid release into the knee joint: Key role in nociception and inflammation. *Pain* **86**, 69–74.

Lawand, N.B., W.D. Willis, and K.N. Westlund (1997). Excitatory amino acid receptor involvement in peripheral nociceptive transmission in rats. *Eur. J. Pharm.* **324**, 169–177.

Lewis, S.J., M.C. Cincotta, A.J.M. Verberne, B. Jarrott, D. Lodge, and P.M. Beart (1987). Receptor autoradiography with [^{3}H]L-glutamate reveals the presence and axonal transport of glutamate receptors in vsagal afferent neurones of the rat. *Eur. J. Pharm.* **144**, 413–415.

Li, H., H. Ohishi, A. Kinoshita, R. Shigemoto, S. Nomura, and N. Mizuno (1997). Localization of a metabotropic glutamate receptor, mGluR7, in axon terminals of presumed nociceptive, primary afferent fibers in the superficial layers of the spinal dorsal horn: An electron microscope study in the rat. *Neurosci. Lett.* **223**, 153–156.

Liu, H., H. Wang, M. Sheng, L.Y. Jan, Y.N. Jan, and A.I. Basbaum (1994). Evidence for presynaptic N-methyl-D-aspartate autoreceptors in the spinal cord dorsal horn. *Proc. Natl. Acad. Sci.* **91**, 8383–8387.

Lovinger, D.M. and F.F. Weight (1988). Glutamate induces a depolarization of adult rat dorsal root ganglion neurons that is mediated predominantly by NMDA receptors. *Neurosci. Lett.* **94**, 314–320.

Ma, Q.-P. and R.J. Hargreaves (2000). Localization of *N*-methyl-D-aspartate NR2B subunits on primary sensory neurons that give rise to small-caliber sciatic nerve fibers in rats. *Neuroscience* **101**, 699–707.

Marvizon, J.C.G., J.A. McRoberts, H.S. Ennes, B. Song, X. Wang, L. Jinton *et al.* (2002). Two N-methyl-D-aspartate receptors in rat dorsal root ganglia with different subunit composition and localization. *J. Comp. Neurol.* **446**, 325–341.

McNearney, T., D. Speegle, N.B. Lawand, J. Lisse, and K.N. Westlund (2000). Excitatory amino acid profiles of synovial fluid from patients with arthritis. *J. Rheumatol.* **27**, 739–745.

McRoberts, J.A., S.V. Coutinho, J.C.G. Marvizon, E.F. Grady, M. Tognetto, J.N. Sengupta *et al.* (2001). Role of peripheral N-methyl-D-aspartate (NMDA) receptors in visceral nociception in rats. *Gastroenterology* **120**, 1737–1748.

Nordlind, K., O. Johansson, S. Liden, and T. Hökfelt (1993). Glutamate-and aspartate-like immunoreactivities in human normal and inflamed skin. *Cell. Path. Mol. Path.* **64**, 75–82.

Ohishi, H., S. Nomura, Y.Q. Ding, R. Shigemoto, E. Wada, A. Kinoshita *et al.* (1995). Presynaptic localization of a metabotropic glutamate receptor, mGluR7, in the primary afferent neurons: An immunohistochemical study in the rat. *Neurosci. Lett.* **202**, 85–88.

Omote, K., T. Kawamata, M. Kawamata, and A. Namiki (1998). Formalin-induced release of excitatory amino acids in the skin of the rat hindpaw. *Brain Res.* **787**, 161–164.

Pin, J.-P. and R. Duvoisin (1995). Review: Neurotransmitter receptors I. The metabotropic glutamate receptors: Structure and functions. *Neuropharmacology* **34**, 1–26.

Sahara, Y., N. Noro, Y. Iida, K. Soma, and Y. Nakamura (1997). Glutamate receptor subunits GluR5 and KA-2 are coexpressed in rat trigeminal ganglion neurons. *J. Neurosci.* **17**, 6611–6620.

Sato, K., H. Kiyama, H.T. Park, and M. Tohyama (1993). AMPA, KA and NMDA receptors are expressed in the rat DRG neurones. *NeuroReport* **4**, 1263–1265.

Shigemoto, R., H. Ohishi, S. Nakanishi, and N. Mizuno (1992). Expression of the mRNA for the rat NMDA receptor (NMDAR1) in the sensory and autonomic ganglion neurons. *Neurosci. Lett.* **144**, 229–232.

Skilling, S.R., D.H. Smullin, and A.A. Larson (1988). Extracellular amino acid concentrations in the dorsal spinal cord of freely moving rats following veratridine and nociceptive stimulation. *J. Neurochem.* **51**, 127–132.

Sorkin, L.S. and J.H. Moore (1996). Evoked release of amino acids and prostanoids in spinal cords of anesthetized rats: Changes during peripheral inflammation and hyperalgesia. *Am. J. Therapeutics* **3**, 268–275.

Ushida, T., T. Tani, M. Kawasaki, O. Iwatsu, and H. Yamamoto (1999). Peripheral administration of an N-methyl-D-aspartate receptor antagonist (MK-801) changes dorsal horn neuronal responses in rats. *Neurosci. Lett.* **260**, 89–92.

Walker, K., A. Reeve, M. Bowes, J. Winter, G. Wotherspoon, A. Davis *et al.* (2001). mGlu5 receptors and nociceptive function II. mGlu5 receptors functionally expressed on peripheral sensory neurones mediate inflammatory hyperalgesia. *Neuropharmacology* **40**, 10–19.

Wang, H., R.J. Liu, R.-X. Zhang, and J.-T. Qiao (1997). Peripheral NMDA receptors contribute to activation of nociceptors: A c-fos expression study in rats. *Neurosci. Lett.* **221**, 101–104.

Wang, H., R.-X. Zhang, R. Wang, and J.-T. Qiao (1999). Decreased expression of N-methyl-D-aspartate (NMDA) receptors in rat dorsal root ganglion following complete Freund's adjuvant-induced inflammation: An immunocytochemical study for NMDA NR1 subunit. *Neurosci. Lett.* **265**, 195–198.

Warncke, T., E. Jorum, and A. Stubhaug (1997). Local treatment with the N-methyl-D-aspartate receptor antagonist ketamine, inhibits development of secondary hyperalgesia in man by a peripheral action. *Neurosci. Lett.* **227**, 1–4.

Westlund, K.N., D.L. McNeill, and R.E. Coggeshall (1989a). Glutamate immunoreactivity in rat dorsal roots. *Neurosci. Lett.* **96**, 13–17.

Westlund, K.N., D.L. McNeill, J.T. Patterson, and R.E. Coggeshall (1989b). Aspartate immunoreactive axons in normal rat L4 dorsal roots. *Brain Res.* **489**, 347–351.

Westlund, K.N., Y.C. Sun, K.A. Sluka, P.M. Dougherty, L.S. Sorkin, and W.D. Willis (1992). Neural changes in acute arthritis in monkeys. II. Increased glutamate immunoreactivity in the medial articular nerve. *Brain Res. Rev.* **17**, 15–27.

Yu, X.-M., B.J. Sessle, D.A. Haas, A. Izzo, H. Vernon, and J.W. Hu (1996). Involvement of NMDA receptor mechanisms in jaw electromyographic activity and plasma extravasation induced by inflammatory irritant application to temporomandibular joint region of rats. *Pain* **68**, 169–178.

Zhou, S., S. Komak, J. Du, and S.M. Carlton (2001). Metabotropic glutamate 1α receptors on peripheral primary afferent fibers: Their role in nociception. *Brain Res.* **913**, 18–26.

Zhou, Z., L. Bonasera, and S.M. Carlton (1996). Peripheral administration of NMDA, AMPA or KA results in pain behaviors in rats. *NeuroReport* **7**, 1–6.

Part II

Specific Target Tissues, Organs, and Systems

6

The Vertebrate Retina

Victoria P. Connaughton

The retina contains a diversity of glutamate receptors and transporters, each localized to different types of cells and mediating different functions in the processing of visual signals. The purpose of this chapter is to review the distribution patterns of ionotropic (iGluR) and metabotropic glutamate receptor (mGluR) types, and glutamate transporters, in retinal neurons. In addition, the functional consequences of these distribution patterns are discussed with regard to signal transduction within the retina, mechanisms of excitotoxicity, and the developmental segregation of ON and OFF processes in the inner retina.

1. Retinal Organization and Cell Types

Light entering the eye passes through the cornea and lens before reaching the retina. The retina is composed of seven cell types that are organized into layers, with nuclear layers containing somata separated by plexiform layers filled with processes (Figure 6.1, see color insert). The distal-most cells are photoreceptors, which absorb light, initiating the signal transduction cascade through the retina. Photoreceptors are presynaptic to horizontal and bipolar cell dendritic processes in the outer plexiform layer (OPL). Bipolar cells relay the signal to the inner retina, where they are presynaptic to amacrine and ganglion cells in the inner plexiform layer (IPL). Ganglion cell axons form the optic nerve and transmit the signal to the brain. It is the connections between photoreceptor, bipolar, and ganglion cells that form the vertical transduction pathway through the retina (Dowling, 1987). Glutamate is the neurotransmitter released between elements in this pathway. The pathway is modulated in the outer retina (i.e., OPL) through electrical (Kamermans *et al.*, 2001) and/or chemical (GABAergic) synapses arising from neighboring horizontal cells (Dowling, 1987). Glycinergic and/or dopaminergic interplexiform cells also provide inputs to the OPL (Dowling and Ehinger, 1978; Kalloniatis and Marc, 1990). In the IPL, inhibitory GABAergic and glycinergic inputs from amacrine cells modify bipolar-to-ganglion cell signaling (Dowling, 1987). In addition to neurons, Muller glial cells are also present. Muller cells extend the width of the retina and play a role in the removal of excess glutamate (Poitry *et al.*, 2000).

Victoria P. Connaughton • Department of Biology, American University, 4400 Massachusetts Ave, NW, Washington, DC 20016.

Glutamate Receptors in Peripheral Tissue, edited by Santokh Gill and Olga Pulido.
Kluwer Academic / Plenum Publishers, New York, 2005.

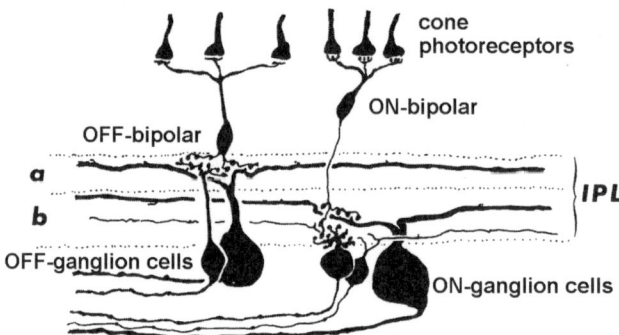

Figure 6.2. ON and OFF retinal neurons are morphologically distinct. Diagram showing the differential distribution of ON and OFF cell processes in the inner plexiform layer (IPL). OFF bipolar cells have axon terminals in the distal sublamina of the IPL, where they contact OFF ganglion cell dendrites. In contrast, ON bipolar cells have synaptic boutons in the proximal half of the IPL (sublamina b), where they are presynaptic to ON amacrine and ON ganglion cells (Modified from Nelson *et al.*, 1978, figure 1.).

Two parallel pathways, ON and OFF, exist in retina. Neurons within each pathway are characterized by their response to light, with ON type cells depolarizing in response to a light stimulus and OFF cells hyperpolarizing to light (or depolarizing in the dark). Signal transduction to both ON and OFF pathways is initiated by photoreceptors, which release glutamate in the *dark* (i.e., in the *absence* of a stimulus). This tonic release of glutamate depolarizes OFF cells and hyperpolarizes ON cells. As physiological evidence suggests, all photoreceptors release glutamate as their neurotransmitter, the distinction between these pathways must occur downstream, at the level of the second-order neurons. In fact, the separation between ON and OFF pathways occurs at the level of the bipolar cells, with ON and OFF bipolar cells expressing different types of postsynaptic glutamate receptors (see below). This differentiation of ON and OFF cell types is maintained in proximal retina, as ON ganglion cells receive inputs from ON bipolar cells and OFF ganglion cells receive inputs from OFF bipolar cells. These pathways are also morphologically distinct with the processes of OFF amacrine, ganglion, and bipolar cells found exclusively in the distal IPL, while ON cell processes are localized to the proximal IPL (Figure 6.2). (Famiglietti and Kolb, 1976; Stell *et al.*, 1977; Nelson *et al.*, 1978; Ishida *et al.*, 1980; Kolb *et al.*, 1981).

2. Glutamate Localization in Retina

Glutamate is believed to be the major excitatory neurotransmitter in retina. In general, glutamate is synthesized from ammonium and α-ketoglutarate (a component of the Krebs Cycle) and is used in the synthesis of proteins, other amino acids, and even other neurotransmitters (such as GABA; Stryer, 1988). Though glutamate is present in all retinal neurons, only a few are glutamatergic, releasing glutamate as their neurotransmitter. Neuroactive glutamate is stored in synaptic vesicles in presynaptic axon terminals (Fykse and Fonnum, 1996). Glutamate is incorporated into the vesicles by a glutamate transporter located in the vesicular membrane. This transporter selectively accumulates glutamate through a sodium-independent, ATP-dependent process (Naito and Ueda, 1983; Tabb and Ueda, 1991; Fykse and Fonnum, 1996), resulting in a high concentration of glutamate in each vesicle.

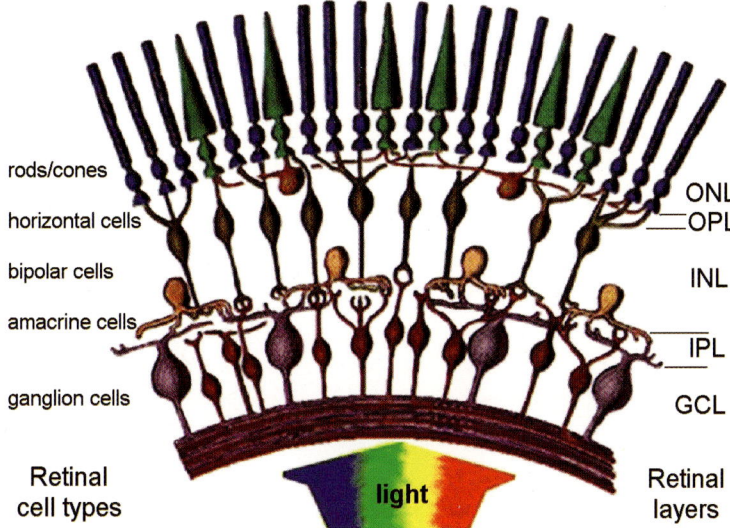

rods/cones

horizontal cells

bipolar cells

amacrine cells

ganglion cells

ONL
OPL

INL

IPL

GCL

Retinal cell types

light

Retinal layers

Figure 6.1. Organization of the vertebrate retina. Schematic diagram showing the different cell types (left) located in different retinal layers (right). The vertical transduction pathway through the retina, which uses glutamate as its neurotransmitter, contains photoreceptors (rods/cones), which initiate the signal, bipolar cells, and ganglion cells. Horizontal cells and amacrine cells provide lateral, inhibitory inputs to this pathway. Photoreceptors are presynaptic to horizontal and bipolar neurons in the outer plexiform layer (OPL). Bipolar cell terminals synapse with dendrites of amacrine and ganglion cells in the inner plexiform layer (IPL). ONL, outer nuclear layer; INL, inner nuclear layer; GL, ganglion cell layer (modified from webvision, www.webvision.utah.med.edu).

Retinal photoreceptors, bipolar cells, and ganglion cells are glutamate immunoreactive (Figure 6.3) (Ehinger *et al.*, 1988; Marc *et al.*, 1990; Van Haesendonck and Missotten, 1990; Kalloniatis and Fletcher, 1993; Yang and Yazulla, 1994; Jojich and Pourcho, 1996; Connaughton *et al.*, 1999). Some horizontal and/or amacrine cells can also display weak labeling with glutamate antibodies (Ehinger *et al.*, 1988; Marc *et al.*, 1990; Jojich and Pourcho, 1996; Yang, 1996; Connaughton *et al.*, 1999). These neurons are believed to release GABA, not glutamate, as their neurotransmitter (Yazulla, 1986), suggesting that the weak glutamate labeling reflects the pool of metabolic glutamate used in the synthesis of GABA. This has been supported by the results from double-labeling studies using antibodies to both GABA and glutamate: glutamate-positive amacrine cells also label with the GABA antibodies (Jojich and Pourcho, 1996; Yang, 1996).

Photoreceptors, which contain glutamate, actively take up radiolabeled glutamate from the extracellular space, as do Muller cells (Marc and Lam, 1981; Yang and Wu, 1997). Glutamate is incorporated into these cell types through a high affinity glutamate transporter

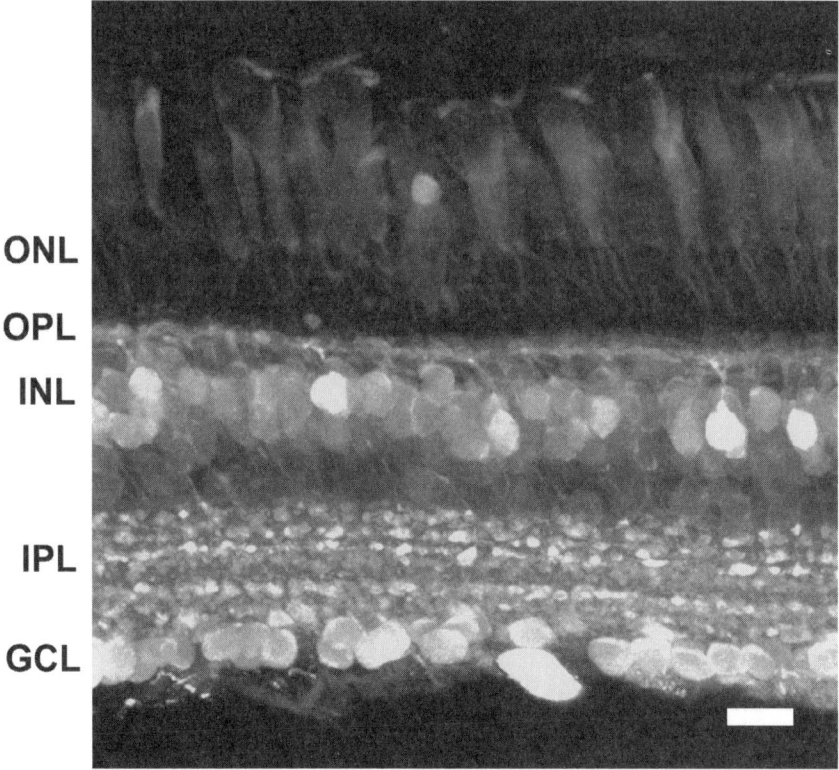

Figure 6.3. Photoreceptors, bipolar cells, and ganglion cells are glutamate immunoreactive. Representative micrograph of the zebrafish retina labeled with an antibody to glutamate. Photoreceptors in the outer nuclear layer (ONL), bipolar cell bodies in the inner nuclear layer (INL), and ganglion cell somata in the ganglion cell layer (GCL) are positively labeled. In addition, process labeling was observed in both the outer (OPL) and inner plexiform layers (IPL), with OPL labeling representing photoreceptor terminals and bipolar cell dendrites. Fluorescent puncta within the IPL represent bipolar cell axon terminals. Bar = $10\mu M$ (Taken from Connaughton *et al.*, 1999, Figure 1.).

located in the plasma membrane. Glutamate transporters (discussed below) maintain the concentration of glutamate within the synaptic cleft at low levels, preventing glutamate-induced cell death (Kanai *et al.*, 1994). Though Muller cells take up glutamate, they do not label with glutamate antibodies (i.e., Jojich and Pourcho, 1996). Glutamate incorporated into Muller cells is rapidly broken down into glutamine (Poitry *et al.*, 2000), which is then exported from glial cells and incorporated into the surrounding neurons (Pow and Crook, 1996). Neurons can then synthesize glutamate from glutamine (Hertz, 1979; Pow and Crook, 1996).

3. Glutamate Receptor Types in Retina

As stated above, photoreceptor, bipolar, and ganglion cells show glutamate immunoreactivity. Glutamate responses have been electrically characterized in horizontal and bipolar cells, which are postsynaptic to photoreceptors, and in amacrine and ganglion cells, which are postsynaptic to bipolar cells (see Figure 6.1). Taken together, these findings suggest that glutamate is the neurotransmitter released by neurons in the vertical transduction pathway. Both *in situ* hybridization and immunocytochemical studies have localized the expression iGluR subunits, mGluR, and glutamate transporter proteins in the retina. The goal here is to interrelate these three components of glutamate signaling to understand functional processes within the retina.

3.1. Ionotropic Glutamate Receptors (iGluRs)

3.1.1. Non-NMDA Receptors

Glutamate binding onto a non-NMDA (N-methyl-D-aspartate) receptor opens non-selective cation channels more permeable to sodium (Na^+) and potassium (K^+) ions than calcium (Ca^{+2}) (Mayer and Westbrook, 1987). However, calcium entry does occur through these channels (i.e., Gilbertson *et al.*, 1991). Glutamate binding elicits a rapidly activating inward current at membrane potentials negative to 0 mV, and an outward current at potentials positive to 0 mV. Kainate, quisqualate, and α-*a*mino-3-hydroxy-5-*m*ethyl-4-isoxazolepropionic *a*cid (AMPA) are the specific agonists at these receptors; 6-*c*yano-7-*n*itroquinoxaline-2,3-dione (CNQX), 1,2,3,4-tetrahydro-6-*n*itro-2,3-dione-*b*enzo[f]quinoxaline-7-sulfonamide (NBQX), and 6,7-*di*nitroquinoxaline-2,3-dione (DNQX) are the antagonists. The benzodiazepine GYKI compounds were recently identified as selective antagonists for AMPA-type receptors (Paternain *et al.*, 1995).

Each non-NMDA receptor is formed from the co-assembly of several subunits (Boulter *et al.*, 1990; Nakanishi *et al.*, 1990; Nakanishi, 1992). To date, seven subunits (named GluR1 through GluR7) have been cloned (Hollmann *et al.*, 1989; Bettler *et al.*, 1990, 1992; Boulter *et al.*, 1990; Keinanen *et al.*, 1990; Nakanishi *et al.*, 1990; Egebjerg *et al.*, 1991). Expression of subunit clones in *Xenopus* oocytes revealed that GluR5, GluR6, and GluR7 (along with subunits KA1 and KA2) co-assemble to form kainate (-preferring) receptors, whereas, GluR1, GluR2, GluR3, and GluR4 are assembled into AMPA (-preferring) receptors (Nakanishi, 1992).

In retina, antibodies to the different non-NMDA receptor subunits differentially label all retinal layers (i.e., Hartveit *et al.*, 1994; Peng *et al.*, 1995; Hughes, 1997; Hack *et al.*, 1999, 2001; Vandenbranden *et al.*, 2000a; Haverkamp *et al.*, 2001a; Grunert *et al.*, 2002) and mRNAs encoding these different subunits are similarly expressed (Hughes *et al.*, 1992; Hamassaki-Britto *et al.*, 1993; Brandstatter *et al.*, 1994). In particular, horizontal cells, OFF bipolar cells, amacrine cells, and ganglion cells express GluR1-4 (Tables 6.1 and 6.2). Briefly, all four subunits have been identified at flat (OFF type) bipolar cell contacts and/or invaginating horizontal cell contacts with photoreceptors (Hack *et al.*, 1999, 2001; Haverkamp *et al.*, 2001a; Grunert *et al.*, 2002). In the inner retina, antibodies to the different subunits are differentially labeled processes postsynaptic to bipolar cell terminals (Qin and Pourcho, 1999; Ghosh *et al.*, 2001), with only one element of the dyad typically immunoreactive for a given subunit (Grunert *et al.*, 2002).

Kainate receptor subunits (GluR5-7) are distributed in a pattern similar to that seen for AMPA subunits (Tables 6.1 and 6.2). In particular, labeling in the OPL has been observed at flat bipolar cell contacts, (Brandstatter *et al.*, 1997; Haverkamp *et al.*, 2001b; Qin and Pourcho, 2001; Grunert *et al.*, 2002), horizontal cell processes postsynaptic to photoreceptor terminals (Haverkamp *et al.*, 2001b; Qin and Pourcho, 2001), and at junctions between horizontal cell processes (Haverkamp *et al.*, 2001b; Grunert *et al.*, 2002). Labeling was also found throughout the IPL (Brandstatter *et al.*, 1997; Vandenbranden *et al.*, 2000a) on ganglion cells and ganglion cell processes postsynaptic to bipolar terminal dyads (Qin and Pourcho, 2001). Cells in the ganglion cell layer (GCL) and some amacrine cells were also positively labeled (Qin and Pourcho, 2001).

Thus, the distribution of non-NMDA receptor subunits has been found on neurons comprising the OFF pathway in retina. In addition, some ON type cells, such as amacrine and ganglion cells also contain these iGluRs. Interestingly, immunocytochemical studies have identified ON bipolar cells that express AMPA receptor subunits (i.e., Morigiwa and Vardi, 1999; Klooster *et al.*, 2001; Schultz *et al.*, 2001), though the function of these subunits is unknown as AMPA-elicited currents have not been recorded in these cells.

3.1.2. NMDA Receptors

Glutamate binding onto an NMDA receptor also opens nonselective cation channels, resulting in a conductance increase. However, the high conductance channel associated with these receptors is more permeable to Ca^{+2} than Na^{+} ions (Mayer and Westbrook, 1987) and NMDA-gated currents are typically more sustained than kainate- and AMPA-gated channels. As the name suggests, NMDA (*N-m*ethyl-*D-a*spartate) is the selective agonist at these receptors. Compounds such as MK-801, AP-5 (2-amino-5-phosphonopentanoic acid), and AP-7 (2-amino-7-phosphoheptanoic acid) are antagonists.

NMDA receptors are structurally complex with separate binding sites for glutamate, glycine, magnesium ions (Mg^{+2}), zinc ions (Zn^{+2}), and a polyamine recognition site. There is also an antagonist-binding site for PCP and MK-801 (Lodge, 1997). The glutamate, glycine, and magnesium-binding sites are important for receptor activation and gating of the ion channel as they confer both ligand- and voltage-gated properties onto NMDA receptors. NMDA receptors require the binding of glutamate, as well as micromolar concentrations of glycine (Johnson and Ascher, 1987; Kleckner and Dingledine, 1988), to activate the channel. The requirement for both glutamate and glycine makes them "co-agonists" (Kleckner and Dingledine, 1988) at these receptors. Mg^{+2} ions provide a voltage-dependent block of NMDA-gated channels (Nowak *et al.*, 1984). Current–voltage relationships plotted from currents recorded in the presence of Mg^{+2} have a characteristic "J-shape," whereas a linear relationship is observed in Mg^{+2}-free solutions. At negative membrane potentials, Mg^{+2} ions occupy a binding site causing less current to flow through the channel. As the membrane

Table 6.1. Localization of Ionotropic Glutamate Receptor (iGluR) Subunits in Identified Retinal Neurons

| Cell type or layer | Non-NMDA receptor subunits | | | | | | | | | NMDA receptor subunits | | Species | References |
| | AMPA | | | | Kainate | | | | | | | | |
	GluR1	GluR2	GluR2/3	GluR4	GluR5	GluR6	GluR6/7	GluR7	KA2	NR1	NR2 (a–d)		
PR	++[2]		++[1]									Goldfish	[1]Peng et al. (1995) [2]Vandenbranden et al. (2000a)
										++	++(b)	Cat	Pourcho et al. (2001)
BC	++[1]	++[1]	++[1]		++[1]							Rat	[1]Hughes et al. (1992)
	++[3]	++[1,2]		++[3,4]								Goldfish	[1]Peng et al. (1995) [2]Klooster et al. (2001) [3]Vandenbranden et al. (2000a) [4]Yazulla and Studholme (1999)
	++[1,3]	++[1]	++[1]	++[1]	++[2,3]		++[2,3]		++[2]			Monkey (and mouse[2])	[1]Haverkamp et al. (2001a) [2]Haverkamp et al. (2001b) [3]Grunert et al. (2002)
	++[1,2]	++[1,2]	++[2]	++[2]								Rodent	[1]Hack et al. (1999) [2]Hack et al. (2001)
					++	++						Cat	Qin and Pourcho (2001)
HC	++[1]	++[1]		++[1]	++[1]							Rat	[1]Hughes et al. (1992)
		++[1]					++(GluR5-7)[2]					Goldfish	[1]Klooster et al. (2001) [2]Vandenbranden et al. (2000a)
		++[1]	++[1]	++[1]			++[2,3]					Monkey	[1]Haverkamp et al. (2001a) [2]Haverkamp et al. (2001b) [3]Grunert et al. (2002)

Cell type							Species	References
AC	++²(rat)	++^{1,2}	++²	++²			Rodent	Hack et al. (1999)[1]; Hack et al. (2001)[2]
		++¹	++¹	++²			Cat	Qin and Pourcho (1999)[1]; Qin and Pourcho (2001)[2]
	++^{2,3}	++²	++³	++²	++¹	++(a–c)[1]	Rat	Brandstatter et al. (1994)[1]; Hughes et al. (1992)[2]; Peng et al. (1995)[3]
		++	++		++		Goldfish	Vandenbranden et al. (2000a)
GC	++¹	++¹	++¹	++²			Cat	Qin and Pourcho (1999)[1]; Qin and Pourcho (2001)[2]
		++	++				Monkey	Ghosh et al. (2001)
	++^{1,2}	++¹	++¹		++¹		Rat	Hughes et al. (1992)[1]; Peng et al. (1995)[2]
		++	++		++		Cat	Qin and Pourcho (2001)
	++	++ (GluR3)	++				Goldfish	Vandenbranden et al. (2000a)
		++	++			++(a)	Monkey	Grunert et al. (2002)
M	++¹	++^{1,2,3}					Rat	Peng et al. (1995)[1]
							Goldfish	Peng et al. (1995)[1]; Vandenbranden et al. (2000a)[2]; Yazulla and Studholme (1999)[3]

Note: Summary table indicating the distribution of non-NMDA and NMDA receptor subunits in the retinas of different vertebrates. Data was obtained through immunocytochemical and *in situ* hybridization techniques. Retinal cell types are listed on the left where PR, photoreceptors; BC, bipolar cells; HC, horizontal cells; AC, amacrine cells; GC, ganglion cells; and M, Muller cells.

Table 6.2. Localization of Ionotropic Glutamate Receptor (iGluR) Subunits to Different Retinal Layers

| Retinal layer | Non-NMDA receptor subunits | | | | | | | | | NMDA receptor subunits | | Species | References |
| | AMPA | | | | | Kainate | | | | NR1 | NR2 (a–d) | | |
	GluR1	GluR2	GluR2/3	GluR4	GluR5	GluR6	GluR6/7	GluR7	KA2				
OPL											++(a)[1] ++(b)[2]	Cat	Hartveit et al. (1994)[1] Pourcho et al. (2001)[2]
		++[1]	++[1]									Rat	Peng et al. (1995)[1]
	++[1,2,3]		++[2]	++[2,4]			++[1]					Goldfish	Peng et al. (1995)[1] Vandenbranden et al. (2000a)[2] Klooster et al. (2001)[3] Yazulla and Studholme (1999)[4]
	++[1,3]	++[1,3]	++[1,3]	++[1,4]	++[2]		++[2,3]		++[2,3]			Monkey	Haverkamp et al. (2001a)[1] Haverkamp et al. (2001b)[2] Grunert et al. (2002)[3] Ghosh et al. (2001)[4]
	++[1,2]	++[1,2]	++[2] (GluR3)	++[2]								Rodent (rat/mouse)	Hack et al. (1999)[1] Hack et al. (2000)[2]
INL	++[2]	++[2]	++[2]	++[2]		++[1,2]		++[1,2]	++[1]	++[1]	++(a–c)[1]	Rat	Brandstatter et al. (1994)[1] Peng et al. (1995)[2]
		++[1]		++[1]						++[2]		Cat	Qin and Pourcho (1999)[1] Pourcho et al. (2001)[2]
	++	++	++ (and GluR3)	++		++(GluR5–7)				++		Goldfish	Vandenbranden et al. (2000a)

Layer							Species	Reference
IPL	++[1]					++(a)	Rat, cat, rabbit, monkey	Hartveit et al. (1994)
	++[1]	++[1]			++[1]		Rat	[1]Peng et al. (1995)
	++[1]	++[1,2]	++[1,2]			++(b)[3]	Cat	[1]Qin and Pourcho (1999) [2]Qin and Pourcho (2001) [3]Pourcho et al. (2001)
	++	++	++				Rodent	Hack et al. (2001)
	++[1,2]	++[1,2]	++[1]	++[1,3]	++[2] ++[1]	++(a)[1]	Monkey	[1]Grunert et al. (2002) [2]Ghosh et al. (2001) [3]Haverkamp et al. (2001b)
	++	++	++	++ (GluR5–7)	++		Goldfish	Vandenbranden et al. (2000a)
GCL	++[3]	++[3]	++[2]	++[2]		++(a)[1] ++(a–c)	Rat	[1]Hartveit et al. (1994) [2]Brandstatter et al. (1994) [3]Peng et al. (1995)
	++[1]	++[1]	++[2]	++[2]	++[3]	++(b)[3]	Cat	[1]Qin and Pourcho (1999) [2]Qin and Pourcho (2001) [3]Pourcho et al. (2001)
	++	++	++	++			Goldfish	Vandenbranden et al. (2000a)

Note: Summary table indicating the distribution of non-NMDA and NMDA receptor subunits through the retinal layers. Data was obtained through immunocytochemical and *in situ* hybridization techniques. Retinal layers are listed on the left where OPL, outer plexiform layer; INL, inner nuclear layer; IPL, inner plexiform layer; and GCL, ganglion cell layer.

depolarizes, the Mg^{+2} block is removed (Nowak *et al.*, 1984). Zinc blocks the channel in a voltage-independent manner (Westbrook and Mayer, 1987). The polyamine site (Ransom and Stec, 1988; Williams *et al.*, 1994) binds compounds such as spermine or spermidine, either potentiating (Ranson and Stec, 1988; Williams *et al.*, 1994) or inhibiting (Williams *et al.*, 1994) the activity of the receptor, depending on the combination of subunits forming each NMDA receptor (Williams *et al.*, 1994).

To date, five subunits (NR1, NR2a, N2b, N2c, and N2d) of NMDA receptors have been cloned (Moriyoshi *et al.*, 1991; Ikeda *et al.*, 1992; Katsuwada *et al.*, 1992; Meguro *et al.*, 1992; Ishii *et al.*, 1993), as well as a splice variant of the NR1 subunit (NR1C2'; antibody to this subunit used in Grunert *et al.*, 2002). NMDA receptor subunits can co-assemble as homomers (i.e., five NR1 subunits; Moriyoshi *et al.*, 1991; Moyner *et al.*, 1992) or heteromers (one NR1 + four NR2 subunits; Katsuwada *et al.*, 1992; Meguro *et al.*, 1992; Moyner *et al.*, 1992; Ishii *et al.*, 1993) to form ion channels. However, all functional NMDA receptors express the NR1 subunit (Moyner *et al.*, 1992; Nakanishi, 1992; Ishii *et al.*, 1993).

NMDA receptor subunits are expressed predominantly in the inner retina (i.e., inner nuclear layer (INL), IPL, GCL; Tables 6.1 and 6.2) (Brandstatter *et al.*, 1994; Hartveit *et al.*, 1994; Vandenbranden *et al.*, 2000a; Pourcho *et al.*, 2001; Grunert *et al.*, 2002). The distribution of NMDA receptors is in striking contrast to that observed for non-NMDA receptors, which are found in all retinal layers. Immunocytochemical (Hughes, 1997; Wenzel *et al.*, 1997; Pourcho *et al.*, 2001) and *in situ* hybridization (Hartveit *et al.*, 1994) studies have identified specific NMDA receptor subunits in the outer retina, though NMDA-gated currents have not been recorded from distal retinal neurons (except in catfish horizontal cells [O'Dell and Christensen, 1989; Eliasof and Jahr, 1997]).

3.2. Metabotropic Glutamate Receptors (mGluRs)

Unlike ionotropic receptors, which are directly linked to an ion channel, metabotropic receptors are coupled to their associated ion channel through a second messenger pathway. Glutamate binding activates a G-protein and initiates an intracellular cascade (Nestler and Duman, 1994). mGluRs are not co-assembled from multiple subunits, but are one polypeptide. To date, eight mGluRs (mGluR1–mGluR8) have been cloned (Houamed *et al.*, 1991; Masu *et al.*, 1991; Abe *et al.*, 1992; Tanabe *et al.*, 1992; Nakajima *et al.*, 1993; Saugstad *et al.*, 1994; Duvoisin *et al.*, 1995). These receptors are classified into three groups (I, II, and III) based on structural homology, agonist selectivity, and their associated second messenger cascade (reviewed in Nakanishi, 1994; Knopfel *et al.*, 1995; Pin and Bockaert, 1995; Pin and Duvoisin, 1995). In brief, Group I mGluRs (mGluR1 and mGluR5) are coupled to the hydrolysis of fatty acids and the release of calcium from internal stores. Quisqualate and *trans*-ACPD are Group I agonists. Group II (mGluR2 and mGluR3) and Group III (mGluRs 4, 6, 7, and 8) receptors are considered inhibitory because they are coupled to the downregulation of cyclic nucleotide synthesis (Pin and Duvoisin, 1995). L-CCG-1 and *trans*-ACPD agonize Group II receptors; L-AP4 (or 2-amino-4-Phosphobutyric acid, APB) selectively agonizes Group III receptors.

All mGluRs, except mGluR3, have been identified in retina either through antibody staining (Peng *et al.*, 1995; Brandstatter *et al.*, 1996; Koulen *et al.*, 1997; Pourcho *et al.*, 1997; Vardi *et al.*, 2000; Kriemborg *et al.*, 2001) or *in situ* hybridization (Nakajima *et al.*, 1993; Duvoisin *et al.*, 1995; Hartveit *et al.*, 1995). mGluRs are differentially expressed throughout the retina, in the OPL, INL, IPL, and the GCL (Table 6.3), with receptors localized both pre- and postsynaptically. Presynaptic mGluRs found on cone pedicles (Hirasawa *et al.*, 2002) and bipolar terminals (Awatramani and Slaughter, 2001) most likely function as autoreceptors (Hirasawa *et al.*, 2002). Evidence for the role of presynaptic mGluRs can be found in a recent study in salamander in which the application of CPPG, a selective antagonist of Group III receptors, blocks glutamate release from bipolar cell terminals (Awatramani and Slaughter, 2001), decreasing glutamate-elicited responses in amacrine and ganglion cells.

Though different patterns of mGluR expression have been observed in the retina, only the APB receptor (Group III) found on ON type bipolar cells has been physiologically examined in detail and will be discussed.

3.2.1. APB Receptor

In contrast to the iGluRs, glutamate binding onto the APB receptor elicits a conductance *decrease* (Slaughter and Miller, 1981; Nawy and Copenhagen, 1987, 1990) due to the *closure* of nonselective cation channels (Nawy and Jahr, 1990; Nawy, 1999; but see Hirano and MacLeish, 1991). This is a unique action as glutamate, which is excitatory, actually inhibits postsynaptic cells containing the APB receptor. In retina, 2-*amino*-4-*phospho*butyric acid (APB, also called DL-AP4, with the L-form being active) application selectively affects the ON pathway (Slaughter and Miller, 1981), that is, ON bipolar cell responses and the ON responses of amacrine (i.e., Taylor and Wassle, 1995) and ganglion cells (i.e., Cohen and Miller, 1994; Kittila and Massey, 1995; Jin and Brunken, 1996) are eliminated by APB. Physiological evidence (i.e., Slaughter and Miller 1981; Massey *et al.*, 1983) suggests that the APB receptor is localized to ON bipolar cell dendrites (though APB-induced hyperpolarizations have been reported in some fish horizontal cells (Takahashi and Copenhagen, 1992)). Thus, the observed blockade of amacrine and ganglion cell ON light responses following APB application is probably due to an elimination of inputs to bipolar cells, rather than a direct effect of APB on postsynaptic amacrine and ganglion cells.

4. Physiology of Retinal Glutamate Receptors

4.1. Ionotropic Receptors

Patch-clamp recordings (i.e., Gilbertson *et al.*, 1991; Boos *et al.*, 1993; Zhou *et al.*, 1993; Cohen and Miller, 1994; Yu and Miller, 1995; Shen *et al.*, 1999; Connaughton and Nelson, 2000) indicate that AMPA and/or kainate application can evoke currents in horizontal, bipolar, amacrine, and ganglion cells. In addition, the kinetics of AMPA- and kainate-elicited currents differ, with AMPA- and quisqualate-elicited currents showing rapid desensitization and resensitization, while, kainate-gated currents resensitize slowly (i.e., Shen *et al.*, 1999). Recently, however, desensitizing kainate-receptor responses have been identified on OFF bipolar cells in squirrel (DeVries and Schwartz, 1999). The desensitization of AMPA receptors can be reduced by adding cyclothiazide (Yamada and Tang, 1993), which stabilizes the receptor in an active (or non-desensitized) state (Yamada and Tang, 1993; Kessler *et al.*, 1996). Studies using AMPA blockers show that GYKI compounds block

Table 6.3. Localization of Metabotropic Glutamate Receptor (mGluR) in Retina

Cell type or layer	Group I		Group II		Group III				Species	References
	mGluR1 (mGluRα)	mGluR5 (mGluR5α)	mGluR2 (mGluR2/3)	mGluR3	mGluR4	mGluR6	mGluR7	mGluR8		
PR	++	++							Chicken	Kriemborg et al. (2001)
OPL						++			Monkey	Vardi et al. (2000)
	++[1]	++[1]				++[2]			Rat	[1]Koulen et al. (1997) [2]Nomura et al. (1994)
	++	++							Chicken	Kriemborg et al. (2001)
Bipolar cells	++[1]	++[1,2]				++[2]	++[2]		Rat	[1]Koulen et al. (1997) [2]Hartveit et al. (1995)
						++			Primate	Vardi et al. (2000)
Horizontal cells		++							Rat	Hartveit et al. (1995)
INL								++	Mouse	Duvoisin et al. (1995)
	++	++							Chicken	Kriemborg et al. (2001)
	++[3]	++[1,3]	++[3]		++[3]	++[2,3]	++[3]		Rat	[1]Koulen et al. (1997) [2]Nakajima et al. (1993) [3]Hartveit et al. (1995)

Cell type/Layer	Species	PR	OPL	INL	IPL	GCL	References
PL	Rat	++[1,3]	++[3]			++[2]	[1]Peng et al. (1995) [2]Brandstatter et al. (1996) [3]Koulen et al. (1997)
	Chicken	++	++				Kriemborg et al. (2001)
Amacrine cells	Rat	++[1,2]	++[3]	++[2]		++[2]	[1]Peng et al. (1995) [2]Hartveit et al. (1995) [3]Koulen et al. (1997)
	Chicken	++		++			Kriemborg et al. (2001)
	Cat		++	++			Pourcho et al. (1997)
Ganglion cells	Rat	++[1,2]	++	++[2]		++[2]	[1]Peng et al. (1995) [2]Koulen et al. (1997)
	Chicken	++					Kriemborg et al. (2001)
	Mouse				++		Duvoisin et al. (1995)
GCL	Cat	++	++	++		++	Pourcho et al. (1997)
	Rat	++	++	++		++	Hartveit et al. (1997)

Note: Results are presented for both retinal cell types and retinal layers. PR, photoreceptors; OPL, outer plexiform layer; INL, inner nuclear layer; IPL, inner plexiform layer; and GCL, ganglion cell layer.

these rapidly activating glutamate-elicited currents (i.e., Shen *et al.*, 1999; Cohen, 2000; Matsui *et al.*, 2001), suggesting that AMPA receptors (GluR1-4) and their associated ion channels underlie the majority of iGluR mediated responses in retina.

Physiologically NMDA receptors are localized to proximal retinal neurons, as ganglion and some amacrine cells (i.e., Massey and Miller, 1988, 1990; Mittman *et al.*, 1990; Dixon and Copenhagen 1992; Boos *et al.*, 1993; Diamond and Copenhagen, 1993; Cohen and Miller, 1994, but see Hartveit and Veruki, 1997) display NMDA-gated currents. As discussed above, these cells also express AMPA/kainate receptors. Both NMDA and non-NMDA receptor types underlie EPSCs recorded in these cells (i.e., Cohen, 2000; Matsui *et al.*, 2001; Chen and Diamond, 2002), with spontaneous EPSPs mediated by AMPA-type receptors (Matsui *et al.*, 2001; Chen and Diamond, 2002). At negative holding potentials (-80 mV), AMPA receptors provide the bulk of the postsynaptic glutamate response in ganglion cells; NMDA-gated currents were revealed at more positive holding potentials (Cohen, 2000; Chen and Diamond, 2002). Pharmacologically, non-NMDA receptor antagonists block a transient component of the ganglion cell light response, whereas NMDA antagonists block a more sustained component (Mittman *et al.*, 1990; Diamond and Copenhagen, 1993; Hensley *et al.*, 1993; Cohen and Miller, 1994). Interestingly, though AMPA and NMDA receptors co-localize on the same post-synaptic process, they may be distributed at different distances from the synapse itself. If a high glutamate concentration is allowed to remain in the cleft either by blocking $GABA_C$ receptors on bipolar cell axon terminals, preventing feedback inhibition from amacrine cells (Matsui *et al.*, 2001), or by blocking glutamate uptake through transporters (Chen and Diamond, 2002), both AMPA and NMDA responses are observed. On the other hand, if glutamate concentration in the synapse is only transiently increased, resulting from *in vivo* mechanisms that remove/reduce glutamate from within the cleft, primarily AMPA-receptor generated responses were recorded (Matsui *et al.*, 2001; Chen and Diamond, 2002). Taken together, these studies suggest that functionally NMDA receptors, which can be stimulated by a lower concentration of glutamate (Chen and Diamond, 2002), may be located farther away from the site of neuro-transmitter release than AMPA receptors (Matsui *et al.*, 2001; Chen and Diamond, 2002).

4.2. Metabotropic (APB) Receptor

The intracellular pathway through which the APB receptor closes ion channels in ON type bipolar cells begins with glutamate binding onto mGluR6, the metabotropic receptor localized to ON bipolar cell dendrites (Vardi *et al.*, 2000). Mice lacking mGluR6 lack ON responses and display abnormal ERG b-waves (Masu *et al.*, 1995). As the b-wave is believed to be generated by ON bipolar cell activity (Stockton and Slaughter, 1989; Tian and Slaughter, 1995), these findings suggest that mGluR6 mutants have a defect in the ON retinal pathway at the level of the bipolar cell response (Masu *et al.*, 1995). The G-protein coupled to this receptor is G_O (Vardi *et al.*, 1993; Vardi, 1998; Nawy, 1999). Knockout mice lacking the α-subunit ($G_{\alpha O}$) also lacked the ERG b-wave (Dhingra *et al.*, 2000), indicating that $G_{\alpha O}$ is an important component regulating ON bipolar cell responses. The ion channel coupled to the APB receptor is cGMP-gated (Nawy and Jahr, 1990), suggesting that the closure of ion channels occurs following glutamate binding onto mGluR6, activating of G_O, and resulting in the increased hydrolysis of cGMP (Nawy and Jahr, 1990).

However, blocking phosphodiesterase (PDE) activity or adding non-hydrolyzable cGMP analogs (and, therefore, preventing the breakdown of cGMP) does not inhibit the glutamate responses generated through APB receptors on ON bipolar cells, suggesting that removal of cGMP is not required for channel closure (Nawy, 1999). So, how do the channels close in response to glutamate application? Recent studies suggest that calcium ions entering through the ion channel closed by glutamate (Yamashita and Wassle, 1991; Nawy, 2000) effect channel function. Increased internal calcium may directly downregulate the channel, resulting in the observed run-down of glutamate-elicited currents in ON bipolar cells (Nawy, 2000). Other studies report that elevated internal calcium levels activate calcium-dependent enzymes, such as CaMKII (Walters et al., 1998; Shiells and Falk, 2001), which modulate ion channel conductance. Thus, mGluR6 on ON bipolar cells is coupled, through G_O, to a cGMP-gated cation channel permeable to Na^+, K^+, and Ca^{+2}, which is modulated by calcium and/or calcium-dependent phosphorylation (see discussion in Nawy, 2000).

5. Glutamate Transporters and Transporter-Like Receptors in Retina

Glutamate transporters have been identified on photoreceptors (Marc and Lam, 1981; Tachibana and Kaneko, 1988; Eliasof and Werblin, 1993) and Muller cells (Marc and Lam, 1981; Yang and Wu, 1997). From glutamate labeling studies, the average concentration of glutamate in photoreceptors, bipolar cells, and ganglion cells is ~5 mM (Marc et al., 1990). Physiological studies using isolated cells indicate that only micromolar levels of glutamate are required to activate glutamate receptors (i.e., Aizenman et al., 1988; Zhou et al., 1993; Sasaki and Kaneko, 1996). The amount of glutamate released into the synaptic cleft is several orders of magnitude higher than the concentration required to activate most postsynaptic receptors. High affinity glutamate transporters located on adjacent neurons and surrounding glial cells rapidly remove glutamate from the synaptic cleft to prevent cell death (Kanai et al., 1994). Five glutamate transporters, EAAT-1 (or GLT-1), EAAT-2 (or L-glutamate/L-aspartate transporter, GLAST), EAAT-3 (or EAAC-1), EAAT-4, and EAAT-5 have been cloned (Kanai and Hediger, 1992; Pines et al., 1992; Fairman et al., 1995; Schultz and Stell, 1996; Arriza et al., 1997).

Glutamate transporters are pharmacologically distinct from both iGluRs and mGluRs. L-glutamate, L-aspartate, and D-aspartate are substrates for the transporters (Brew and Attwell, 1987; Tachibana and Kaneko, 1988; Eliasof and Werblin, 1993); other glutamate receptor agonists (Brew and Attwell, 1987; Tachibana and Kaneko, 1988; Schwartz and Tachibana, 1990; Eliasof and Werblin, 1993) and antagonists (Barbour et al., 1991; Eliasof and Werblin, 1993) are not. Glutamate uptake can be blocked by the transporter blockers dihydrokainate (DHKA), DL-threo-beta-hydroxyaspartate (beta-HA) (Barbour et al., 1991; Eliasof and Werblin, 1993), and DL-threo-beta-benzyloxyaspartate (DL-TBOA; Shimamoto et al., 1998).

Glutamate transporters incorporate glutamate into Muller cells along with the co-transport of three Na^+ ions (Brew and Attwell, 1987; Barbour et al., 1988) and the antiport of one K^+ ion (Barbour et al., 1988; Bouvier et al., 1992) and either one OH^- or one HCO_3^- ion (Bouvier et al., 1992). The excess sodium ions generate a net positive inward current (Brew and Attwell, 1987; Barbour et al., 1988). More recent findings indicate that a glutamate-elicited chloride current is also associated with some transporters (Eliasof and Jahr, 1996; Arriza et al., 1997).

Glutamate receptors with transporter-like pharmacology have been described in photoreceptors (Picaud et al., 1995a, b; Grant and Werblin, 1996) and ON bipolar cells (Grant and Dowling, 1995, 1996; Connaughton and Nelson, 2000). These receptors are coupled to a chloride current. The pharmacology of these receptors is similar to that described for glutamate transporters, as the glutamate-elicited current is (a) dependent upon external Na^+, (b) reduced by transporter blockers, and (c) insensitive to glutamate agonists and antagonists. However, altering internal Na^+ concentration does not change the reversal potential (Picaud *et al.*, 1995b) or the amplitude (Grant and Werblin, 1996; Grant and Dowling, 1996) of the glutamate-elicited current, suggesting that the receptor is distinct from glutamate transporters. At the photoreceptor terminals, the glutamate-elicited chloride current may regulate membrane potential and subsequent voltage-gated channel activity (i.e., Picaud *et al.*, 1995a). Postsynaptically, this receptor is believed to mediate conductance changes underlying photoreceptor input to ON cone bipolar cells in teleosts (Grant and Dowling, 1995; Connaughton and Nelson, 2000).

5.1. Neurons Expressing Glutamate Transporters

The glutamate transporters GLAST, EAAC-1, and GLT-1 have been identified in retina (Table 6.4). GLAST immunoreactivity is found in all retinal layers, but not in neurons. GLAST is localized to Muller cell membranes (Derouiche and Rauen, 1995; Rauen *et al.*, 1996; Lehre *et al.*, 1997; Rauen, 2000). In contrast, *excitatory amino acid carrier-1* (EAAC-1) antibodies do not label Muller cells (but see Zhao and Yang, 2001). EAAC-1 immunoreactivity in both lower and higher vertebrates is observed in all retinal layers, and in all neuronal types, except photoreceptors (Table 6.3). *Glutamate transporter-1* (GLT-1) proteins have been identified in monkey (Grunert *et al.*, 1994), rat (Rauen *et al.*, 1996; Rauen, 2000), rabbit (Massey *et al.*, 1997), and goldfish (Vandenbranden *et al.*, 2000b) bipolar cells. In addition, a few amacrine cells were weakly labeled with the GLT-1 antibody in rat (Rauen *et al.*, 1996), as were photoreceptor terminals in rabbit (Massey *et al.*, 1997), rat (Rauen, 2000), and goldfish (Vandenbranden *et al.*, 2000b).

6. Role of Glutamate and Glutamate Receptors in Retina

6.1. Signal Transduction

As shown above, the retina contains a diversity of glutamate receptor and transporter types, each localized to different classes of cells and mediating different functions in the processing of visual signals. The importance of glutamate receptors to retinal processing occurs in the transmission of information within the ON and OFF pathways, in the stimulation of neurons providing lateral inhibitory inputs to the vertical transduction pathway, and in the overall understanding and formation of light responses.

Visual signals are relayed through the retina via two pathways. Glutamate is released from photoreceptors in the dark. Photoreceptors synapse onto ON and OFF type bipolar cells, generating the parallel retinal pathways. Thus, ON and OFF bipolar cells can be classified experimentally by their response to glutamate, as well as by morphological differences in stratification patterns in the IPL (see Figure 6.2). All OFF type cells (i.e., bipolar, horizontal, amacrine, and ganglion cells) are depolarized by glutamate release through the activity of iGluRs. The different iGluR types have distinct distributions within the retina: AMPA/kainate (non-NMDA) receptors are distributed throughout the retina, while NMDA receptors are found predominantly in the IPL. The colocalization of AMPA/kainate and NMDA receptors in proximal retina underlies the responses of amacrine and ganglion cells to a light stimulus.

Table 6.4. Localization of Glutamate Transporters in Retina

Cell type/layer	EAAC-1	GLAST	GLT-1	Species	References
Photoreceptors			+ (cone soma to pedicles)	Rabbit	Massey et al. (1997)
			++	Rat	Rauen (2000)
OPL	++			Rat	Raun et al. (1996)
			++ (cone pedicles)	Goldfish	Vandenbranden et al. (2000b)
				Carp, bullfrog	Zhao and Yang (2001)
	++		++ (rod spherules > cone pedicles)	Rabbit	Massey et al. (1997)
Horizontal cells	++			Rat	Schultz and Stell (1996), Rauen et al. (1996), Rauen (2000)
	++			Bullfrog	Zhao and Yang (2001)
			++ (2 types of CBCs)	Rabbit	Massey et al. (1997)
Bipolar cells	++ (faint)		++	Rat	Rauen et al. (1996), Rauen (2000)
			++ (Mb, cone bipolars)	Goldfish	Vandenbranden et al. (2000b)
	++			Turtle, salamander	Schultz and Stell (1996)
			++ (DB2, flat midget bipolar cells)	Monkey	Grunert et al. (1994)

Table 6.4. Continued

Cell type/layer	EAAC-1	GLAST	GLT-1	Species	References
IPL			++ (diffuse)	Rabbit	Massey et al. (1997)
	++		++	Goldfish	Vandenbranden et al. (2000b)
	++[1]		++[1,2]	Carp, bullfrog	Zhao and yang (2001)
				Rat	[1]Rauen et al. (1996)
					[2]Rauen (2000)
	++			Goldfish, salamander, turtle, chicken, rat	Schultz and Stell (1996)
Amacrine cells	++[1,2]		++[1]	Rat	[1]Rauen et al. (1996)
					[2]Rauen (2000)
	++			Carp, bullfrog	Zhao and Yang (2001)
	++			Chicken, rat,	Schultz and Stell (1996)
Ganglion cells	++			goldfish, turtle	Schultz and Stell (1996)
	++			Carp	Zhao and Yang (2001)
	++			Rat	Rauen et al. (1996), Rauen (2000)
					Derouiche and Rauen (1995), Rauen et al. (1996),
					Lehre et al. (1997), Rauen (2000)
Muller cells		++		Rat	Zhao and Yang (2001)
	++			Carp, bullfrog	Vandenbranden et al. (2000b)
			++	Goldfish	

Note: Data is presented for retinal cell types and retinal layers, listed on the left. OPL, outer plexiform layer and IPL, inner plexiform layer.

In the ON pathway, neurons are hyperpolarized by glutamate release from photorecep-tors. This pathway is initiated at the level of ON bipolar cells where glutamate binding onto mGluR6 forms a sign-inverting synapse in which the second-order neurons are hyperpolar-ized due to the closure of nonselective cation channels. Hyperpolarization of ON bipolar cells decreases glutamate release from bipolar synaptic terminals, resulting in a decrease in stimu-lation of postsynaptic ON amacrine and ON ganglion cells. Though ON bipolar cells express a mGluR, ON type amacrine and ganglion cells exhibit primarily iGluR-mediated responses to glutamate. Thus, all proximal neuronal cell types express similar glutamate receptors, regardless of their presynaptic ON or OFF type inputs.

Though glutamate release governs direct (photoreceptor to bipolar to ganglion cell) signal propagation through the vertical transduction pathway, it is important to note that glu-tamatergic synapses also occur onto neurons providing lateral inhibitory inputs to this path-way. Horizontal cells and amacrine cells, which contain inhibitory neurotransmitters, express postsynaptic iGluRs. When stimulated, these cell types release GABA (horizontal and amacrine cells) and/or glycine (amacrine cells), altering neurotransmitter release from presy-naptic photoreceptors (horizontal cell connections) and/or bipolar cells (amacrine cell synapses), contributing to the surround component of recorded light responses (reviewed in Kamermans and Spekreijse, 1999). For example, recordings of bipolar cell responses identify two components: a center response due to direct glutamatergic inputs from photoreceptors, and a surround response, of opposite polarity to the center, generated when the peripheral receptive field is stimulated (Werblin and Dowling, 1969). Stimulation of the surround decreases a neuron's response to center stimulation, contributing to differences and/or changes in recorded light responses.

6.2. Excitotoxicity

Under normal circumstances, external glutamate concentrations are low. However, dur-ing synaptic transmission, a high (mM) concentration of glutamate is released into the synap-tic cleft. Excess glutamate is quickly removed by transporters located on presynaptic terminals and/or surrounding glial processes (see above). Rapid removal of glutamate is important as a sustained increase in extracellular glutamate (Vorwerk et al., 1996)—in the low micromolar range (Otori et al., 1998)—is toxic to neurons (reviewed in Choi, 1988) both in culture condi-tions (i.e., Sucher et al., 1991; Kashii et al., 1996; Otori et al., 1998) and in vivo (i.e., Olney, 1969; Sisk and Kuwabara, 1985; Siliprandi et al., 1992). High intraocular glutamate levels, measured in the aqueous or vitreous humor, have been associated with damage to the optic nerve (Yoles and Schwartz, 1998), ischemic conditions (Mosinger et al., 1991), and glaucoma (Dreyer et al., 1996; Brooks et al., 1997). Elevated glutamate levels in the vitreous follow a concentration gradient, with higher glutamate concentrations observed posterior, adjacent to the GCL (Dreyer et al., 1996; Brooks et al., 1997). Consequently, glutamate-induced cell death has been identified predominantly in proximal retinal neurons, such as amacrine and ganglion cells (Abrams et al., 1989; Mosinger et al., 1991; Wygnanski et al., 1995; Vorwerk et al., 1996; Luo et al., 2001). Of these cell types, ganglion cells are more susceptible to excitotoxicity due to (a) their location next to the vitreous, where high glutamate levels are found, and (b) the expression of iGluRs on ganglion cell dendrites (see above). Interestingly, high levels of glutamate are not toxic to all types of ganglion cells, as cells with a larger somal size appear to be more vulnerable to glutamate- and/or kainate-induced excitotoxicity (Dreyer et al., 1994; Vorwerk et al., 1999; Luo et al., 2001).

Excitotoxicity is characterized as a two-stage process. Initially, neuronal swelling, due to the movement of Na^+ and Cl^- ions, is observed in the damaged area, followed by neuronal degeneration as a result of calcium entry (Choi, 1988, 1994) through both NMDA receptor-gated (Choi, 1994) and voltage-gated calcium (Sucher *et al.*, 1991) channels. Recent work has suggested that AMPA-type receptors are also stimulated (Choi, 1988, 1994; Mosinger *et al.*, 1991; Otori *et al.*, 1998; Schuettaf *et al.*, 2000; Luo *et al.*, 2001; Santos *et al.*, 2001). The excess increase in internal calcium levels leads to the activation of nitric oxide (NO) synthase (Dawson *et al.*, 1991; Kashii *et al.*, 1996) and the formation of NO. The subsequent effects of NO are not known (Dawson *et al.*, 1991), though damaged cells ultimately die by apoptosis (Villani *et al.*, 1997; Glazner *et al.*, 2000; Santos *et al.*, 2001), possibly through a mechanism involving the loss of AMPA receptor subunits following caspase activation (Glazner *et al.*, 2000).

Protection from neurotoxicity can be achieved experimentally through the application of iGluR antagonists, such as CNQX and/or MK-801 (i.e., Mosinger *et al.*, 1991; Schuettaf *et al.*, 2000; Luo *et al.*, 2001). *In vivo*, Muller cells provide protection from both glutamate (Kanai *et al.*, 1994; Vorwerk *et al.*, 2000) and NO neurotoxicity (Kawasaki *et al.*, 2000). Ganglion cell cultures containing Muller cells displayed increased neuronal survival in response to high glutamate levels compared to ganglion cell-only cultures (Kawasaki *et al.*, 2000). In addition, a downregulation of transporter protein synthesis has been correlated with increased glutamate levels in the vitreous (Naskar *et al.*, 2000; Vorwerk *et al.*, 2000) and a decrease in ganglion cell viability (Vorwerk *et al.*, 2000). This suggests that transporters play a major role in the prevention of neurotoxicity as increased extracellular levels of glutamate may result from either the inability of existing transporters to reduce glutamate levels and/or a decrease in overall transporter function (Dreyer *et al.*, 1996).

6.3. ON/OFF Stratification Patterns in the IPL

During neuronal development, it is important that each neuron establishes connections with the correct pre- and postsynaptic partners. This is exemplified in retina with the segregation of ON and OFF processes in the IPL, resulting in morphologically and physiologically distinct classes of neurons. Glutamate and glutamate receptors have been shown to play a role in this stratification (Bodnarenko and Chalupa, 1993; Bodnarenko *et al.*, 1995), as well as in the generation of spontaneous waves of retinal activity (Bansal *et al.*, 2000; Wong *et al.*, 2000; Zhou and Zhao, 2000). Both of these events are indicative of and occur during circuit formation within the retina.

Glutamate immunoreactivity has been observed throughout the developing mammalian retina (Pow and Barnett, 1999). As development proceeds, observed immunoreactivity patterns become confined to photoreceptors, bipolar cells, and ganglion cells as in adults (Redburn and Rowe-Rendelman, 1996; Pow and Barnett, 1999). The change in glutamate immunoreactivity patterns has been correlated with increased GLAST expression in Muller cells, suggesting that glial cell transporters remove the excess external glutamate (Pow and Barnett, 1999).

Glutamatergic activity is important in generating spontaneous waves of retinal activity (Bansal *et al.*, 2000; Wong *et al.*, 2000; Zhou and Zhao, 2000). Initially, cholinergic inputs to ganglion cells (from amacrine cells) underlie wave generation, as both waves (Bansal *et al.*, 2000; Zhou and Zhao, 2000) and ganglion cell activity (Wong *et al.*, 2000) can be blocked with the application of nicotinic receptor antagonists. The contribution of glutamate to these waves increases with developmental time (Wong *et al.*, 2000), as later in development (i.e., >P13 in ferret, Wong *et al.*, 2000; >P12–P14 in mice, Bansal *et al.*, 2000; P0–P1 in rabbit, Zhou and Zhao, 2000), the waves are selectively blocked by iGluR antagonists (Wong *et al.*, 2000; Bansal *et al.*, 2000; Zhou and Zhao, 2000). At these later times, synapses between bipolar and ganglion cells have formed, suggesting that direct glutamatergic inputs to ganglion cells underlie wave generation (Wong *et al.*, 2000).

Functional connections between bipolar and ganglion cells have also been shown to be responsible for segregation of ON and OFF ganglion cell dendrites (Bodnarenko and Chalupa, 1993; Bodnarenko *et al.*, 1995). Initially, ganglion cell processes are broadly stratified. Over time, the dendrites are restructured becoming confined to either the distal (OFF) or proximal (ON) regions of the IPL (Bodnarenko *et al.*, 1995, 1999). The pruning of ganglion cell dendrites is believed to occur in an activity-dependent manner (Bodnarenko and Chalupa, 1993; Bodnarenko *et al.*, 1995) requiring active synapses between presynaptic bipolar cell terminals and postsynaptic ganglion cell dendrites. Evidence for this comes from experiments in which intraocular injection of APB, acting on ON bipolar dendrites, inhibits process retraction, resulting in the maintenance of ganglion cells stratified in both ON and OFF zones of the IPL (Bodnarenko and Chalupa, 1993). The effect of APB, if administered over a short term, is transient, resulting in only a delay of ganglion cell stratification (Bodnarenko *et al.*, 1995). However, long-term (for the first postnatal month) application of APB resulted in much more severe effects as ganglion cell physiology (specifically center-surround responses) was altered by prolonged APB treatment (Bisti *et al.*, 1998). Interestingly, knockout mice lacking mGluR6, the ON bipolar receptor, displayed no abnormalities in either bipolar, amacrine, or ganglion cell organization or stratification patterns in the IPL (Tagawa *et al.*, 1999), suggesting that the mechanism involved in the establishment of ON and OFF processes is complex.

7. Summary and Conclusion

Histological analyses of presynaptic neurons and physiological recordings from postsynaptic cells suggest that photoreceptor, bipolar, and ganglion cells release glutamate as their neurotransmitter. Multiple glutamate receptor types are present and differentially distributed in the retina. iGluRs, found on neurons within the OFF pathway, directly gate ion channels and mediate rapid synaptic transmission. IGluRs are also present on ON-type amacrine and ganglion cells in proximal retina, while ON-bipolar cells express the metabotropic APB receptor. Signaling through the APB receptor is important during development as bipolar cell input to ganglion cells mediate the formation of ON and OFF processes in the IPL. Glutamate transporters are also present at retinal glutamatergic synapses. Transporters remove excess glutamate from the synaptic cleft to prevent neurotoxicity. Thus, postsynaptic responses to glutamate are determined by the distribution of receptors and transporters at glutamatergic synapses which, in retina, determine the conductance mechanisms underlying visual information processing and development within the ON and OFF pathways.

References

Abe, T., H. Sugihara, H., Nawa, R. Shigemoto, N. Mizuno, and S. Nakanishi (1992). Molecular characterization of a novel metabotropic glutamate receptor mGluR5 coupled to inositol phosphate/Ca^{+2} signal transduction. *J. Biol. Chem.* **267**, 13361–13368.

Abrams, L., L.E. Politi, and R. Adler (1989). Differential susceptibility of isolated mouse retinal neurons and photoreceptors to kainic acid toxicity. *Invest. Ophthal. Vis. Sci.* **30**, 2300–2308.

Aizenman, E., M.P. Frosch, and S.A. Lipton (1988). Responses mediated by excitatory amino acid receptors in solitary retinal ganglion cells from rat. *J. Physiol.* **396**, 75–91.

Arriza, J.L., S. Eliasof, M.P. Kavanaugh, and S.G. Amara (1997). Excitatory amino acid transporter 5, a retinal glutamate transporter coupled to a chloride conductance. *Proc. Natl. Acad. Sci.* **94**, 4155–4160.

Awatramani, G.B. and M.M. Slaughter (2001). Intensity-dependent, rapid activation of presynaptic metabotropic glutamate receptors at a central synapse. *J. Neurosci.* **21**, 741–749.

Bansal, A., J.H. Singer, B.J. Hwang, W. Zu, A. Beaudet, and M.B. Feller (2000). Mice lacking specific nicotinic acethycholine receptor subunits exhibit dramatically altered spontaneous activity patterns and reveal a limited role for retinal waves in forming ON and OFF circuits in the inner retina. *J. Neurosci.* **2000**, 7672–7681.

Barbour, B., H. Brew, and D. Attwell (1988). Electrogenic glutamate uptake in glial cells is activated by intracellular potassium. *Nature* **335**, 433–435.

Barbour, B., H. Brew, and D. Attwell (1991). Electrogenic uptake of glutamate and aspartate into glial cells isolated from the salamander (Ambystoma) retina. *J. Physiol.* **436**, 169–193.

Bettler, B., J. Boulter, I. Hermans-Borgmeyer, O.A. Shea-Greenfield, E.S. Deneris, C. Moll *et al.* (1990). Cloning of a novel glutamate receptor subunit, GluR5: Expression in the nervous system during development. *Neuron* **5**, 583–595.

Bettler, B., J. Egebjerg, G. Sharma, G. Pecht, I. Hermans-Borgmeyer, C. Moll *et al.* (1992). Cloning of a putative glutamate receptor: a low affinity kainate-binding subunit. *Neuron* **8**, 257–265.

Bisti, S., C. Gargini, and L.M. Chalupa (1998). Blockade of glutamate-mediated activity in the developing retina perturbs the functional segregation of ON and OFF pathways. *J. Neurosci.* **18**, 5019–5025.

Bodnarenko, S.R. and L.M. Chalupa (1993). Stratification of ON and OFF ganglion cell dendrites depends on glutamate-mediated afferent activity in the developing retina. *Nature* **364**, 144–146.

Bodnarenko, S.R., G. Jeyarasasingam, and L.M. Chalupa (1995). Developmental and regulation of dendritic stratification in retinal ganglion cells by glutamate-mediated afferent activity. *J. Neurosci.* **15**, 7037–7045.

Bodnarenko, S.R., G. Yeung, L. Thomas, and M. McCarthy (1999). The development of retinal ganglion cell dendritic stratification in ferrets. *Neuroreport* **10**, 2955–2959.

Boos, R., H. Schneider, and H. Wassle (1993). Voltage- and transmitter-gated currents of AII amacrine cells in a slice preparation of the rat retina. *J. Neurosci.* **13**, 2874–2888.

Boulter, J., M. Hollmann, A. O'Shea-Greenfield, M. Hartley, E. Deneris, C. Maron *et al.* (1990). Molecular cloning and functional expression of glutamate receptor subunit genes. *Science* **249**, 1033–1037.

Bouvier, M., M. Szatkowski, A. Amato, and D. Attwell (1992) The glial cell glutamate uptake carrier countertransports pH-changing ions. *Nature* **360**, 471–474.

Brandstatter, J.H., E. Hartveit, M. Sassoe-Pognetto, and H. Wassle (1994). Expression of NMDA and high-affinity kainate receptor subunit mRNAs in the adult rat retina. *Eur. J. Neurosci.* **6**, 1100–1112.

Brandstatter, J.H., P. Koulen, R. Kuhn, H. van der Putten, and H. Wassle (1996). Compartmental localization of a metabotropic glutamate receptor (mGluR7): Two different active sites at a retinal synapse. *J. Neurosci.* **16**, 4749–4756.

Brandstatter, J.H., P. Koulen, and H. Wassle (1997). Selective synaptic distribution of kainate receptor subunits in the two plexiform layers of the rat retina. *J. Neurosci.* **17**, 9298–9307.

Brew, H. and D. Attwell (1987). Electrogenic glutamate uptake is a major current carrier in the membrane of axolotl retinal glial cells. *Nature* **327**, 707–709.

Brooks, D.E., G.A. Garcia, E.D. Dreyer, D. Zurakowski, and R.E. Franco-Bourland (1997). Vitreous body glutamate concentration in dogs with glaucoma. *Am. J. Vet. Res.* **58**, 864–867.

Chen, S. and J.S. Diamond (2002). Synaptically released glutamate activates extrasynaptic NMDA receptors on cells in the ganglion cell layer of rat retina. *J. Neurosci.* **22**, 2165–2173.

Choi, D.W. (1988). Glutamate neurotoxicity and diseases of the nervous system. *Neuron* **1**, 623–634.

Choi, D.W. (1994). Glutamate receptors and the induction of excitotoxic neuronal death. *Progr. Brain. Res.* **100**, 47–51.

Cohen, E.D. (2000). Light-evoked excitatory synaptic currents of Z-type retinal ganglion cells. *J. Neurophysiol.* **83**, 3217–3229.

Cohen, E.D. and R.F. Miller (1994). The role of NMDA and non-NMDA excitatory amino acid receptors in the functional organization of primate retinal ganglion cells. *Vis. Neurosci.* **11**, 317–332.

Connaughton, V.P., T.N. Behar, W.-L.S. Liu, and S. Massey (1999). Immunocytochemical localization of excitatory and inhibitory neurotransmitters in the zebrafish retina. *Vis. Neurosci.* **16**, 483–490.

Connaughton, V.P and R. Nelson (2000). Axonal stratification patterns and glutamate-gated conductance mechanisms in zebrafish retinal bipolar cells. *J. Physiol.* **524**, 135–146.

Dawson, V.L., T.M. Dawson, E.D. London, D.S. Bredt, and S.N. Snyder (1991). Nitric oxide mediates glutamate neurotoxicity in primary cortical cultures. *Proc. Natl. Acad. Sci.* **88**, 6368–6371.

Derouiche, A. and T. Rauen (1995). Coincidence of L-glutamate/L-aspartate transporter (GLAST) and glutamine synthetase (GS) immunoreactions in retinal glia: Evidence for coupling of GLAST and GS in transmitter clearance. *J. Neurosci. Res.* **42**, 131–143.

DeVries, S.H. and E.A. Schwartz (1999). Kainate receptors mediate synaptic transmission between cones and OFF bipolar cells in a mammalian retina. *Nature* **397**, 157–160.

Dhingra, A., A. Lyubarsky, M. Jiang, E.N. Pugh, Jr., L. Birnbaumer, P. Sterling *et al.* (2000). The light response of ON bipolar neurons requires G(o. *J. Neurosci.* **20**, 9053–9058.

Diamond, J.A. and D.R. Copenhagen (1993). The contribution of NMDA and non-NMDA receptors to the light-evoked input-output characteristics of retinal ganglion cells. *Neuron* **11**, 725–738.

Dixon, D.B. and D.R. Copenhagen (1992). Two types of glutamate receptors differentially excite amacrine cells in the tiger salamander retina. *J. Physiol.* **449**, 589–606.

Dowling, J.E. (1987). The Retina, an Approachable Part of the Brain. The Belknap Press of Harvard University Press, Cambridge, MA.

Dowling, J.E. and B. Ehinger (1978). The interplexiform cell system I. synapses of the dopaminergic neurons of the goldfish retina. *Proc. R. Soc. Lond. B.* **201**, 7–26.

Dreyer, E.B., Z.-H. Pan, S. Storm, and S.A. Lipton (1994). Greater sensitivity of larger retinal ganglion cells to NMDA-mediated cell death. *Neuroreport.* **5**, 629–631.

Dreyer, E.B., D. Zurakowski, R.A. Schumer, S.M. Podos, and S.A. Lipton (1996). Elevated glutamate levels in the vitreous body of humans and monkeys with glaucoma. *Arch. Ophthalmol.* **114**, 299–305.

Duvoisin, R.M., C. Zhang, and K. Ramonell (1995). A novel metabotropic glutamate receptor expressed in the retina and olfactory bulb. *J. Neurosci.* **15**, 3075–3083.

Egebjerg, J., B. Bettler, I. Hermans-Borgmeyer, and S. Heinemann (1991). Cloning of a cDNA for a glutamate receptor subunit activated by kainate but not AMPA. *Nature* **351**, 745–748.

Ehinger, B., O.P. Ottersen, J. Storm-Mathisen, and J.E. Dowling (1988). Bipolar cells in the turtle retina are strongly immunoreactive for glutamate. *Proc. Natl. Acad. Sci.* **85**, 8321–8325.

Eliasof, S. and C.E. Jahr (1996). Retinal glial cell glutamate transporter is coupled to an anionic conductance. *Proc. Natl. Acad. Sci.* **93**, 4153–4158.

Eliasof, S. and C.E. Jahr (1997). Rapid AMPA receptor desensitization in catfish cone horizontal cells. *Vis. Neurosci.* **14**, 13–18.

Eliasof, S. and F. Werblin (1993). Characterization of the glutamate transporter in retinal cones of the tiger salamander. *J. Neurosci.* **13**, 402–411.

Fairman, W.A., R.J. Vandengerg, J.L. Arriza, M.P. Kavanaugh, and S.G. Amara (1995). An excitatory amino-acid transporter with properties of a ligand-gated chloride channel. *Nature* **375**, 599–603.

Famiglietti, Jr., E.V. and H. Kolb (1976). Structural basis for ON- and OFF-center responses in retinal ganglion cells. *Science* **194**, 193–195.

Fykse, E.M. and F. Fonnum (1996). Amino acid neurotransmission: Dynamics of vesicular uptake. *Neurochem. Res.* **21**, 1053–1060.

Ghosh, K.K., S. Haverkamp, and H. Wassle (2001). Glutamate receptors in the rod pathway of the mammalian retina. *J. Neurosci.* **21**, 8636–8647.

Gilbertson, T.A., R. Scobey, and M. Wilson (1991). Permeation of calcium ions through non-NMDA glutamate channels in retinal bipolar cells. *Science* **251**, 1613–1615.

Glazner, G.W., S.L. Chan, C. Lu, and M.P. Mattson (2000). Caspase-mediated degradation of AMPA receptor subunits: A mechanism for preventing excitotoxic necrosis and ensuring apoptosis. *J. Neurosci.* **20**, 3641–3649.

Grant, G.B. and J.E. Dowling (1995). A glutamate-activated chloride current in cone-driven ON bipolar cells of the white perch retina. *J. Neurosci.* **15**, 3852–3862.

Grant, G.B. and J.E. Dowling (1996). ON bipolar cell responses in the teleost retina are generated by two distinct mechanisms. *J. Neurophysiol.* **76**, 3842–3849.

Grant, G.B. and F.S. Werblin (1996). A glutamate-elicited chloride current with transporter-like properties in rod photoreceptors of the tiger salamander. *Vis. Neurosci.* **13**, 135–144.

Grunert, U., S. Haverkamp, E.L. Fletcher, and H. Wassle (2002). Synaptic distribution of ionotropic glutamate receptors in the inner plexiform layer of the primate retina. *J. Comp. Neurol.* **447**, 138–151.

Grunert, U., P.R. Martin, and H. Wassle (1994). Immunocytochemical analysis of bipolar cells in the Macaque monkey retina. *J. Comp. Neurol.* **348**, 607–627.

Hack, I., M. Frech, O. Dick, L. Peichl, and J.H. Brandstatter (2001). Heterogeneous distribution of AMPA glutamate receptor subunits at the photoreceptor synapses of rodent retina. *Eur. J. Neurosci.* **13**, 15–24.

Hack, I., L. Peichl, and J.H. Brandstatter (1999). An alternative pathway for rod signals in the rodent retina: Rod photoreceptors, cone bipolar cells, and the localization of glutamate receptors. *Proc. Natl. Acad. Sci.* **96**, 14130–14135.

Hamassaki-Britto, D.E., I. Hermans-Borgmeyer, S. Heinemann, and T.E. Hughes (1993). Expression of glutamate receptor genes in the mammalian retina: The localization of GluR1 through GluR7 mRNAs. *J. Neurosci.* **13**, 1888–1898.

Hartveit, E., J.H. Brandstatter, R. Enz, and H. Wassle (1995). Expression of the mRNA of seven metabotropic glutamate receptors (mGluR1 to 7) in the rat retina. An in situ hybridization study on tissue section and isolated cells. *Eur. J. Neurosci.* **7**, 1472–1483.

Hartveit, E., J.H. Brandstatter, M. Sasso-Pognetto, D.J. Laurie, P.H. Seeburgh, and H. Wassle (1994). Localization and developmental expression of the NMDA receptor subunit NR2A in the mammalian retina. *J. Comp. Neurol.* **348**, 570–582.

Hartveit, E. and M.L. Veruki (1997). All amacrine cells express functional NMDA receptors. *Neuroreport* **8**, 1219–1223.

Haverkamp, S., U. Grunert, and H. Wassle (2001a). The synaptic architecture of AMPA receptors at the cone pedicle of the primate retina. *J. Neurosci.* **21**, 2488–2500.

Haverkamp, S., U. Grunert, and H. Wassle (2001b). Localization of kainate receptors at the cone pedicles of the primate retina. *J. Comp. Neurol.* **436**, 471–186.

Hensley, S.H., X.-L. Yang, and S.M. Wu (1993). Identification of glutamate receptor subtypes mediating inputs to bipolar cells and ganglion cells in the tiger salamander retina. *J. Neurophysiol.* **69**, 2099–2107.

Hertz, L. (1979). Functional interactions between neurons and astrocytes I. Turnover and metabolism of putative amino acid transmitters. *Progr. Neurobiol.* **13**, 277–323.

Hirano, A.A. and P.R. MacLeish (1991). Glutamate and 2-amino-4-phosphobutyric acid evoke an increase in potassium conductance in retinal bipolar cells. *Proc. Natl. Acad. Sci.* **88**, 805–809.

Hirasawa, H., R. Shiells, and M. Yamada (2002). A metabotropic glutamate receptor regulates transmitter release from cone presynaptic terminals in carp retinal slices. *J. Gen. Physiol.* **119**, 55–68.

Hollmann, M., A. O'Shea-Greenfield, S.W. Rogers, and S. Heinemann (1989). Cloning by functional expression of a member of the glutamate receptor family. *Nature* **342**, 643–648.

Houamed, K.M., J.L. Kuijper, T.L. Gilbert, B.A. Haldeman, P.J. O'Hara, E.R. Mulvihill *et al.* (1991). Cloning, expression, and gene structure of a G protein-coupled glutamate receptor from rat brain. *Science* **252**, 1318–1321.

Hughes, T.E. (1997). Are there ionotropic glutamate receptors on the rod bipolar cell of the mouse retina? *Vis. Neurosci.* **14**, 103–109.

Hughes, T.E., I. Hermans-Borgmeyer, and S. Heinemann (1992). Differential expression of glutamate receptor genes (GluR1-5) in the rat retina. *Vis. Neurosci.* **8**, 49–55.

Ikeda, K., M. Nagasawa, H. Mori, K. Araki, K. Sakimura, M. Watanabe *et al.* (1992). Cloning and expression of the ε4 subunit of the NMDA receptor channel. *FEBS Lett.* **313**, 34–38.

Ishida, A.T., W.K. Stell, and D.O. Lightfoot (1980). Rod and cone inputs to bipolar cells in goldfish retina. *J. Comp. Neurol.* **191**, 315–335

Ishii, T., K. Moriyoshi, H. Sugihara, K. Sakurada, H. Kadotani, M. Yokoi *et al.* (1993). Molecular characterization of the family of N-methyl-D-aspartate receptor subunits. *J. Biol. Chem.* **268**, 2836–2843.

Jin, X.T. and W.J. Brunken (1996). A differential effect of APB on ON- and OFF-center ganglion cells in the dark adapted rabbit retina. *Brain Res.* **708**, 191–196.

Johnson, J.W. and P. Ascher (1987). Glycine potentiates the NMDA response in cultured mouse brain neurons. *Nature* **325**, 529–531.

Jojich, L. and R.G. Pourcho (1996). Glutamate immunoreactivity in the cat retina: A Quantitative study. *Vis. Neurosci.* **13**, 117–133.

Kalloniatis, M. and E.L. Fletcher (1993). Immunocytochemical localization of the amino acid neurotransmitters in the chicken retina. *J. Comp. Neurol.* **336**, 174–193.

Kalloniatis, M. and R.E. Marc (1990). Interplexiform cells of the goldfish retina. *J. Comp. Neurol.* **297**, 340–358.

Kamermans, M. I. Fahrenfort, K. Schultz, U. Janssen-Bienhold, T. Sjoerdsma, and R. Weiler (2001). Hemichannel-meidated inhibition in the outer retina. *Science* **292**, 1178–1180.

Kamermans, M. and H. Spekreijse (1999). The feedback pathway from horizontal cells to cones. A mini review with a look ahead. *Vision Res.* **39**, 2449–2468.

Kanai, Y. and M.A. Hediger (1992). Primary structure and functional characterization of a high-affinity glutamate transporter. *Nature* **360**, 467–471.

Kanai, Y., C.P. Smith, and M.A. Hediger (1994). A new family of neurotransmitter transporters: The high-affinity glutamate transporters. *FASEB J.* **8**, 1450–1459.

Kashii, S., M. Mandai, M. Kikuchi, Y. Honda, Y. Tamura, K. Kaneda *et al.* (1996). Dual actions of nitric oxide in N-methyl-D-aspartate receptor-mediated neurotoxicity in cultured retinal neurons. *Brain Res.* **711**, 93–101.

Katsuwada, T., N. Kashiwabuchi, H. Mori, K. Sakimura, E. Kushiya, K. Araki *et al.* (1992). Molecular diversity of the NMDA receptor channel. *Nature* **358**, 36–41.

Kawasaki, A., Y. Otori, and C.J. Barnstable (2000). Muller cell protection of rat retinal ganglion cells from glutamate and nitric oxide neurotoxicity. *Invest. Ophthal. Vis. Sci.* **41**, 3444–3450.

Keinanen, K., W. Wisden, B. Sommer, P. Werner, A. Herb, T.A. Versoorn *et al.* (1990). A family of AMPA-selective glutamate receptors. *Science* **249**, 556–560.

Kessler, M., A. Arai, A. Quan, and G. Lynch (1996). Effect of cyclothiazide on binding properties of AMPA-type glutamate receptors: Lack of competition between cyclothiazide and GYKI 52466. *Molec. Pharmacol.* **49**, 123–131.

Kittila, C.A. and S.C. Massey (1995). Effect of ON pathway blockade on directional selectively in the rabbit retina. *J. Neurophysiol.* **73**, 703–712.

Kleckner, N.W. and R. Dingledine (1988). Requirement for glycine activation of NMDA-receptors expressed in Xenopus oocytes. *Science* **241**, 835–837.

Klooster, J., K.M. Studholme, and S. Yazulla (2001). Localization of the AMPA subunit GluR2 in the outer plexiform layer of goldfish retina. *J. Comp. Neurol.* **441**, 155–167.

Knopfel,T., R. Kuhn, and H. Allgeier (1995). Metabotropic glutamate receptors: Novel targets for drug development. *J. Med Chem* **38**, 1417–1426.

Kolb, H., R. Nelson, and A. Mariani (1981). Amacrine cells, bipolar cells, and ganglion cells of the cat retina: A Golgi study. *Vision Res.* **21**, 1081–1114.

Koulen, P., R. Kuhn, H. Wassle, and J.H. Brandstatter (1997). Group I metabotropic glutamate receptors mGluR1alpha and mGluR5a: Localization in both synaptic layers of the rat retina. *J. Neurosci.* **17**, 2200–2211.

Kriemborg, K.M., M.L. Lester, K.F. Medler, and E.L. Gleason (2001). Group I metabotropic glutamate receptors are expressed in the chicken retina and by cultured retinal amacrine cells. *J. Neurochem.* **77**, 452–465.

Lehre, K.P., S. Davanger, and N.C. Danbolt (1997). Localization of the glutamate transporter protein GLAST in rat retina. *Brain Res.* **744**, 129–137.

Lodge, D. (1997). Subtypes of glutamate receptors. Historical perspectives on their pharmacological differentiation. In D.T. Monaghan and R.J. Weinhold (eds), *The Ionotropic Glutamate Receptors*. Humana Press, NJ. pp. 1–38.

Luo, X., V. Heidinger, S. Picaud, G. Lambrou, H. Dreyfus, J. Sahel *et al.* (2001). Selective excitotoxic degeneration of adult pig retinal ganglion cells in vitro. *Invest. Ophthal. Vis. Sci.* **42**, 1096–1106.

Marc, R.E. and D.M.K. Lam (1981). Uptake of aspartic and glutamic acid by photoreceptors in goldfish retina. *Proc. Natl. Acad. Sci.* **78**, 7185–7189.

Marc, R.E, W.-L.S. Liu, M. Kalloniatis, S.F. Raiguel, and E. Van Haesendonck (1990). Patterns of glutamate immunoreactivity in the goldfish retina. *J. Neurosci.* **10**, 4006–4034.

Massey, S.C., J.M. Koomen, S. Liu, K.P. Lehre, and N.C. Danbolt (1997). Distribution of the glutamate transporter GLT-1 in the rabbit retina. *Invest. Ophthal. Vis. Sci.* **38**, S689.

Massey, S.C. and R.F. Miller (1990). N-Methyl-D-Aspartate receptors of ganglion cells in rabbit retina. *J. Neurophysiol.* **63**, 16–30.

Massey, S.C. and R.F. Miller (1988). Glutamate receptors of ganglion cells in the rabbit retina: Evidence for glutamate as a bipolar cell transmitter. *J. Physiol.* **405**, 635–655.

Massey, S.C., D.A. Redburn, and M.L.J. Crawford (1983). The effects of 2-amino-4-phosphobutyric acid (APB) on the ERG and ganglion cell discharge of rabbit retina. *Vision Res.* **23**, 1607–1613.

Masu, M., H. Iwakabe, Y. Tagawa, T. Miyoshi, M. Yamashita, Y. Fukuda *et al.* (1995). Specific deficit of the ON response in visual transmission by targeted disruption of the mGluR6 gene. *Cell* **80**, 757–765.

Masu, M., Y. Tanabe, K. Tsuchida, R. Shigemoto, and S. Nakanishi (1991). Sequence and expression of a metabotropic glutamate receptor. *Nature* **349**, 760–765.

Matsui, K., J. Hasegawa, and M. Tachibana (2001). Modulation of excitatory synaptic transmission by GABAC receptor-mediated feedback in the mouse inner retina. *J. Neurophysiol.* **86**, 2285–2298.

Mayer, M.L. and G.L. Westbrook (1987). Permeation and block of N-Methyl-D-Aspartic acid receptor channels by divalent cations in mouse cultured central neurones. *J. Physiol.* **394**, 501–527.

Meguro, H., H. Mori, K. Araki, E. Kushiya, T. Kutsuwada, M. Yamazaki *et al.* (1992). Functional characterization of a heteromeric NMDA receptor channel expressed from cloned cDNAs. *Nature* **357**, 70–74.

Mittman, S., W.R. Taylor, and D.R. Copenhagen (1990). Concomitant activation of two types of glutamate receptor mediates excitation of salamander retinal ganglion cells. *J. Physiol.* **428**, 175–197.

Morigiwa, K. and N. Vardi (1999). Differential expression of ionotropic glutamate receptor subunits in the outer retina. *J. Comp. Neurol.* **405**, 173–184.

Moriyoshi, K., M. Masu, T. Ishii, R. Shigemoto, N. Mizuno, and S. Nakanishi (1991). Molecular cloning and characterization of the rat NMDA receptor. *Nature* **354**, 31–37.

Mosinger, J.L., M.T. Price, H.Y. Bai, H. Xiao, D.F. Wozniak, and J.W. Olncy (1991). Blockade of both NMDA and non-NMDA receptors is required for optimal protection against ischemic neuronal degeneration in the in vivo adult mammalian retina. *Exp. Neurol.* **113**, 10–17.

Moyner, H., R. Sprengel, R. Schoepfer, A. Herb, M. Higuchi, H. Lorneli *et al.* (1992). Heteromeric NMDA receptors: molecular and functional distinction of subtypes. *Science* **256**, 1217–1221.

Naito, S. and T. Ueda (1983). Adenosine triphosphate-dependent uptake of glutamate into Protein I-associated synaptic vesicles. *J. Biol. Chem.* **258**, 696–699.

Nakajima, Y., H. Iwakabe, C. Akazawa, H. Nawa, R. Shigemoto, N. Mizuno *et al.* (1993). Molecular characterization of a novel retinal metabotropic glutamate receptor mGluR6 with a high agonist selectively for L-2-amino-4-phosphobutyrate. *J. Biol. Chem.* **268**, 11868–11873.

Nakanishi, S. (1994). Metabotropic glutamate receptors: Synaptic transmission, modulation, and plasticity. *Neuron* **13**, 1031–1037.

Nakanishi, S. (1992). Molecular diversity of glutamate receptors and implications for brain function. *Science* **258**, 597–603.

Nakanishi, N., N.A. Schneider, and R. Axel (1990). A family of glutamate receptor genes: evidence for the formation of heteromultimeric receptors with distinct channel properties. *Neuron* **5**, 569–581.

Naskar, R., C.K. Vorwerk, and E.B. Dreyer (2000). Concurrent downregulation of a glutamate transporter and receptor in glaucoma. *Invest. Ophthal. Vis. Sci.* **41**, 1940–1944.

Nawy, S. (2000). Regulation of the ON bipolar cell mGluR6 pathway by Ca^{+2}. *J. Neurosci.* **20**, 4471–4479.

Nawy, S. (1999). The metabotropic receptor mGluR6 may signal through Go, but not phosphodiesterase, in retinal bipolar cells. *J. Neurosci.* **19**, 2938–2944.

Nawy, S. and D.R. Copenhagen (1990). Intracellular cesium separates two glutamate conductances in retinal bipolar cells of goldfish. *Vision Res.* **30**, 967–972.

Nawy, S. and D.R. Copenhagen (1987). Multiple classes of glutamate receptor on depolarizing bipolar cells in retina. *Nature* **325**, 56–58.

Nawy, S. and C.E. Jahr (1990). Suppression by glutamate of cGMP-activated conductance in retinal bipolar cells. *Nature* **346**, 269–271.

Nelson, R., E.V. Famiglietti Jr., and H. Kolb (1978). Intracellular staining reveals different levels of stratification for ON- and OFF-center ganglion cells in cat retina. *J. Neurophysiol.* **41**, 472–483.

Nestler, E.J. and E.S. Duman (1994). G proteins and cyclic nucleotides in the nervous system. In G.J. Siegel, B.W. Agranoff, R.W. Albers, and P.B. Molinoff, (eds). *Basic Neurochemistry*, 5th edn. Raven Press, New York, pp. 429–448.

Nomura, A., R. Shigemoto, Y. Nakamura, N. Okamoto, N. Mizuno, and S. Nakanishi (1994). Developmentally regulated postsynaptic localization of a metabotropic glutamate receptor in rat bipolar cells. *Cell* **77**, 361–369.

Nowak, L., P. Bregestovski, P. Ascher, A. Herbet, and A. Prochiantz (1984). Magnesium gates glutamate-activated channels in mouse central neurones. *Nature* **307**, 462–465.

O'Dell, T.J. and B.N. Christensen (1989). Horizontal cells isolated from catfish retina contain two types of excitatory amino acid receptors. *J. Neurophysiol.* **61**, 1097–1109.

Olney, J.W. (1969). Glutamate-induced retinal degeneration in neonatal mice. Electron microscopy of the acutely evolving lesion. *J. Neuropathol. Exp. Neurol.* **28**, 455–474.

Otori, Y., J.-Y. Wei, and C.J. Barnstable (1998). Neurotoxic effects of low doses of glutamate on purified rat retinal ganglion cells. *Invest. Ophthal. Vis. Sci.* **39**, 972–981.

Paternain, A.V., M. Morales, and J. Lerma (1995). Selective antagonism of AMPA receptors unmasks kainate receptor-mediated responses in hippocampal neurons. *Neuron* **14**, 185–189.

Peng, Y.W., C.D. Blackstone, R.L. Huganir, and K.W. Yau (1995). Distribution of glutamate receptor subtypes in the vertebrate retina. *Neuroscience.* **66**, 483–497.

Picaud, S., H.P. Larsson, D.P. Wellis, H. Lecar, and F. Werblin (1995a). Cone photoreceptors respond to their own glutamate release in the tiger salamander. *Proc. Natl. Acad. Sci.* **92**, 9417–9421.

Picaud, S.A., H.P. Larsson, G.B. Grant, H. Lecar, and F.S. Werblin (1995b). Glutamate-gated chloride channel with glutamate-transporter-like properties in cone photoreceptors of the tiger salamander. *J. Neurophysiol.* **74**, 1760–1771.

Pin, J.P. and J. Bockaert (1995). Get receptive to metabotropic glutamate receptors. *Curr. Opin. Neurobiol.* **5**, 342–349.

Pin, J.P. and R. Duvoisin (1995). Review: Neurotransmitter receptors I: The metabotropic glutamate receptors: Structure and functions. *Neuropharmacology* **34**, 1–26.

Pines, G., N.C. Danbolt, M. Bjoras, Y. Zhang, A. Bendahan, L. Eide *et al.* (1992). Cloning and expression of a rat brain L-glutamate transporter. *Nature* **360**, 464–467.

Poitry, S., C. Poitry-Yamate, J. Ueberfeld, P.R. MacLeish, and M. Tsacopoulos (2000). Mechanisms of glutamate metabolic signalling in retinal glial (Muller) cells. *J. Neurosci.* **20**, 1809–1821.

Pourcho, R.G., W. Cai, and P. Qin (1997). Glutamate receptor subunits in cat retina: light and electron microscopic observations. *Invest. Ophthal. Vis. Sci.* **38**, S46.

Pourcho, R.G., P. Qin, and D.J. Goebel (2001). Cellular and subcellular distribution of NMDA receptor subunit NR2B in the retina. *J. Comp. Neurol.* **433**, 75–85.

Pow, D.V. and N.L. Barnett (1999). Changing patterns of spatial buffering of glutamate in developing rat retinae are mediated by the Muller cell glutamate transporter GLAST. *Cell Tissue Res.* **297**, 57–66.

Pow, D.V. and D.R. Crook (1996). Direct immunocytochemical evidence for the transfer of glutamine from glial cells to neurons: Use of specific antibodies directed against the D-stereoisomers of glutamate and glutamine. *Neuroscience.* **70**, 295–302.

Qin, P. and R.G. Pourcho (2001). Immunocytochemical localization of kainate-selective glutamate receptor subunits GluR5, GluR6, GluR7 in the cat retina. *Brain Res.* **890**, 211–221.

Qin P. and R.G. Pourcho (1999). AMPA-selective glutamate receptors subunits GluR2 and GluR4 in the cat retina: an immunocytochemical study. *Vis. Neurosci.* **16**, 1105–1114.

Ransom, R.W. and N.L. Stec (1988). Cooperative modulation of [3H]MK-801 binding to the N-Methyl-D-Aspartate receptor-ion channel complex by L-glutamate, glycine, and polyamines. *J. Neurochem.* **51**, 830–836.

Rauen, T. (2000). Diversity of glutamate transporter expression and function in the mammalian retina. *Amino Acids.* **19**, 53–62.

Rauen, T., J.F. Rothstein, and H. Wassle (1996). Differential expression of three glutamate transporter subtypes in the rat retina. *Cell Tissue Res.* **286**, 325–336.

Redburn, D.A. and Rowe-Rendelman, C. (1996). Development neurotransmitters Signals for shaping neuronal circuitry. *Invest. Ophthal. Vis. Sci.* **37**, 1479–1482.

Santos, A.E., A.L. Carvalho, M.C. Lopes, and A.P. Carvalho (2001). Differential postreceptor signaling events triggered by excitotoxic stimulation of different ionotropic glutamate receptors in retinal neurons. *J. Neurosci. Res.* **66**, 643–655.

Sasaki, T. and A. Kaneko (1996). L-glutamate-induced responses in OFF-type bipolar cells of the cat retina. *Vision Res.* **36**, 787–795.

Saugstad, J.A., J.M. Kinzie, E.R. Mulvihill, T.P. Segerson, and G.L. Westbrook (1994). Cloning and expression of a new member of the L-2-amino-4-phosphobutyric acid-sensitive class of metabotropic glutamate receptors. *Mol. Pharmacol.* **45**, 367–372.

Schuettaf, F., R. Naskar, C.K. Vorwerk, D. Zurakowski, and E.B. Dreyer (2000). Ganglion cell loss after optic nerve crush mediated through AMPA-kainate and NMDA receptors. *Invest. Ophthal. Vis. Sci.* **41**, 4313–4316.

Schultz, K., D.J. Goldman, T. Ohtsuka, J. Hirano, L. Barton, and W.K. Stell (1997). Identification and localization of an immunoreactive AMPA-type glutamate receptor subunit (GluR4) with respect to identified photoreceptor synapses in the outer plexiform layer of goldfish retina. *J. Neurocytol.* **26**, 651–666.

Schultz, K., U. Janssen-Bienhold, and R. Weiler (2001). Selective synaptic distribution of AMPA and kainate receptor subunits in the outer plexiform layer of the carp retina. *J. Comp. Neurol.* **435**, 433–449.

Schultz, K. and W.K. Stell (1996). Immunocytochemical localization of the high-affinity glutamate transporter, EAAC1, in the retina of representative vertebrate species. *Neurosci. Lett.* **211**, 191–194.

Schwartz, E.A. and M. Tachibana (1990). Electrophysiology of glutamate and sodium co-transport in a glial cell of the salamander retina. *J. Physiol.* **426**, 43–80.

Shen, Y., Y. Zhou, and Yang, Z.-L. (1999). Characterization of AMPA receptors on isolated amacrine-like cells in carp retina. *Eur. J. Neurosci.* **11**, 4233–4240.

Shiells, R.A. and G. Falk (2001). Rectification of cGMP-activated channels induced by phosphorylation in dogfish retinal "on" bipolar cells. *J. Physiol.* **535**, 697–702.

Shimamoto, K., B. Lebrun, Y. Yasuda-Kamatani, M. Sakaitani, Y. Shigeri, N. Yumoto *et al.* (1998). DL-threo-beta-benzyloxyaspartate, a potent blocker of excitatory amino acid transporters. *Mol. Pharmacol.* **53**, 195–201.

Siliprandi, R., R. Canella, G. Carmignoto, N. Schiavo, A. Zanellato, R. Zanoni *et al.* (1992). N-methyl-D-aspartate induced neurotoxicity in the adult rat retina. *Vis. Neurosci.* **8**, 567–573.

Sisk, D.R. and T. Kuwabara (1985). Histologic changes in the inner retina of albino rats following intravitreal injection of monosodium L-glutamate. *Graefe's Arch. Clin. Exp. Ophthalmol.* **223**, 250–258.

Slaughter, M.M. and R.F. Miller (1981). 2-amino-4-phosphobutyric acid: A new pharmacological tool for retina research. *Science* **211**, 182–184.

Stell, W.K., A.T. Ishida, and D.O. Lightfoot (1977). Structural basis for ON- and OFF-center responses in retinal bipolar cells. *Science* **198**, 1269–1271.

Stockton, R.A. and M.M. Slaughter (1989). B-wave of the electroretinogram a reflection of ON bipolar cell activity. *J. Gen. Physiol.* **93**, 101–122

Stryer, L. (1988). *Biochemistry*, 3rd edn. W.H. Freeman & Co., New York.

Sucher, N.J., S.Z. Lei, and S.A. Lipton (1991). Calcium channel antagonists attenuate NMDA-receptor mediated neurotoxicity of retinal ganglion cells in culture. *Brain Res.* **297**, 297–302.

Tabb, J.S. and T. Ueda (1991). Phylogenetic studies on the synaptic vesicle glutamate transporter. *J. Neurosci.* **11**, 1822–1828.

Tachibana, M. and A. Kaneko (1988). L-glutamate-induced depolarization in solitary photoreceptors: A process that may contribute to the interaction between photoreceptors in situ. *Proc. Natl. Acad. Sci.* **85**, 5315–5319.

Tagawa, Y., H. Sawai, Y. Ueda, M. Tauchi, and S. Nakanishi (1999). Immunohistological studies of metabotropic glutamate receptor subtype 6-deficient mice show no abnormality of retinal cell organization and ganglion cell maturation. *J. Neurosci.* **19**, 2568–2579.

Takahashi, K.-I. and D.R. Copenhagen (1992). APB suppresses synaptic input to retinal horizontal cells modulated by intracellular pH. *J. Neurophysiol.* **67**, 1633–1642.

Tanabe, Y., M. Masu, T. Ishii, R. Shigemoto, and S. Nakanishi (1992). A family of metabotropic glutamate receptors. *Neuron.* **8**, 169–179.

Taylor, W.R. and H. Wassle (1995). Receptive field properties of starburst cholinergic amacrine cells in the rabbit retina. *Eur. J. Neurosci.* **7**, 2308–2321.

Tian, N. and M.M. Slaughter (1995). Correlation of dynamic responses in the ON bipolar neuron and the b-wave of the electroretinogram. *Vision Res.* **35**, 1359–1364.

Vandenbranden, C.A.V., W. Kamphuis, B. Nunes Cardozo, and M. Kamermans (2000a). Expression and localization of ionotropic glutamate receptor subunits in the goldfish retina—an in situ hybridization and immunocyto-chemical study. *J. Neurocytol.* **29**, 729–742.

Vandenbranden, C.A.V., S. Yazulla, K.M. Studholme, W. Kamphuis, and M. Kamermans (2000b). Immunocytochemical localization of the glutamate transporter GLT-1 in the goldfish (Carassius auratus) retina. *J. Comp Neurol.* **423**, 440–451.

Van Haesendonck, E. and L. Missotten (1990). Glutamate-like immunoreactivity in the retina of a marine teleost, the dragonet. *Neurosci. Lett.* **111**, 281–286.

Vardi, N. (1998). Alpha subunit of Go localizes in the dendritic tips of ON bipolar cells. *J. Comp. Neurol.* **395**, 43–52.

Vardi, N., R. Duvoisin, G. Wu, and P. Sterling (2000). Localization of mGluR6 to dendrites of ON bipolar cells in primate retina. *J. Comp Neurol.* **423**, 402–412.

Vardi, N., D.F. Matesic, D.R. Manning, P.A. Liebman, and P. Sterling (1993). Identification of a G-protein in depolarizing rod bipolar cells. *Vis. NeuroSci.* **10**, 473–478.

Villani, L., T. Guarnieri, and G. Dell'Erba (1997). Apoptosis is induced by excitotoxicity in goldfish retina. *J. Brain Res.* **38**, 481–486.

Vorwerk, C.K., M.R. Kreutz, T.M. Bockers, M. Brosz, E.B. Dreyer, and B.A. Sabel (1999). Susceptibility of retinal ganglion cells to excitotoxicity depends on soma size and retinal eccentricity. *Curr. Eye Res.* **19**, 59–65.

Vorwerk, C.K., S.A. Lipton, D. Zurakowski, B.T. Hyman, B.A. Sabel, and E.B. Dreyer (1996). Chronic low-dose glutamate is toxic to retinal ganglion cells Toxicity blocked by memantine. *Invest. Ophthal. Vis. Sci.* **37**, 1618–1624.

Vorwerk, C.K., R. Naskar, F. Schuettauf, K. Quinto, D. Zurakowski, G. Gochenauer *et al.* (2000). Depression of retinal glutamate transporter function leads to elevated intravitreal glutamate levels and ganglion cell death. *Invest. Ophthal. Vis. Sci.* **41**, 3615–3621.

Walters, R.J., R.H. Kramer, and S. Nawy (1998). Regulation of cGMP-dependent current in ON bipolar cells by calcium/calmodulin-dependent kinase. *Vis. Neurosci.* **15**, 257–261.

Wenzel, A., D. Benke, H. Mohler, and Fritschy, J.-M. (1997). N-methyl-D-aspartate receptors containing the NR2D subunit in the retina are selectively expressed in rod bipolar cells. *Neuroscience* **78**, 1105–1112.

Werblin, F.S and J.E. Dowling (1969). Organization of the retina of the mudpuppy, Necturus Maculosus. II. Intracellular recording. *J. Neurophysiol.* **32**, 339–355.

Westbrook, G.L. and M.L. Mayer (1987). Micromolar concentrations of Zn^{+2} antagonize NMDA and GABA responses of hippocampal neurons. *Nature* **328**, 640–643.

Williams, K., A.M. Zappia, D.B. Pritchett, Y.M. Shen, and P.B. Molinoff (1994). Sensitivity of the N-Methyl-D-Aspartate receptor to polyamines is controlled by NR2 subunits. *Mol. Pharmacol.* **45**, 803–809.

Wong, W.T., K.L Myhr, E.D. Miller, and R.O.L. Wong (2000). Developmental changes in the neurotransmitter regulation of correlated spontaneous retinal activity. *J. Neurosci.* **20**, 351–360.

Wygnanski, T., H. Desatnik, H.A. Quigley, and Y. Glovinsky (1995). Comparison of ganglion cell loss and cone loss in experimental glaucoma. *Am. J. Ophthal.* **120**, 184–189.

Yamada, K.A. and C.-M. Tang (1993). Benzothiadiazides inhibit rapid glutamate receptor desensitization and enhance glutamatergic synaptic currents. *J. Neurosci.* **13**, 3904–3915.

Yamashita, M. and H. Wassle (1991). Responses of rod bipolar cells isolated from the rat retina to the glutamate agonist 2-amino-4-phosphonobutyric acid (APB). *J. NeuroSci.* **11**, 2372–2382.

Yang, C.-Y. (1996). Glutamate immunoreactivity in the tiger salamander retina differentiates between GABA-immunoreactive and glycine-immunoreactive amacrine cells. *J. Neurocytol.* **25**, 391–403.

Yang, C.-Y. and S. Yazulla (1994). Glutamate-, GABA-, and GAD-immunoreactivities co-localize in bipolar cells of tiger salamander retina. *Vis. Neurosci.* **11**, 1193–1203.

Yang, J.H. and S.M. Wu (1997). Characterization of glutamate transporter function in the tiger salamander retina. *Vision Res.* **37**, 827–838.

Yazulla, S. (1986). GABAergic neurons in the retina. *Progr. Retinal Res.* **5**, 1–52.

Yazulla, S. and K.M. Studholme (1999). Co-localization of Shaker A-type K+ channel (Kv1.4) and AMPA-glutamate receptor (GluR4) immunoreactivities to dendrites of OFF-bipolar cells of goldfish retina. *J. Neurocytol.* **28**, 63–73.

Yoles, E. and M. Schwartz (1998). Elevation of intraocular glutamate levels in rats with partial lesion of the optic nerve. *Arch. Ophthal.* **116**, 906–910.

Yu, W. and R.F. Miller (1995). NBQX, an improved non-NMDA antagonist studied in retinal ganglion cells. *Brain Res.* **692**, 190–194.

Zhao, J.-W. and Yang, X.-L. (2001). Glutamate transporter EAAC1 is expressed on Muller cells of lower vertebrate retinas. *J. Neurosci. Res.* **66**, 89–95.

Zhou, Z.J., G.L. Fain, and J.E. Dowling (1993). The excitatory and inhibitory amino acid receptors on horizontal cells isolated from the white perch retina. *J. Neurophysiol.* **70**, 8–19.

Zhou, Z.J. and D. Zhao (2000). Coordinated transitions in neurotransmitter systems for the initiation and propagation of spontaneous retinal waves. *J. Neurosci.* **20**, 6570–6577.

7

Glutamate Receptors in Taste Receptor Cells

Albertino Bigiani

1. Introduction

Taste receptor cells (*TRCs*) are specialized epithelial cells that are clustered together into ovoid end organs (*taste buds*: Figure 7.1) located in the epithelium of the tongue, the soft palate, and the epiglottis (Finger and Simon, 2000). On the tongue, taste buds are found in structures called *taste papillae* (fungiform, foliate, and vallate; Figure 7.1). TRCs detect chemical substances occurring in food and relay sensory information to the brain through the activation of specific synapses on gustatory nerve fibers (Figure 7.1). Sensory information built up at the level of TRCs is essential for proper selection of nutrients and for energy and material homeostasis (Breslin, 2000; Scott and Verhagen, 2000).

As peripheral detectors of chemical stimuli, TRCs are polarized elements: tight junctions between adjacent TRCs partition the cell membrane into an *apical membrane* (bathed by saliva or analogous fluid) that interacts with the chemicals occurring in food, and a *basolateral membrane* (bathed by interstitial fluid) that interacts with adjacent excitable elements, such as nerve endings through chemical synapses (Figure 7.1). Both the detection of chemical stimuli and the synaptic activity rely on the presence of specific membrane proteins (reviewed in Lindemann, 1996, 2001). Among these proteins, *glutamate receptors* play a key role in TRCs' physiology, acting as molecular sensors for food L-glutamate and as synaptic receptors in glutamatergic interactions with nerve endings (Figure 7.1).

2. Glutamate Receptors as Taste Receptors ("umami" Receptors)

Glutamate is a relevant taste stimulus in vertebrates. Glutamate (occurring as free amino acid, not bound in proteins) is a key molecule in determining the flavor of many protein-rich foods, including meat, cheese, fish, and certain vegetables (Fuke and Konosu,

---

Albertino Bigiani • Dipartimento di Scienze Biomediche, Sezione di Fisiologia, Università di Modena e Reggio Emilia, via Campi 287, 41100 Modena, Italy.

Glutamate Receptors in Peripheral Tissue, edited by Santokh Gill and Olga Pulido.
Kluwer Academic / Plenum Publishers, New York, 2005.

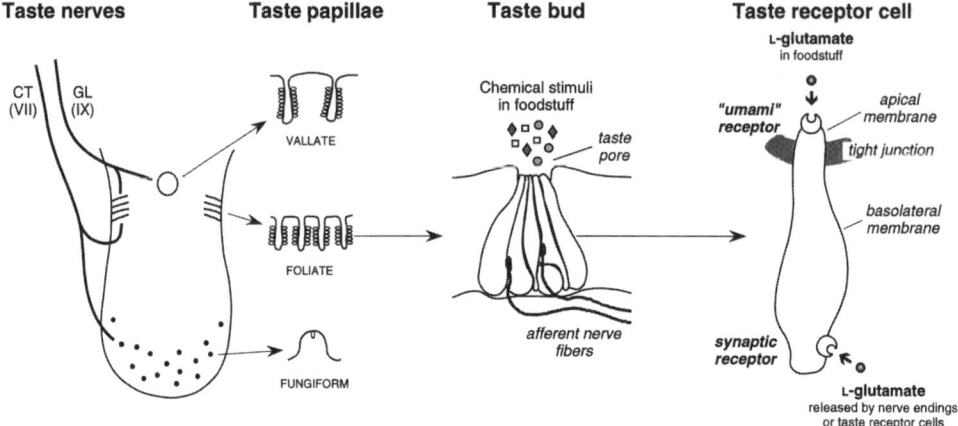

Figure 7.1. Schematic drawing of the anatomical organization of the peripheral taste system. Taste receptor cells are clustered in taste buds, onion-shaped sensory organs that can be found in three kinds of specialized taste papillae on the tongue surface. These papillae (fungiform, foliate, vallate) are distinguished on the basis of their morphology, location, and innervation. The main taste nerves to the tongue are the chorda tympani (CT) nerve, a branch of the VII nerve that innervates the fungiform papillae and the anterior part of the foliate papillae, and the glossopharyngeal (GL) nerve (IX nerve) innervating the vallate papilla and the posterior part of the foliate papilla. At the level of single TRCs, tight junctions partition the cell membrane in two regions: the apical membrane, above the junctions, that interacts directly with chemical stimuli like glutamate in foodstuff; the basolateral membrane, below the junctions, where functional, synaptic interactions with other cells can occur through glutamate signaling. "Umami receptor" refers to the glutamate receptor that detects food glutamate, whereas "synaptic receptor" refers to the glutamate receptor that detects glutamate as neurotransmitter.

1991; Yamaguchi and Ninomiya, 2000). Glutamate signals the presence of protein sources, and it is now believed that the taste of L-glutamate, called *umami* (a Japanese word meaning "delicious") is a separate and independent taste quality on par with sweet, sour, salty, and bitter (reviewed in Bellisle, 1999). Indeed, a new specific ageusia (lack of taste) for L-glutamate has been recently reported in human subjects (Lugaz *et al.*, 2002). Detection of glutamate by TRCs seems to be important for the regulation of ingestive behavior, metabolism, and protein balance (Rogers and Blundell, 1990; Mori *et al.*, 1991; Nijima, 1991; Viarouge *et al.*, 1991, 1992; Horio, 2000; Kondoh *et al.*, 2000; for review, see also Bellisle, 1999). In addition to its natural occurrence in many foods, L-glutamate is added to a wide range of packaged foods worldwide, mostly in the form of *monosodium glutamate* (*MSG*), as taste enhancer.

A characteristic taste-enhancing effect, or *synergism*, between L-glutamate and some *5′-ribonucleotides* (IMP and GMP), is a hallmark of umami taste. Synergism refers to the observation that low concentrations of the nucleotides, which by themselves have virtually no effect, potentiate the glutamate responses. This synergy has been documented in humans (Yamaguchi, 1991; Schiffman *et al.*, 1994), mice (Ninomiya, 1991, 1992), rats (Yoshii *et al.*, 1986; Yamamoto *et al.*, 1991; Sako and Yamamoto, 1999; Delay *et al.*, 2000), dogs (Kumazawa and Kurihara, 1990; Kumazawa *et al.*, 1991; Kurihara and Kashiwayanagi, 2000), and monkeys (Hellekant and Ninomiya, 1991). The molecular/cellular mechanisms underlying glutamate taste synergism are not fully clarified.

Given the relevance of glutamate in nutrition and human health, in recent years, a lot of experimental effort has been devoted to understand how food glutamate acts on TRCs.

Although preliminary biochemical data indicated that glutamate likely binds to the apical membrane receptors on TRCs (e.g., Hayashi *et al.*, 1994), only the recent application of molecular cloning techniques has allowed to identify candidate receptor molecules for glutamate as taste stimulus (*umami receptor*, Figure 7.1). These include the taste-specific GPCRs, "taste-mGluR4" and "T1R1/T1R3 heteromer."

2.1. Taste-mGluR4

Before the application of molecular methods, electrophysiological and behavioral studies suggested that taste receptors for glutamate could be somehow related to the glutamate receptors expressed in the brain (e.g., see Faurion, 1991). Using RT-PCR, Chaudhari and collaborators found that indeed several ionotropic (ion channels) and metabotropic (GPCRs) glutamate receptors were present in the rat lingual tissues (Chaudhari *et al.*, 1996; Yang *et al.*, 1999). However, they were able to demonstrate that only a metabotropic glutamate receptor, mGluR4, was specifically expressed by TRCs in foliate and vallate papillae, suggesting that this molecule could be the candidate taste receptor for glutamate. Behavioral studies supported this view by indicating that MSG and L-AP4, a mGluR4-selective ligand, elicited similar tastes in the rat (Chaudhari *et al.*, 1996). In subsequent studies, Chaudhari *et al.* (2000) were able to clone the mGluR4 from taste tissues and found that it differed substantially from the brain-mGluR4 by exhibiting a truncated extracellular domain (about 50% of the N-terminus was lacking). They called the taste-specific mGluR4 as *taste-mGluR4*. The extracellular N-terminus of brain-mGluR4 is supposed to be the glutamate-binding domain. Thus, Chaudhari and collaborators reasoned that the affinity of taste-mGluR4 for glutamate had to be altered, perhaps lower than in the brain isoform. Functional expression of taste-mGluR4 in CHO cells revealed that indeed activation of taste-mGluR4 by glutamate (evaluated by the reduction in intracellular cAMP) occurred at much higher concentration (about 100-fold) than that required to activate brain-mGluR4 (Table 7.1). This is what one would expect for a "taste" glutamate receptor: 0.1–3 mM is in fact the concentration range necessary to trigger a neural or behavioral taste response in juvenile (lower concentration) and adult (higher concentration) rodents (Ninomiya *et al.*, 1991; Yamamoto *et al.*, 1991; Chaudhari *et al.*, 1996; Stapleton *et al.*, 1999). In adult humans, detection threshold for glutamate is about 1 mM (Yamaguchi, 1991).

L-AP4 mimics the taste of L-glutamate in conditioned taste aversion in rats (Chaudhari *et al.*, 1996) and mice (Nakashima *et al.*, 2001), and in human psychophysical measurements (Kurihara and Kashiwayanagi, 1998). Consistently, Chaudhari *et al.* (2000) found that CHO cells expressing taste-mGluR4 responded to L-AP4 in a concentration range effective as taste stimulus (100 μM to 1 mM). On the contrary, in cell expressing brain-mGluR4, the response saturated for concentration >10 μM (Table 7.1).

How about the localization of taste-mGluR4 *in situ*? As a taste receptor, it is expected to be located at the apical ends of TRCs (Figure 7.1). Recent immunocytochemical observations have confirmed that in rat taste buds, taste-mGluR4 localizes exclusively at the level of taste pore (Toyono *et al.*, 2002), namely where the apical membrane of TRCs occurs (Figure 7.1). This finding further supports the hypothesis for a role of taste-mGluR4 as umami receptor.

The comprehensive studies by Chaudhari and collaborators also elucidated some aspects of the signal transduction pathway activated by food glutamate in TRCs. The cytoplasmic C terminus of taste-mGluR4, which participates in the interaction with G proteins, was found to be 100% identical with one of the two isoforms of mGluR4 found in the brain, namely

Table 7.1. G-Protein-Coupled Receptors (GPCRs) for Glutamate in Mammalian Taste Receptor Cells

	Taste-mGluR4	Brain-mGluR4a	T1R1/T1R3 (heteromer)		
Species	Rat	Rat	Mouse	Human	Rat
Papilla	Foliate Vallate	Foliate Vallate	Fungiform	ND	ND
Ligand	L-glutamate	L-glutamate	L-amino acids	L-glutamate	L-glutamate L-aspartate
L-glutamate efficacy (EC$_{50}$ in mM)	0.28	0.002	ND	~3-4	ND
N-terminal extracellular domain	Short	Long	Long	Long	Long
Effector system	↓cAMP	↓cAMP	ND	ND	ND
L-AP4 as agonist (effective concentration in μM)[b]	Yes (100–1,000)	Yes (<10)	Yes in the presence of IMP	Yes in the presence of IMP	No
Membrane localization in TRCs	Apical	Apical	T1R1: apical	ND	T1R1: apical in fungiform TRCs (Hoon et al. 1999; Kitagawa et al., 2001; Max et al., 2001; Montmayeur et al., 2001; Nelson et al., 2001, 2002; Sainz et al., 2001)
Proposed function	Umami receptor	?	Amino acid receptor Umami receptor	Umami receptor	Umami receptor
Purine ribonucleotide synergism	Yes[a]	ND	Yes for many AA, including L-Glu[b]	Yes for L-Glu[b]	Yes for L-Glu and L-Asp[b]
References	Chaudhari et al. (1996, 2000), Zhou and Chaudhari (1997), and Toyono et al. (2002)	Chaudhari et al. (1996, 2000), Zhou and Chaudhari (1997), and Toyono et al. (2002)	Hoon et al. (1999), Kitagawa et al. (2001), Max et al. (2001), Montmayeur et al. (2001), Nelson et al. (2001, 2002), and Sainz et al. (2001)	Li et al. (2002)	Li et al. (2002)

Note: ND: not determined.

[a]Evaluated on the basis of the effect on cAMP concentration in taste tissue from the rat foliate/vallate papillae.

[b]Tested in heterologous cell lines functionally expressing the relevant receptors.

brain-mGluR4a that couples to a decrease in cAMP (Thomsen *et al.*, 1997). When expressed in heterologous cells, taste-mGluR4 also responded to millimolar level of glutamate with decreases in cAMP (Chaudhari *et al.*, 2000). Also, stimulating rat taste buds with glutamate decreases cellular cAMP, and IMP potentiates this effect (Zhou and Chaudhari, 1997). Thus, taste-mGluR4 is a GPCR that binds food glutamate and signals through a G protein to regulate the level of intracellular cAMP (Figure 7.2A). Recent behavioral data in mice (He *et al.*, 2002; Ruiz *et al.*, 2003) indicate a role for gustducin in mediating umami transduction. Gustducin is a taste-specific Gα subunit expressed by subsets of TRCs in all taste organs (McLaughlin *et al.*, 1992; Boughter *et al.*, 1997). Gustducin activates cAMP-specific phosphodiesterase to decrease intracellular cAMP in TRCs (McLaughlin *et al.*, 1994; Ruiz-Avila

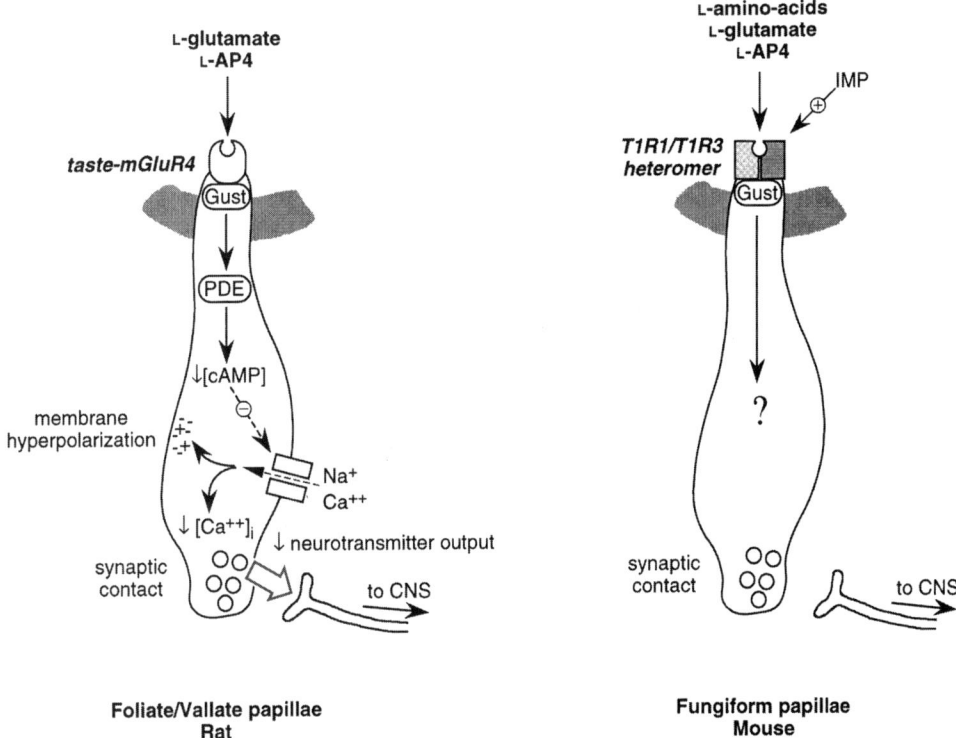

Figure 7.2. Candidate transduction mechanisms involved in the detection of glutamate as taste stimulus. (A) In TRCs from rat foliate/vallate papillae, food glutamate is supposed to interact with a taste-specific isoform of the mGluR4 called taste-mGluR4. Receptor activation leads to a decrease in the intracellular concentration of cAMP ([cAMP]$_i$) probably via gustducin (Gust)-mediated activation of a phosphodiesterase (PDE). Reduction in [cAMP]$_i$ should induce the closure of cation channels in the plasma membrane. It is not known whether the effect of a decrease of cAMP on the membrane channel is direct or mediated through intervening steps (dashed arrows). The reduction in the inflow of Na$^+$ and Ca^{2+} is expected to have two functional consequences: membrane hyperpolarization and decrease in [Ca^{2+}]$_i$, which ultimately should lead to a reduction in synaptic signaling to the nerve endings. However, it is not yet known whether actually hyperpolarization and decrease in [Ca^{2+}]$_i$ in TRCs can modulate a preexisting tonic release of neurotransmitter. (B) In TRCs from mouse fungiform papillae, food glutamate is believed to interact with the heteromer (T1R1/T1R3) of two GPCRs, T1R1 and T1R3. T1R1/T1R3 is actually a broadly tuned amino acid receptor. However, it displays properties consistent with being an umami receptor, such as activation by L-AP4 and synergism with IMP. It is likely that T1R1/T1R3 couples to gustducin (Gust) and also to α-transducin (not shown). The intracellular signaling cascade for this receptor has not been elucidated yet.

et al., 1995). Thus, gustducin represents a possible candidate for coupling activation of taste-mGluR4 with the intracellular effector system.

How the decrease in cAMP causes TRCs to signal the presence of glutamate to nerve endings is not known yet. Physiological studies on isolated TRCs have provided some cues. Using dye-imaging techniques, Hayashi *et al.* (1996) reported that L-AP4 primarily decreases $[Ca^{2+}]_i$ in TRCs from mouse foliate/vallate papillae. Patch-clamp studies showed that L-AP4 mimics glutamate by modulating membrane conductances in TRCs isolated from rat (Bigiani *et al.*, 1997; Lin and Kinnamon, 1999), and mouse (Iseki *et al.*, 2001). In rat foliate/vallate TRCs, L-AP4 (and glutamate) shuts down nonselective cation channels (Bigiani *et al.*, 1997). Consequently, a model has been proposed in which food glutamate triggers a decrease in cAMP, resulting in the closure of cation channels and hyperpolarization of TRCs, at least in rat foliate/vallate papillae (Brand, 2000; Lindemann, 2000) (for details, see Figure 7.2A). According to this hypothesis, TRCs would operate like photoreceptors (Rodieck, 1998). Interestingly, in rat fungiform TRCs, L-AP4 induces membrane hyperpolarization by the closure of chloride channels rather than of cation channels (Lin and Kinnamon, 1999). Also in this case, however, it seems that a reduction in intracellular cAMP is responsible for the effect on membrane channels (Lin and Kinnamon, 1999). Moreover, in some fungiform TRCs, L-AP4 elicits membrane depolarizations, suggesting the presence of either two forms of mGluR-4 or other receptors for glutamate (Lin and Kinnamon, 1999). It is then clear that more studies are necessary to fully clarify the sequence of events downstream of the activation of taste-mGluR4 and that eventually signals the presence of food glutamate to nerve endings.

Taste tissues also express brain-mGluR4a (Chaudhari *et al.*, 2000; Toyono *et al.*, 2002). The role of this receptor in glutamate reception (umami taste), if any, has to be elucidated. As demonstrated by Chaudhari *et al.* (2000), brain-mGluR4a has a sensitivity for L-glutamate that is more compatible with a neurotransmitter receptor, rather than with an umami receptor.

2.2. T1R1/T1R3 Heteromer

In addition to taste-mGluR4, recent molecular studies have provided evidence for the expression in TRCs of other GPCRs that may be involved in the detection of alimentary glutamate. By using an expression screening strategy, Nelson *et al.* (2002) identified and characterized an amino acid taste receptor in mice. This receptor, called *T1R1/T1R3*, is a heteromer of two taste-specific GPCRs, namely T1R1 and T1R3. T1R1 and T1R3 are distantly related to mGluRs (22–30% amino acid identity) and are distinguished from other GPCRs by the presence of a very long N-terminal extracellular domain (Hoon *et al.*, 1999; Kitagawa *et al.*, 2001; Max *et al.*, 2001; Montmayeur *et al.*, 2001; Nelson *et al.*, 2001; Sainz *et al.*, 2001). Nelson and collaborators (2002) transfected heterologous cell lines (HEK cells, expressing "promiscuous" Gα subunits that couple to a variety of receptors) with mouse T1R1 and T1R3 genes, singly or in combination, and tested for their abilities to respond to tastants by monitoring variation in $[Ca^{2+}]_i$ with calcium-imaging techniques. Quantification of amino acid responses revealed that only the combination of T1R1 with T1R3 functioned as a broadly tuned L-amino acid receptor (Nelson *et al.*, 2002). Thus, like the structurally related GPCR, GABA$_B$ receptor (Marshall *et al.*, 1999), T1R1 is expressed as a heterodimer with T1R3 in heterologous cell lines. *In situ*, T1R1 is coexpressed with T1R3 prominently in mouse fungiform TRCs (Nelson *et al.*, 2001), which are innervated by the chorda tympani fibers (Figure 7.1). In addition, it has been demonstrated that T1R1 is specifically located at the level of taste pores in rat taste buds (Hoon *et al.*, 1999), namely where a taste receptor is supposed to be (Figure 7.1).

Is T1R1/T1R3 an umami receptor? Although the combination of mouse T1R1 and T1R3 functions as a receptor that responds to most of the 20 L-amino acids commonly found in proteins, in their study, Nelson and collaborators (2002) showed that T1R1/T1R3 responds also to L-glutamate and to L-AP4, and that the responses are potentiated by IMP (synergistic effect), which is a basic characteristic of umami taste. Thus, they proposed that T1R1/T1R3 is involved in the umami response (Figure 7.2B).

Soon after the identification and functional characterization of mouse T1R1/T1R3 as a candidate umami receptor, another study reported that human T1R1/T1R3 and rat T1R1/T1R3 also may function as umami receptors (Li *et al.*, 2002). After cloning human and rat T1Rs, Li and collaborators transfected T1R genes, singly or in combination, into heterologous cell lines that stably expressed $G_{\alpha 15}$, a promiscuous phospholipase C-linked G protein that couple receptor activation to increases in intracellular calcium. Then they tested these cells for their abilities to respond to L-glutamate and other substances by monitoring increases in $[Ca^{2+}]_i$ in calcium-imaging experiments. They found that L-glutamate elicited transient increases in cells transfected with human T1R1/T1R3, but not in cells transfected with T1R1 or T1R3 alone. In addition, they showed that the 5'-ribonucleotides, IMP and GMP, potentiated the T1R1/T1R3 response to L-glutamate. Finally, T1R1/T1R3-transfected cells responded to L-AP4 in the presence of IMP or GMP, consistently with psychophysical findings (Kurihara and Kashiwayanagi, 1998). Similar results were also obtained with heterologous cells transfected with rat T1R genes (Li *et al.*, 2002). Interestingly, rat T1R1/T1R3-transfected cells did not respond to L-AP4 either alone or in the presence of IMP (see Section 2.4).

Unlike mouse, the human T1R1+3 heteromer apparently shows a strong preference in activation by L-glutamate and L-aspartate among the L-amino acids tested (Li *et al.*, 2002). This could be due to the difference in the T1R1 sequence between man and mouse, as suggested by the finding that the combination of human T1R1 with mouse T1R3 alters the sensitivity to L-glutamate of the mouse amino acid receptor (Nelson *et al.*, 2002).

The G protein or G proteins coupled to the T1R1/T1R3 in TRCs, as well as their intracellular effector are not known. TRCs express a multitude of G proteins, including α-gustducin (McLaughlin *et al.*, 1992), α-transducin (McLaughlin *et al.*, 1994), a number of other Gα subunits (Kusakabe *et al.*, 1996, 2000), several Gβ subunits, and a taste-specific Gγ subunit (Huang *et al.*, 1999). cAMP phosphodiesterases (McLaughlin *et al.*, 1994; Ruiz-Avila *et al.*, 1995) and phospholipaseβ2 (Rössler *et al.*, 1998) cloned from mammalian taste buds could potentially participate in sensory transduction pathways downstream of T1R1/T1R3 activation (see e.g., Asano-Miyoshi *et al.*, 2001).

Gustducin is expressed by subsets of TRCs in all taste organs (McLaughlin *et al.*, 1992; Boughter *et al.*, 1997). Double-labeling *in situ* hybridizations, however, revealed that T1R1 does not colocalize with gustducin in rodent fungiform TRCs (Hoon *et al.*, 1999). Similar results were obtained for T1R3 (Montmayeur *et al.*, 2001). On the contrary, by profiling the pattern of gene expression for T1R3 and α-gustducin in single mouse TRCs, Max and collaborators (2001) found that their expression overlapped considerably. Consistently with these results, double labeling for T1R3 and α-gustducin showed that many cells were positive for both in human taste buds (Max *et al.*, 2001). More recent studies using two-color fluorescent *in situ* hybridization indicate that T1R1 and T1R3 are coexpressed with gustducin in mice (Kim *et al.*, 2003). The discrepancy among the above studies could be due to differences in the sensitivity of the methods used. In any case, molecular data from Max *et al.* (2001) suggest a possible functional link between gustducin and T1R3. In agreement with this hypothesis, recent behavioral studies have shown that α-gustducin knockout (KO) mice have diminished responses to MSG (He et al., 2002; Ruiz et al., 2003). Thus, it seems that the T1R1/T1R3–gustducin linkage may play a role for umami reception in the fungiform papillae, at least in mice (Figure 7.2B). Interestingly, mice KO for both α-gustducin and α-transducin is insensitive to MSG (He et al., 2002), suggesting a role also for α-transducin in the transduction of MSG signal.

2.3. iGluRs

The first evidence for the presence of ionotropic glutamate receptors (iGluRs) in TRCs was provided by bilayer studies, in which epithelial membranes from foliate/vallate taste regions were incorporated (Brand et al., 1991; Teeter et al., 1992). Both MSG (10 mM) and NMDA (50–100 μM) activated a cation conductance. With the use of optical methods to monitor membrane potentials and $[Ca^{2+}]_i$, Hayashi et al. (1996) showed that there may be multiple receptors (metabotropic and ionotropic) for glutamate in mouse TRCs. By applying an RT-PCR strategy to search for iGluRs, Chaudhari et al. (1996) identified several iGluRs in rat lingual tissues, including NMDA (NMDAR1 and NMDAR2d) and non-NMDA (KA-2) receptors. These receptors (Table 7.2) showed 100% identity to published amino acid sequences. Interestingly, iGluRs seemed not to be associated specifically with TRCs, but were also found in nontaste lingual tissue. In any case, these molecular studies as well as more recent functional findings with the patch-clamp technique (Bigiani et al., 1997; Lin and Kinnamon, 1999; Oh et al., 2001) clearly indicated the occurrence of iGluRs in TRCs. In patch-clamp experiments, however, both apical and basolateral membranes were exposed to glutamate and to iGluR agonists. In situ, only the apical ends of TRCs interact with taste stimuli (Figure 7.1). The obvious question is then: Are iGluRs involved in the detection of food glutamate (umami taste)? To address this issue, different groups have used both behavioral assays and nerve recordings in animal models. The rationale underlying these studies is that any involvement of iGluRs as apical, umami receptors should be demonstrated in experiments in which only the apical ends of TRCs are stimulated by glutamate or relevant GluR agonists. With conditioned taste aversion experiments, it has been shown that the activation of iGluRs (receptors for NMDA, AMPA, or KA) does not mimic the taste of MSG in rats (Chaudhari et al., 1996; Stapleton et al., 1999). A hallmark of umami taste is the synergism between L-glutamate and the 5′ ribonucleotides IMP and GMP. Thus, Sako and Yamamoto (1999) studied the effect of IMP on the taste responses of CT nerve to NMDA, KA, and AMPA (iGluRs agonists), and to L-AP4 (a mGluR4 agonist). They found synergistic responses only for L-AP4, but not for NMDA, KA, or AMPA, and concluded that iGluRs may not be involved in umami transduction, at least in rats. Similar findings were obtained by Delay et al. (2000) by studying the taste preference synergy in behavioral tests in rats. In mice, conditioned taste aversion to monosodium L-glutamate does not generalize to NMDA but to L-AP4 (Nakashima et al., 2001). In humans, psychophysical measurements ruled out a role for ionotropic receptors as umami receptors (Kurihara and Kashiwayanagi, 1998). In conclusion, behavioral assays and nerve recordings in which only the apical membrane of TRCs is stimulated indicate a minimal role for ionotropic-like glutamate receptors in taste transduction for L-glutamate. Instead, recent studies have provided increasing evidence for a role of iGluRs as synaptic receptors in TRCs (see Section 3).

Table 7.2. Ionotropic Glutamate Receptors (iGluRs) in Mammalian Taste Receptor Cells

	NMDA			Non-NMDA (kainate)		Non-NMDA (AMPA)	
Species	Rat		Mouse	Rat		Rat	
Papilla	Fungiform	Foliate Vallate	Foliate Vallate	Foliate Vallate	Fungiform	Foliate Vallate	Fungiform
Molecular evidence		Yes (Chaudhari et al., 1996)		Yes (Chaudhari et al., 1996)		No (Chaudhari et al., 1996)	
Functional evidence	Yes (Lin and Kinnamon, 1999)	Yes (Caicedo et al., 2000a)	Yes (Teeter et al., 1992)	Yes (Caicedo et al., 2000a)		No (Caicedo et al., 2000a)	Yes (Lin and Kinnamon, 1999)
Evidence from uptake studies				Yes (Caicedo et al., 2000b; Kim et al., 2001)			
Ion permeability	Na^+/Ca^{2+} (Lin and Kinnamon, 1999)						
Antagonist	D-AP 5 (Lin and Kinnamon, 1999)	D-AP 5 (Caicedo et al., 2000a)		CNQX and GYKY 52466 (Caicedo et al., 2000a)			
Mg^{2+} block	Weak (Lin and Kinnamon, 1999)						
Purine ribonucleotide synergism tested in electrophysiological nerve recording experiments	No (Sako and Yamamoto, 1999)			No (Sako and Yamamoto, 1999)		No (Sako and Yamamoto, 1999)	
Purine ribonucleotide synergism tested in behavioral assays		No (Stapleton et al., 1999)		No (Delay et al., 2000)			
Proposed function		Synaptic receptor		Synaptic receptor			

2.4. Multiple Receptor Mechanisms for Sensing Food Glutamate?

The identification of different receptor molecules (taste-mGluR4, T1R1/T1R3) as candidate umami receptors raises questions about their relative importance on the sensory mechanisms for the detection of glutamate in foodstuff. In general, both taste-mGluR4 and T1R1/T1R3 meet several criteria as umami receptors (see Table 7.1). For example, the effective concentration of L-glutamate to activate the receptor molecule is in the concentration range required to induce neural or behavioral responses. In addition, both taste-mGluR4 and T1R1/T1R3 are activated by L-AP4, a substance that mimics the taste of L-glutamate. Finally, receptor activation by L-glutamate (or L-AP4) shows synergism with 5′-ribonucleotides, a hallmark of umami taste.

On the other hand, if one considers the specific topographic organization of the peripheral taste system (Figure 7.1) and the relative umami sensitivity of the taste nerves, then discrepancies with the molecular findings come out. Studies examining the responses to umami substances of taste nerves have revealed significant diversity among species and, in a given species, among nerves. The GL nerve (IX, which innervates mainly the vallate papillae: Figure 7.1) seems to be more important than the CT nerve for transmitting the taste of umami substances in mice (Ninomiya and Funakoshi, 1989a; Ninomiya et al., 1991, 1993), monkeys (Hellekant et al., 1997), and humans (Maruyama and Yamaguchi, 1995). For example, mice with bilateral sectioning of the GL nerve could not discriminate MSG from NaCl (Ninomiya and Funakoshi, 1989b). In rats, on the contrary, the umami taste is conveyed predominantly by the CT nerve (a branch of VII, which innervates mainly the fungiform papillae: Figure 7.1) (Sako et al., 2000). Indeed, rats with transection of the CT nerve could not acquire conditioned taste aversions to umami substances, whereas transection of the GL nerve did not impair the aversions (Sako et al., 2000). These patterns of umami sensitivity among taste nerves are not fully consistent with the expression pattern of candidate umami receptors among taste papillae. Taste-mGluR4 has been localized in rat foliate/vallate papillae, that is, a taste region innervated mainly by the GL nerve that does not seem to play a role in mediating taste information on umami substances in rats (Sako et al., 2000).[1] In mice, T1R1/T1R3 seems to be restricted to the fungiform papillae, which are innervated by the CT nerve. This is not consistent with the observation that in mice the CT nerve is less sensitive to umami substances than the GL nerve (Ninomiya et al., 1991, 1993). Finally, the lack of responses of rat T1R1/T1R3-transfected cells to L-AP4 (Li et al., 2002) is in contrast with the observation that L-AP4 mimics umami taste in rats (Chaudhari et al., 1996; Delay et al., 2000).

On the basis of the above observations, it is unlikely that a single receptor molecule may represent "the" umami receptor. Instead, the experimental data suggest that the molecular system devoted to the detection of food glutamate is quite complex and diverse in terms of both the receptor entities (taste-mGluR4 and T1R1/T1R3) and of their distribution among taste papillae in different species. It is reasonable to conceive that either taste-mGluR4 or T1R1/T1R3, or both could play a predominant/accessory role depending, for example, on the developmental stages or the aging of the animal. Postnatal developmental changes in the detection threshold for MSG as well as in the synergistic effect by GMP have been documented for the GL nerve but not for the CT nerve in mice (Ninomiya et al., 1991).

---

[1]Recent immunohistochemical studies indicate the occurrence of mGluR4a (and perhaps taste-mGluR4) also in the rat fungiform papillae (Toyono et al., 2002).

In particular, the sensitivity of the GL nerve to MSG shows a marked decrease in developing mice (Ninomiya *et al.*, 1991). Consistent with this observation is the finding that the mRNA for taste-mGluR4 is expressed at 2- to 3-fold higher levels in juvenile than in adult rats (Chaudhari *et al.*, 2000). Finally, the predominant role of taste-mGluR4 or T1R1/T1R3 in umami taste could also depend on the physiological conditions of the animals, such as their nutritional status. Indeed, taste sensitivity is partly dependent on what has been eaten (Bellisle, 1999), and the relative expression of taste-mGluR4 and T1R1/T1R3 could then reflect nutritional needs.

The molecular diversity of the umami receptors is consistent with physiological observations in single TRCs. For example, in rats, L-glutamate and L-AP4 elicit different kind of membrane ion currents according to the papillar location of TRCs (Bigiani *et al.*, 1997; Lin and Kinnamon, 1999). Similar findings have also been obtained by studying changes in $[Ca^{2+}]_i$, IP_3, and cAMP in response to umami substances in mice (Hayashi *et al.*, 1996; Ninomiya *et al.*, 2000). Further, it is possible that different transduction mechanisms may be present in the same papilla as shown by Ninomiya *et al.* (2000).

Taste detection mechanisms likely reflect the different alimentary habits among mammals, and one should expect variability even in the detection of a given taste stimulus. There are several examples that this is, indeed, the case (reviewed in Kinnamon and Margolskee, 1996; Lindemann, 1996, 2001; Stewart *et al.*, 1997; Herness and Gilbertson, 1999; Gilbertson *et al.*, 2000). Thus, the existence of multiple receptors mechanisms for the detection of alimentary glutamate is fully compatible with our current view of the biology of the taste system, in which several redundant mechanisms for chemical detection have been evolved. A future challenge in umami taste research will be, therefore, to decipher how the expression pattern of umami receptors on the tongue is linked to the pattern of activity in the network of the gustatory nerves that signals the presence of food glutamate to the brain.[2]

3. Glutamate Receptors as Synaptic Receptors

Activation of chemical synapses between TRCs and afferent nerve fibers is thought to be necessary for transmitting gustatory information to the brain (Figure 7.1). Interestingly, morphological studies have shown the existence of synaptic vesicles in some nerve terminals, suggesting that reciprocal synapses may occur at this level (Yoshie *et al.*, 1990, 1996). SNAP-25, a presynaptic protein involved in transmitter release, has been localized in both TRCs and nerve endings (Yang *et al.*, 2000).Whatever the role (afferent or efferent) of chemical synapses between TRCs and nerve endings, the nature of the neurotransmitters and their receptors has been elusive over the past decades. Recent findings have raised the possibility that glutamate and iGluRs might act as a signaling system at the peripheral synapses inside taste buds.

Glutamate is a ubiquitous neurotransmitter in the vertebrate CNS and sensory organs, such as the retina and the cochlea. In taste, first indications that glutamate could act as peripheral neurotransmitter came from studies on the amphibian *Necturus maculosus*.

---

[2]There is increasing evidence that food L-glutamate interacts with the signaling pathway involved in sucrose detection (Yamamoto *et al.*, 1991; Sako and Yamamoto, 1999; Ninomiya *et al.*, 2000; Nakashima *et al.*, 2001; Stapleton *et al.*, 2002). Thus, the generation of umami sensory signals at the peripheral level might involve several parallel and possibly interacting processes, some of which rely on the glutamate sensors, taste-mGluR4 and T1R1/T1R3.

Immunocytochemistry showed that glutamate was contained in nerve fibers innervating the taste buds (Jain and Roper, 1991; Lu and Roper, 1993). Uptake studies on radiolabeled glutamate showed that [^{3}H]-glutamate was taken up by several cell types in the lingual epithelium of *Necturus*, including some TRCs (Nagai *et al.*, 1998). However, [^{3}H]-glutamate that had been accumulated was not released upon depolarization by K$^+$ (Nagai *et al.*, 1998). On the basis of these studies, it has been proposed that glutamate may represent an efferent neurotransmitter in taste buds, and that it is transported into some TRCs as metabolically important amino acid. In this respect, certain TRCs could regulate the extracellular concentration of glutamate inside the taste buds, similarly to glial cells in other systems (e.g., the Müller cells in the retina: Newman and Reichenbach, 1996). Support to this hypothesis has been recently provided by an immunolabeling study on the glutamate-aspartate transporter (GLAST) in rat taste buds. GLAST was found in some TRCs of the vallate taste buds, and it has been proposed that GLAST-positive cells play a glia-like role in the uptake of glutamate following its release at synapses within the taste bud (Lawton *et al.*, 2000).

Physiological studies with optical methods (Hayashi *et al.*, 1996) and patch-clamp techniques (Bigiani *et al.*, 1997; Lin and Kinnamon, 1999; Iseki *et al.*, 2001; Oh *et al.*, 2001) provided evidence for iGluR and mGluR in isolated mammalian TRCs (Table 7.2). However, in those studies, glutamate and receptor agonists were applied to the entire taste cell membrane, thus raising the possibility that the applied substances could activate receptors located either on the basolateral, synaptic membrane, or in the apical, chemosensitive membrane, or both. Although molecular, biochemical, nerve recording, and behavioral studies indicated a role for metabotropic receptors as glutamate "taste" receptors (Chaudhari *et al.*, 1996, 2000; Chaudhari and Roper, 1998; Kurihara and Kashiwayanagi, 1998; Sako and Yamamoto, 1999; Stapleton *et al.*, 1999; Yang *et al.*, 1999; Delay *et al.*, 2000; Nakashima *et al.*, 2001), yet evidence that iGluRs were functional synaptic receptors was lacking.

Direct evidence for the existence of "synaptic" glutamate receptors on the membrane of mammalian TRCs *in situ* has been provided recently by an elegant Ca^{2+} microfluorometric study by Caicedo *et al.* (2000a). CaGD-labeled TRCs in lingual slices were imaged with a scanning confocal microscope to measure [Ca^{2+}]$_i$ and stimulated by superfusion with glutamate, kainate, AMPA, or NMDA. In cells sensitive to glutamate, the response ($\uparrow$[Ca^{2+}]$_i$) was localized to the basal processes and cell bodies, which are the synaptic regions for TRCs (Figure 7.1). The effective concentration of glutamate was as low as 30 μM (compare with the effective concentration for umami taste in Table 7.1). Both NMDA and kainate, but non-AMPA, elicited Ca^{2+} responses. On the basis of their findings, Caicedo *et al.* (2000a) concluded that glutamate functions as a neurotransmitter at the TRC level, and that iGluRs in TRCs might be presynaptic autoreceptors or postsynaptic receptors at afferent or efferent synapses (Figure 7.3). Consistent with the functional findings, Caicedo *et al.* (2000b) showed the occurrence of non-NMDA (kainate type) receptors in rat TRCs by monitoring the glutamate-induced cobalt uptake. Again, the concentration-response relation for cobalt uptake (range from 10 μM to 1 mM glutamate) suggested that synaptic glutamate receptors, not umami receptors, were activated. Data on postnatal development of non-NMDA glutamate receptors in rat foliate taste buds further support the hypothesis of a role as synaptic receptor (Kim *et al.*, 2001).

An interesting finding from the functional studies by Caicedo *et al.* (2000a) was that TRCs could be subdivided in two subsets, one with NMDA receptors and the other with non-NMDA receptors. This observation is consistent with the hypothesis that a complex system of signaling is present in taste buds (Nagai *et al.*, 1996; Bigiani, 2002).

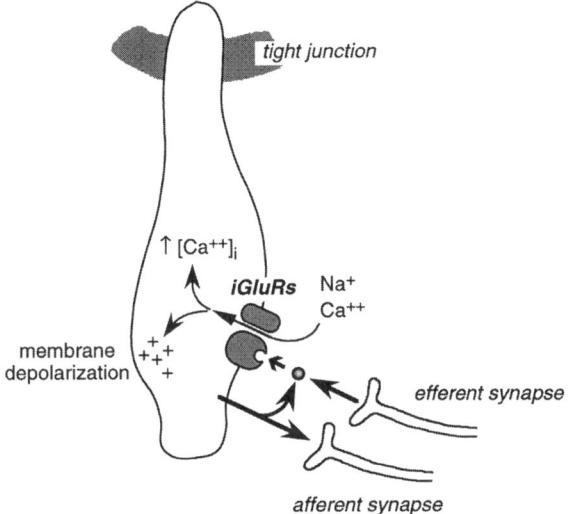

Figure 7.3. Glutamate signaling at the basolateral membrane of TRCs involved ionotropic glutamate receptors (iGluRs). iGluRs (either NMDA or non-NMDA receptors) may play a role in efferent synapses (messages from the brain and directed to the TRCs) and/or as autoreceptors in afferent synapses with the nerve endings. In either cases, activation of iGluRs by L-glutamate would modulate the activity of TRCs by increasing $[Ca^{2+}]_i$, and through membrane depolarization.

As to the AMPA receptors, at the moment there are only very preliminary electrophysiological data suggesting the possible occurrence of these receptors in rat fungiform TRCs (Lin and Kinnamon, 1999). Molecular and Ca^{2+}-imaging studies, on the contrary, indicate that AMPA receptors are not present in the rat foliate/vallate TRCs (Chaudhari *et al.*, 1996; Caicedo *et al.*, 2000a).

Patch-clamp experiments have clearly shown that in a few cases both L-AP4 and NMDA elicit responses in the same TRC (Lin and Kinnamon, 1999). Thus, it is likely that some TRCs are able to sense glutamate as a "taste" stimulus (through umami receptors, which are activated by L-AP4) and as a neurotransmitter (through synaptic receptors, which include NMDA receptors). This, of course, adds complexity to the physiology of TRCs. For example, as demonstrated by Hayashi *et al.* (1996), stimulation of L-AP4 or NMDA receptors elicits opposite changes in $[Ca^{2+}]_i$ in mouse TRCs.

4. Concluding Remarks

Glutamate plays a double role in the physiology of TRCs. As a free-occurring component of some foodstuff, glutamate is detected by TRCs and conveys information about the presence of protein-rich source. As a substance released by TRC and/or nerve endings, it is involved in cell-to-cell communication at chemical synapses in taste organs. In both cases, glutamate is sensed by TRC membrane through specific receptors, some of which (iGluRs) are similar to those found in the CNS, whereas others (taste-mGluR4 and T1R1/T1R3) are specifically expressed by TRCs. One interesting aspect of the biology of glutamate receptors in TRCs is that their expression seems to depend on the specific papillary localization of

TRCs. For example, T1R1/T1R3 receptors are found predominantly in fungiform papillae but not in the foliate/vallate papillae of the mouse. Evidence for taste-mGluR4 has been obtained, on the other hand, by analyzing foliate/vallate papillae in rat. It is then reasonable to conceive that food glutamate may give rise to different patterns of TRC activation depending on their localization on the tongue surface. This may have a profound impact on the sensory coding for glutamate compared to other taste stimuli, including the central representation of this substance as "umami" taste.

Acknowledgments

I thank Drs. Nirupa Chaudhari (University of Miami School of Medicine, Miami, USA), Robert F. Margolskee (Mount Sinai School of Medicine, New York, USA), Yuzo Ninomiya (Kyushu University, Fukuoka, Japan), and Stephen D. Roper (University of Miami School of Medicine, Miami, USA) for helpful discussions and suggestions. This work was supported in part by the Italian Board of Education and University (MURST, Cofin).

References

Asano-Miyoshi, M., K. Abe, and Y. Emori (2001). IP3 receptor type 3 and PLCβ2 are co-expressed with taste receptors T1R and T2R in rat taste bud cells. *Chem. Senses* **26**, 259–265.

Bellisle, F. (1999). Glutamate and the UMAMI taste: Sensory, metabolic, nutritional and behavioral considerations. A review of the literature published in the last 10 years. *Neurosci. Biobehav. Rev.* **23**, 423–438.

Bigiani, A. (2002). Electrophysiology of *Necturus* taste cells. *Prog. Neurobiol.* **66**, 123–159.

Bigiani, A., R.J. Delay, N. Chaudhari, S.C. Kinnamon, and S.D. Roper (1997). Responses to glutamate in rat taste cells. *J. Neurophysiol.* **77**, 3048–3059.

Boughter, J.D. Jr., D.W. Pumplin, C. Yu, R.C. Christy, and D.V. Smith (1997). Differential expression of α-gustducin in taste bud populations of the rat and hamster. *J. Neurosci.* **17**, 2852–2858.

Brand, J.G. (2000). Receptor and transduction processes for umami taste. *J. Nutr.* **130**, 942S–945S.

Brand, J.G., J.H. Teeter, T. Kumazawa, T. Huque, and D.L. Bayley (1991). Transduction mechanisms for the taste of amino acids. *Physiol. Behav.* **49**, 899–904.

Breslin, P.A.S. (2000). Human gestation. In T.E. Finger, W.L. Silver, and D. Restrepo (eds.), *The Neurobiology of Taste and Smell*. Wiley, New York, pp. 423–461.

Caicedo, A., M.S. Jafri, and S.D. Roper (2000a). In situ Ca^{2+} imaging reveals neurotransmitter receptors for glutamate in taste receptor cells. *J. Neurosci.* **20**, 7978–7985.

Caicedo, A., K.-N. Kim, and S.D. Roper (2000b). Glutamate-induced cobalt uptake reveals non-NMDA receptors in rat taste cells. *J. Comp. Neurol.* **417**, 315–324.

Chaudhari, N. and S.D. Roper (1998). Molecular and physiological evidence for glutamate (Umami) taste transduction via a G protein-coupled receptor. *Ann. NY. Acad. Sci.* **855**, 398–406.

Chaudhari, N., H. Yang, C. Lamp, E. Delay, C. Cartford, T. Than et al. (1996). The taste of monosodium glutamate: Membrane receptors in taste buds. *J. Neurosci.* **16**, 3817–3826.

Chaudhari, N., A.M. Landin, and S.D. Roper (2000). A metabotropic glutamate receptor variant functions as a taste receptor. *Nat. Neurosci.* **3**, 113–119.

Delay, E.R., A.J. Beaver, K.A. Wagner, J.R. Stapleton, J.O. Harbaugh, K.D. Catron et al. (2000). Taste preference synergy between glutamate receptor agonists and inosine monophosphate in rats. *Chem. Senses* **25**, 507–515.

Faurion, A. (1991). Are umami taste receptor sites structurally related to glutamate CNS receptor sites? *Physiol. Behav.* **49**, 905–912.

Finger, T.E. and S.A. Simon (2000). Cell biology of taste epithelium. In T.E. Finger, W.L. Silver, and D. Restrepo (eds.), *The Neurobiology of Taste and Smell*. Wiley, New York, pp. 423–461.

Fuke, S. and S. Konosu (1991). Taste-active components in some foods: A review of Japanese research. *Physiol. Behav.* **49**, 863–868.

Gilbertson, T.A., S. Damak, and R.F. Margolskee (2000). The molecular physiology of taste transduction. *Curr. Op. Neurobiol.* **10**, 519–527.

Hayashi, Y., T. Tsunenari, and T. Mori (1994). Binding of umami substance to plasma membrane isolated from bovine circumvallate papillae. *Biosci. Biotech. Biochem.* **58**, 1403–1406.

Hayashi, Y., M.M. Zviman, J.G. Brand, J.H. Teeter, and D. Restrepo (1996). Measurement of membrane potential and $[Ca^{2+}]_i$ in cell ensembles: Application to the study of glutamate taste in mice. *Biophys. J.* **71**, 1057–1070.

He, W., R.F. Margolskee, and S. Damak (2002). Signal transduction of umami taste by α-gustducin and α-transducin. *Chem. Senses* **27**, A8 (Abstract).

Hellekant, G., and Y. Ninomiya (1991). On the taste of umami in chimpanzee. *Physiol. Behav.* **49**, 927–934.

Hellekant, G., V. Danilova, and Y. Ninomiya (1997). Primate sense of taste: Behavioral and single chorda tympani and glossopharyngeal nerve fiber recordings in the rhesus monkey, *Macaca mulatto*. *J. Neurophysiol.* **77**, 978–993.

Herness, M.S. and T.A. Gilbertson (1999). Cellular mechanisms of taste transduction. *Annu. Rev. Physiol.* **61**, 873–900.

Hoon, M.A., E. Adler, J. Lindemeier, J.F. Battey, N.J.P. Ryba, and C.S. Zuker (1999). Putative mammalian taste receptors: A class of taste-specific GPCRs with distinct topographic selectivity. *Cell* **96**, 541–551.

Horio, T. (2000). Effects of various taste stimuli on heart rate in humans. *Chem. Senses* **25**, 149–153.

Huang, L., Y.G. Shanker, J. Dubauskaite, J.Z. Zheng, W. Yan, S. Rosenzweig *et al.* (1999). Gγ13 colocalizes with gustducin in taste receptor cells and mediates IP_3 responses to bitter denatonium. *Nat. Neurosci.* **2**, 1055–1062.

Iseki, K., Y. Hayashi, S.-H. Oh, J. Teeter, D. Restrepo, and T. Mori (2001). Umami taste: electrophysiological recordings of synergism in mouse taste cells. *Sens. Neuron* **3**, 155–167.

Jain, S. and S.D. Roper (1991). Immunocytochemistry of gamma-aminobutyric acid, glutamate, serotonin, and histamine in *Necturus* taste buds. *J. Comp. Neurol.* **307**, 675–682.

Kim, K.-N., A. Caicedo, and S.D. Roper (2001). Glutamate-induced cobalt uptake reveals non-NMDA receptors in developing rat taste buds. *NeuroReport* **12**, 1715–1718.

Kim, M.-R., Y. Kusakabe, H. Miura, Y. Shindo, and A. Hino (2003). Expression patterns of T1R3 and various taste perception related genes. *Chem. Senses* **28**, E29 (Abstract).

Kinnamon, S.C., and R.F. Margolskee (1996). Mechanisms of taste transduction. *Curr. Opin. Neurobiol.* **6**, 506–513.

Kitagawa, M., Y. Kusakabe, H. Miura, Y. Ninomiya, and A. Hino (2001). Molecular genetic identification of a candidate receptor gene for sweet taste. *Biochem. Biophys. Res. Commun.* **283**, 236–242.

Kondoh, T., M. Mori, T. Ono, and K. Torii (2000). Mechanisms of umami taste preference and aversion in rats. *J. Nutr.* **130**, 966S–970S.

Kumazawa, T. and K. Kurihara (1990). Large synergism between monosodium glutamate and 5′-nucleotides in canine taste nerve responses. *Am. J. Physiol.* **259**, R420–R426.

Kumazawa, T., M. Nakamura, and K. Kurihara (1991). Canine taste nerve responses to umami substances. *Physiol. Behav.* **49**, 875–881.

Kurihara, K. and M. Kashiwayanagi (1998). Introductory remarks on umami taste. *Ann. NY. Acad. Sci.* **855**, 393–397.

Kurihara, K. and M. Kashiwayanagi (2000). Physiological studies on umami taste. *J. Nutr.* **130**, 931S–934S.

Kusakabe, Y., K. Abe, K. Tanemura, Y. Emori, and S. Arai (1996). GUST27 and closely related G-protein-coupled receptors are localized in taste buds together with Gi-protein α-subunit. *Chem. Senses* **21**, 335–340.

Kusakabe, Y., A. Yasuoka, M. Asano-Miyoshi, K. Iwabuchi, I. Matsumoto, S. Arai *et al.* (2000). Comprehensive study on G protein α-subunits in taste bud cells, with special reference to the occurrence of Gαi2 as a major Gα species. *Chem. Senses* **25**, 525–531.

Lawton, D.M., D.N. Furness, B. Lindemann, and C.M. Hackney (2000). Localization of the glutamate-aspartate transporter, GLAST, in rat taste buds. *Eur. J. Neurosci.* **12**, 3163–3171.

Li, X., L. Staszewski, H. Xu, K. Durick, M. Zoller, and E. Adler (2002). Human receptors for sweet and umami taste. *Proc. Natl. Acad. Sci. USA* **99**, 4692–4696.

Lin, W. and S.C. Kinnamon (1999). Physiological evidence for ionotropic and metabotropic glutamate receptors in rat taste cells. *J. Neurophysiol.* **82**, 2061–2069.

Lindemann, B. (1996). Taste reception. *Physiol. Rev.* **76**, 719–766.

Lindemann, B. (2000). A taste for umami. *Nat. Neurosci.* **3**, 99–100.

Lindemann, B. (2001). Receptors and transduction in taste. *Nature* **413**, 219–225.

Lu, K.-S., and S.D. Roper (1993). Electron microscopic immunocytochemistry of glutamate-containing nerve fibers in the taste bud of mudpuppy (*Necturus maculosus*). *Micr. Res. Tech.* **26**, 225–230.

Lugaz, O., A.-M. Pillias, and A. Faurion (2002). A new specific ageusia: Some humans cannot taste L-glutamate. *Chem. Senses* **27**, 105–115.

Marshall, F.H., K.A. Jones, K. Kaupmann, and B. Bettler (1999). GABA$_B$ receptors – The first 7TM heterodimers. *Trends Pharmacol. Sci.* **20**, 396–399.

Maruyama, I., and S. Yamaguchi (1995). Effects of locus of stimulation on the tongue and umami perception. *Chem. Senses* **20**, 367 (Abstract).

Max, M., Y.G. Shanker, L. Huang, M. Rong, Z. Liu, F. Campagne *et al.* (2001). *Tas1r3*, encoding a new candidate taste receptor, is allelic to the sweet responsiveness locus *Sac. Nat. Genet.* **28**, 58–63.

McLaughlin, S.K., P.J. McKinnon, and R.F. Margolskee (1992). Gustducin is a taste-cell-specific G protein closely related to the transducins. *Nature* **357**, 563–569.

McLaughlin, S.K., P.J. McKinnon, N. Spickofsky, W. Danho, and R.F. Margolskee (1994). Molecular cloning of G proteins and phosphodiesterases from rat taste cells. *Physiol. Behav.* **56**, 1157–1164.

Montmayeur, J.-P., S.D. Liberles, H. Matsunami, and L.B. Buck (2001). A candidate taste receptor gene near a sweet taste locus. *Nat. Neurosci.* **4**, 492–498.

Mori, M., T. Kawada, T. Ono, and K. Torii (1991). Taste preferences and protein nutrition and L amino acid homeostasis in male Sprague-Dawley rats. *Physiol. Behav.* **49**, 987–995.

Nagai, T., D.-J. Kim, R.J. Delay, and S.D. Roper (1996). Neuromodulation of transduction and signal processing in the end organs of taste. *Chem. Senses* **21**, 353–365.

Nagai, T., R.J. Delay, J. Welton, and S.D. Roper (1998). Uptake and release of neurotransmitter candidates, [^{3}H]serotonin, [^{3}H]glutamate, and [^{3}H]γ-aminobutyric acid, in taste buds of the mudpuppy, *Necturus maculosu. J. Comp. Neurol.* **393**, 199–208.

Nakashima, K., H. Katsukawa, K. Sasamoto, and Y. Ninomiya (2001). Behavioral taste similarities and differences among monosodium L-glutamate and glutamate receptor agonists in C57BL mice. *J. Nutr. Sci. Vitaminol.* **47**, 161–166.

Nelson, G., M.A. Hoon, J. Chandrashekar, Y. Zhang, N.J.P. Ryba, and C.S. Zuker (2001). Mammalian sweet taste receptors. *Cell* **106**, 381–390.

Nelson, G., J. Chandrashekar, M.A. Hoon, L. Feng, G. Zhao, N.J.P. Ryba *et al.* (2002). An amino-acid taste receptor. *Nature* **416**, 199–202.

Newman, E. and A. Reichenbach (1996). The Müller cell: A functional element of the retina. *Trends Neurosci.* **19**, 307–312.

Nijima, A. (1991). Effects of oral and intestinal stimulation with umami substance on gastric vagus activity. *Physiol. Behav.* **49**, 1025–1028.

Ninomiya, Y. and M. Funakoshi (1989a). Peripheral neural basis for behavioral discrimination between glutamate and the four basic taste substances in mice. *Comp. Biochem. Physiol. A* **92**, 371–376.

Ninomiya, Y. and M. Funakoshi (1989b). Behavioral discrimination between glutamate and the four basic taste substances in mice. *Comp. Biochem. Physiol. A* **92**, 365–370.

Ninomiya, Y., T. Tanimukai, S. Yoshida, and M. Funakoshi (1991). Gustatory neural responses in preweanling mice. *Physiol. Behav.* **49**, 913–918.

Ninomiya, Y., S. Kurenuma, T. Nomura, H. Uebayashi, and H. Kawamura (1992). Taste synergism between monosodium glutamate and 5′-ribonucleotide in mice. *Comp. Biochem. Physiol. A* **101**, 97–102.

Ninomiya, Y., H. Kajiura, and K. Mochizuki (1993). Differential taste responses of mouse chorda tympani and glossopharyngeal nerves to sugars and amino acids. *Neurosci. Lett.* **163**, 197–200.

Ninomiya, Y., K. Nakashima, A. Fukuda, H. Nishino, T. Sugimura, A. Hino *et al.* (2000). Responses to umami substances in taste bud cells innervated by the chorda tympani and glossopharyngeal nerves. *J. Nutr.* **130**, 950S–953S.

Oh, S.-H., Y. Hayashi, K. Iseki, D. Restrepo, J. Teeter, and T. Mori (2001). Participation of ionotropic and metabotropic glutamate receptors in taste cell responses to MSG. *Sens. Neuron* **3**, 169–183.

Rodieck, R.W. (1998). *The First Steps in Seeing*. Sinauer Associates, Sunderland, pp. 158–186.

Rogers, P.J. and J.E. Blundell (1990). Umami and appetite: Effects of monosodium glutamate on hunger and food intake in human subjects. *Physiol. Behav.* **48**, 801–804.

Rössler, P., C. Kroner, J. Freitag, J. Noè, and H. Breer (1998). Identification of a phospholipase C β subtype in rat taste cells. *Eur. J. Cell Biol.* **77**, 253–261.

Ruiz, C., E. Delay, R. Margolskee, and S.C. Kinnamon (2003). Behavioral evidence for a role of gustducin in umami taste. *Chem. Senses* **27**, A105 (Abstract).

Ruiz-Avila, L., S.K. McLaughlin, D. Wildman, P.J. McKinnon, A. Robichon, N. Spickofsky *et al.* (1995). Coupling of bitter receptor to phosphodiesterase through transducin in taste receptor cells. *Nature* **376**, 80–85.

Sainz, E., J.N. Korley, J.F. Battey, and S.L. Sullivan (2001). Identification of a novel member of the T1R family of putative taste receptors. *J. Neurochem.* **77**, 896–903.

Sako, N., S. Harada, and T. Yamamoto (2000). Gustatory information of umami substances in three major taste nerves. *Physiol. Behav.* **71**, 193–198.

Sako, N. and T. Yamamoto (1999). Analyses of taste nerve responses with special reference to possible receptor mechanisms of umami taste in the rat. *Neurosci. Lett.* **261**, 109–112.

Scott, T.R. and J.V. Verhagen (2000). Taste as a factor in the management of nutrition. *Nutrition* **16**, 874–885.

Schiffman, S.S., E.A. Sattely-Miller, I.A. Zimmerman, B.G. Graham, and R.P. Erickson (1994). Taste perception of monosodium glutamate (MSG) in foods in young and elderly subjects. *Physiol. Behav.* **56**, 265–275.

Stapleton, J.R., S.D. Roper, and E.R. Delay (1999). The taste of monosodium glutamate (MSG), L-aspartic acid, and N-methyl-D-aspartate (NMDA) in rats: Are NMDA receptors involved in MSG taste? *Chem. Senses* **24**, 449–457.

Stapleton, J.R., M. Luellig, S.D. Roper, and E.R. Delay (2002). Discrimination between the tastes of sucrose and monosodium glutamate in rats. *Chem. Senses* **27**, 375–382.

Stewart, R.E., J.A. DeSimone, and D.L. Hill (1997). New perspectives in gustatory physiology: Transduction, development, and plasticity. *Am. J. Physiol.* **272**, C1–C26.

Teeter, J.H., T. Kumazawa, J.G. Brand, D.L. Kalinoski, E. Honda, and G. Smutzer (1992). Amino acid receptor channels in taste cells. In D.P. Corey and S.D. Roper (eds.), *Sensory Transduction*. Rockefeller University Press, New York, pp. 291–306.

Thomsen, C., R. Pekhletski, B. Haldenman, T.A. Gilbert, P. O'Hara, and D.R. Hampson (1997). Cloning and characterization of a metabotropic glutamate receptor, mGluR4b. *Neuropharmacology* **36**, 21–30.

Toyono, T., Y. Seta, S. Kataoka, H. Harada, T. Morotomi, S. Kawano *et al.* (2002). Expression of the metabotropic glutamate receptor, mGluR4a, in the taste hairs of taste buds in rat gustatory papillae. *Arch. Histol. Cytol.* **65**, 91–96.

Viarouge, C., P. Even, C. Rougeot, and S. Nicolaidis (1991). Effects on metabolic and hormonal parameters of monosodium glutamate (Umami taste) ingestion in the rat. *Physiol. Behav.* **49**, 1013–1018.

Viarouge, C., R. Caulliez, and S. Nicolaidis (1992). Umami taste of monosodium glutamate enhances the thermal effect of food and affects the respiratory quotient in the rat. *Physiol. Behav.* **52**, 879–884.

Yamaguchi, S. (1991). Basic properties of umami and effect on humans. *Physiol. Behav.* **49**, 833–841.

Yamaguchi, S. and K. Ninomiya (2000). Umami and food palatability. *J. Nutr.* **130**, 921S–926S.

Yamamoto, T., R. Matsuo, Y. Fujimoto, I. Fukunaga, A. Miyasaka, and T. Imoto (1991). Electrophysiological and behavioral studies on the taste of umami substances in the rat. *Physiol. Behav.* **49**, 919–925.

Yang, H., I.B. Wanner, S.D. Roper, and N. Chaudhari (1999). An optimized method for in situ hybridization with signal amplification that allows the detection of rare mRNAs. *J. Histochem. Cytochem.* **47**, 431–445.

Yang, R., H.H. Crowley, M.E. Rock, and J.C. Kinnamon (2000). Taste cells with synapses in rat circumvallate papillae display SNAP-25-like immunoreactivity. *J. Comp. Neurol.* **424**, 205–215.

Yoshie, S., C. Wakasugi, Y. Teraki, and T. Fujita (1990). Fine structure of the taste bud in guinea pigs. Part I. Cell characterization and innervation patterns. *Arch. Histol. Cytol.* **53**, 103–119.

Yoshie, S., H. Kanazawa, and T. Fujita (1996). A possibility of efferent innervation of the gustatory cell in the rat circumvallate taste bud. *Arch. Histol. Cytol.* **59**, 479–484.

Yoshii, K., C. Yokouchi, and K. Kurihara (1986). Synergistic effects of 5'-nucleotides on rat taste responses to various amino acids. *Brain Res.* **367**, 45–51.

Zhou, X. and N. Chaudhari (1997). Modulation of cAMP levels in rat taste epithelia following exposure to monosodium glutamate. *Chem. Senses* **22**, 834 (Abstract).

8

Glutamate Receptors in Endocrine Tissues

Tania F. Gendron and Paul Morley

1. Introduction

L-glutamate is the major excitatory amino acid in the brain of vertebrates. Glutamate plays important roles in fast synaptic transmission, neural development, neuronal plasticity, learning, and memory (Ozawa *et al.*, 1998; Dingledine *et al.*, 1999). However, the over stimulation of glutamate receptors is believed to initiate cellular processes leading to neurodegeneration (Rothman and Olney, 1986; Choi, 1988; Meldrum and Garthwaite, 1990). Glutamate mediates its physiological effects via ionotropic (iGluRs) and metabotropic receptors (mGluRs) (Dingledine *et al.*, 1999). Postsynaptic ionotropic receptors are ligand-gated ion channels that are subdivided into the N-methyl-D-aspartate (NMDA), kainate (KA), and α-amino-3-hydroxy-5-methyl-4-isoxazolepropionate (AMPA) receptor subtypes according to the preferred agonist that activates the receptor (Dingledine *et al.*, 1999). All ionotropic receptors are permeable to Na^+ and K^+, but only some are permeable to Ca^{2+}. Each receptor type is composed of subunits encoded by at least six gene families. NMDA receptors are composed of NR1, NR2A-D, or NR3A and NR3B subunits; AMPA receptors by GluR1-4; and KA by KA1, KA2, and GluR5-7 subunits (Dingledine *et al.*, 1999). Coassembly of subunits produces channels with different biophysical properties. mGluRs, which are localized both pre- and postsynaptically, are G-protein linked receptors that activate second messenger systems rather than gating ion channels (Sugiyama *et al.*, 1987; Conn and Pin, 1997). Molecular cloning has identified at least eight subtypes (mGluR1-8) (Conn and Pin, 1997). Antagonists to the NMDA and AMPA receptors are neuroprotective in some animal models of neurodegenerative diseases including cerebral ischemia, epilepsy, Parkinson's, and Alzheimer's diseases (Small *et al.*, 1999). While glutamate receptor antagonists have not yet been successful in the clinic due to difficulties with side-effects, the development of new glutamate receptor antagonists, used either alone or as part of combination therapy, may prove to be more successful.

Tania F. Gendron and Paul Morley • Receptors & Ion Channels Group, Institute for Biological Sciences, National Research Council of Canada, 1200 Montreal Road Bldg. M-54, Ottawa, Ontario, Canada K1A 0R6.

Glutamate Receptors in Peripheral Tissue, edited by Santokh Gill and Olga Pulido.
Kluwer Academic / Plenum Publishers, New York, 2005.

In addition to glutamate receptors in the central nervous system (CNS) there is an ever-growing number of reports of glutamate receptors in the periphery suggesting that glutamate may also function as a neurotransmitter in the peripheral nervous system (Erdo, 1991; Gill et al., 1998; Gill and Pulido, 2001; Skerry and Genever, 2001). Various types of iGluRs and mGluRs have been reported in bone (Chenu et al., 1998; Patton et al., 1998; Laketic-Ljubojevic et al., 1999; Gu and Publicover, 2000), liver (Sureda et al., 1997; Gill et al., 2000), spleen (Gill et al., 2000), heart (Lin et al., 1996; Gill et al., 1999), skin (Morhenn et al., 1994; Genever et al., 1999), thymus (Storto et al., 2000), skeletal muscle (Lin et al., 1996), larynx (Robertson et al., 1998), kidney (Gill et al., 2000), esophagus (Robertson et al., 1998), lung and bronchi (Said et al., 1996, 2001; Said, 1999; Gill et al., 2000), myenteric plexus (Moroni et al., 1986; Shannon and Sawyer, 1989; Ren et al., 2000), taste buds (Lin and Kinnamon, 1999; Dingledine and Conn, 2000), urogenital tract (Gill et al., 2000; Gonzalez-Cadavid et al., 2000), adrenal gland (Yoneda and Ogita, 1986; O'Shea et al., 1992; Kristensen, 1993; Watanabe et al., 1994), and pancreas (Miyazaki et al., 1990; Inagaki et al., 1995; Molnar et al., 1995; Lin et al., 1996; Satin and Kinard, 1998; Weaver et al., 1998; Morley et al., 2000). The localization and function of glutamate receptors at these diverse sites are discussed throughout this book. The topic of this chapter is the potential role of glutamate receptors in the function of the endocrine pancreas and adrenal gland. We will examine receptor subunit expression, signaling, and functional studies in isolated cells and tissues and in whole animal experiments.

2. Glutamate Receptors in the Pancreas

2.1. The Endocrine Pancreas

The pancreas has two functional parts. The exocrine pancreas is responsible for the secretion of fluid and enzymes involved in the process of digestion. The endocrine pancreas, which is the subject of this chapter, secretes hormones that regulate energy metabolism and fuel homeostasis (Munger, 1981; Pipeleers et al., 1992). The endocrine pancreas is composed of groups of cells known as the islets of Langerhans that are contained within the exocrine part of the pancreas. The islets make up only 2% of the weight of the pancreas and they are richly supplied by blood vessels and nerves. Islets are composed of four cell types that each secrete a different hormone. Insulin and glucagon produced by β-cells and α-cells, respectively, regulate blood glucose concentration. Somatostatin produced by δ-cells regulates hormone secretion by α- and β-cells. The last cell type, the F-cell, produces pancreatic polypeptide, the function of which is not known. β-Cells contain numerous secretory granules that represent an intracellular insulin store. Insulin increases the uptake of glucose and amino acids into most cells and stimulates glycogen, fat, and protein synthesis. Insulin secretion is regulated by a plethora of signals including metabolites, hormones, nutrients, ions, nerve activity, and intraislet communication.

While at first it may appear odd that the insulin-secreting β-cells of the pancreas possess glutamate receptors typically found in neurons, β-cells share many common characteristics with neurons. Namely, they are electrically excitable and contain tyrosine hydroxylase (Teitelman and Lee, 1987), neuron-specific enolase (Polack et al., 1984), glutamic acid decarboxylase, the presynaptic protein synaptotagmin associated with vesicular secretion, and high levels of the neurotransmitter GABA (Okada, 1986; Bertrand et al., 1992).

The release of the inhibitory neurotransmitter GABA from β-cells may activate $GABA_A$ receptors on glucagon-secreting α-cells and attenuate glucagon secretion when glucose levels are elevated (Rorsman *et al.*, 1989). Understanding the localization, function, and pharmacology of glutamate receptors in the pancreas may lead to a new understanding of the physiology of this tissue and the pathophysiology of diseases like diabetes. In addition, if glutamate receptor antagonists are ever used as therapeutic agents for neurodegenerative disease, their potential side-effects at peripheral sites must be understood.

2.2. *In Vitro* Studies—Pancreatic β-Cell Lines

Numerous studies have shown the presence of glutamate receptors in pancreatic β-cell lines. These include the rat RINm5F (Inagaki *et al.*, 1995; Molnar *et al.*, 1995), hamster HIT T15 (Molnar *et al.*, 1995), mouse insulinoma GK-P3 (Weaver *et al.*, 1998), and mouse insulinoma MIN6 (Miyazaki *et al.*, 1990; Gonoi *et al.*, 1994; Molnar *et al.*, 1995; Morley *et al.*, 2000). While no complete study of the expression of all glutamate receptor subunits in a particular cell line has been reported, some expression data using radioligand binding, reverse-transcriptase polymerase chain reaction (RT-PCR), and Western analysis is available. In addition, functional studies using electrophysiology, Ca^{2+} imaging, and insulin release have been reported.

2.2.1. Binding and Expression Studies

Radioligand binding studies using $[^3H]AMPA$ or $[^3H]KA$ showed the presence of binding in membrane samples prepared from the MIN6 or RINm5F cell lines (Molnar *et al.*, 1995). Significant binding was detected in both cell types, but at a significantly lower level than that seen in brain extracts (Molnar *et al.*, 1995). More recently, RT-PCR has been used to assay for the expression of mRNA encoding glutamate receptor subtypes in pancreatic β-cell lines. The expression patterns reported differ significantly between cell lines and even between similar cell lines in different studies (Table 8.1). Gonoi *et al.* (1994) reported that MIN6 cells express GluR3, KA2, NR1, NR2D, and very low levels of GluR2 and NR2C subunits (Gonoi *et al.*, 1994). On the contrary, another RT-PCR study showed MIN6 cells to express NR1, GluR2, GluR3, and GluR4, but not GluR1 or any of the NR2 subunits (Morley *et al.*, 2000). GluR2, GluR5, KA1, KA2, NR1, and NR2D mRNAs were shown to be expressed in RINm5F cells (Inagaki *et al.*, 1995). GK-P3 cells showed mRNA encoding GluR2, and GluR3 (Weaver *et al.*, 1998). The different expression patterns could be related to the different species of origin of the cell lines or to differences due to the culture conditions or the transformation of the cell lines. Glutamate receptor subunit mRNA expression does not necessarily indicate expression of receptor proteins. Immunocytochemistry was also used to detect subunits of NMDA and AMPA receptors in MIN6 cultures. MIN6 cells lacked immunoreactivity for NR1, NR2A, and GluR1 subunits, however, they expressed weak immunofluorescent staining for the GluR4 subunit and strong staining for the GluR2/3 subunits. Western blot analysis of glutamate receptor subunit expression in HIT T15, MIN6, and RINm5F cells found NR1 and KA2, but no GluR1 (Molnar *et al.*, 1995). Consistent with the radioligand binding data, less protein was present in β-cells compared to brain extracts (Molnar *et al.*, 1995). Western analysis in GK-P3 cells using anti-GluR2/3 antibody detected protein expression, but no GluR1 or GluR4 was seen, consistent with the RT-PCR data

Table 8.1.   Glutamate Receptor Subtypes in β-Cells and Islets

Cell type	Study type	NMDA	AMPA	Kainate	mGluR	References
MIN6	RT-PCR	NR1 NR2C NR2D	GluR3	KA2		Gonoi et al. (1994)
MIN6	RT-PCR	NR1	GluR2 GluR3 GluR4			Morley et al. (2000)
MIN6	RT-PCR				mGluR3 mGluR4 mGluR5 mGluR8	Brice et al. (2002)
MIN6	Western				mGluR3 mGluR5 mGluR8	Brice et al. (2002)
MIN6	Western	NR1		KA2		Molnar et al. (1995)
MIN6	Immuno		GluR2/3 GluR4			Morley et al. (2000)
RINm5F	RT-PCR	NR1 NR2D	GluR2	KA1 KA2 GluR5		Inagaki et al. (1995)
RINm5F	RT-PCR/ Western				mGluR3 mGluR5 mGluR8	Brice et al. (2002)
RINm5F	Western	NR1		KA2		Molnar et al. (1995)
GK-P3	RT-PCR		GluR2 GluR3			Weaver et al. (1998)
GK-P3	Western		GluR2/3			Weaver et al. (1998)
HIT T15	Western	NR1		KA2		Molnar et al. (1995)
Rat islets	RT-PCR	NR1 NR2A NR2C NR2D	GluR1 GluR2 GluR3	GluR6 GluR7 KA2		Inagaki et al. (1995)
Rat islets	RT-PCR/ Western				mGluR3 mGluR4 mGluR5 mGluR8	Brice et al. (2002)
Rat islets	Western	NR1		KA2		Molnar et al. (1995)
Rat islets	Immuno				mGluR8	Weaver et al. (1996)
Human islets	RT-PCR/ Western				mGluR3 mGluR5	Brice et al. (2002)

(Weaver *et al.*, 1998). The expression of the mGluR subunits mGluR1–mGluR7 was not detected in MIN6 cells based on Western and RT-PCR analyses (Wang *et al.*, 1997). In another study, mGluR3, mGluR5, and mGluR8 mRNA and protein were expressed in MIN6 and RINm5F cells (Brice *et al.*, 2002). Very low levels of mGluR4 was also detected in MIN6 cells (Brice *et al.*, 2002).

2.2.2. Functional Studies—Electrophysiology, $[Ca^{2+}]_i$ Imaging, and Insulin Release

In both neurons and β-cells, membrane potential changes are important in the secretion of neurotransmitters and hormones, respectively. Depolarizations open voltage-dependent Ca^{2+} channels (VDCCs), which allow Ca^{2+} entry that is essential for hormone and neurotransmitter release. The resting membrane potential in pancreatic β-cells is regulated by K^+ conductance through ATP-sensitive K^+ channels (Rajan et al., 1990). Closure of these channels by glucose metabolites leads to β-cell depolarization and Ca^{2+} influx through VDCCs. The rise in $[Ca^{2+}]_i$ triggers insulin secretion. The elevated $[Ca^{2+}]_i$ also activates Ca^{2+}-activated K^+ channels to enhance K^+ efflux and inactivate Ca^{2+} channels which results in cell repolarization and return to the resting state.

AMPA, KA, and NMDA depolarized MIN6 cells by 5–15 mV under current clamp conditions (Gonoi et al., 1994). In voltage clamp, AMPA, KA, and NMDA induced inward currents of 20–90 pA in amplitude (Gonoi et al., 1994). As with central NMDA receptors, the currents required glycine as a co-agonist and were blocked by Mg^{2+} and the NMDA receptor antagonists (±)-2-amino-5-phosphonopentanoic acid (AP-5), MK-801 or 7-chlorokynurenic acid. The KA and AMPA currents were blocked by 6-cyano-7-nitroquinoxaline-2,3-dione (CNQX) (Gonoi et al., 1994). In GK-P3 cells, 60% of cells showed glutamate, KA, or AMPA currents, but no NMDA currents were observed (Weaver et al., 1998). The currents in GK-P3 cells were typical AMPA currents, that is, they rapidly desensitized with a single exponential, were potentiated by cyclothiazide (Yamada and Tang, 1993), and were blocked by CNQX (Weaver et al., 1998). As expected, KA-evoked currents in GK-P3 cells did not desensitize (Weaver et al., 1998). Weaver et al. (1998) also looked at GK-P3 cell AMPA receptor permeability ratios (P_{Ca}/P_{Na}) as calculated from reversal potential measurements in the presence of 147 mM external Na^+ or 92 mM external Ca^{2+}. While most cells showed AMPA receptors with low Ca^{2+} permeability, a small subpopulation of cells appeared to contain AMPA receptors with a greater than average Ca^{2+} permeability (Weaver et al., 1998). Perforated-patch recordings of membrane potential in GK-P3 cells showed glutamate depolarized cells an average of 27 mV from a resting level of -50 mV (Weaver et al., 1998). This depolarization led to the firing of action potentials, similar to those seen in β-cells exposed to high glucose (Rorsman and Trube, 1986).

Ionotropic glutamate receptor activation in neurons increases $[Ca^{2+}]_i$. The Ca^{2+} permeability of glutamate receptors depends on their subunit composition. In addition, for AMPA receptors, RNA editing regulates Ca^{2+} permeability (Sommer et al., 1991). In Fura-2-loaded MIN6 cells, KA, AMPA, and NMDA increased $[Ca^{2+}]_i$ and the responses to KA and NMDA were inhibited by CNQX and MK-801, respectively (Gonoi et al., 1994). The NMDA-induced $[Ca^{2+}]_i$ response was also inhibited by Mg^{2+} (Gonoi et al., 1994). In other studies, MIN6 cells responded to glutamate, AMPA, and KA, but not NMDA or 1S, 3R-trans-ACPD, with increases in $[Ca^{2+}]_i$ (Wang et al., 1997; Morley et al., 2000). To maximize the probability of opening of NMDA-gated channels, Ca^{2+} experiments were conducted in Mg^{2+}-free buffer containing glycine (Morley et al., 2000). The lack of $[Ca^{2+}]_i$ responses to NMDA stimulation correlated with the lack of expression of the NMDA NR2 receptor subunits in MIN6 cells, as determined by RT-PCR and immunocytochemistry. In neurons, functional NMDA receptors require a combination of both NMDA NR1 and NR2 receptor subunits (Ishii et al., 1993).

The AMPA-triggered $[Ca^{2+}]_i$ responses in MIN6 cells required cyclothiazide and were blocked by GYKI 52466, CNQX, DNQX, spermine (Isa *et al.*, 1996), and pentobarbital (Taverna *et al.*, 1994; Morley *et al.*, 2000). AMPA-triggered $[Ca^{2+}]_i$ responses could be mediated directly by Ca^{2+} influx through Ca^{2+}-permeable AMPA receptors or indirectly through VDCCs following Na^+ entry-induced depolarization of the MIN6 cells. However, the MIN6 cell AMPA-triggered $[Ca^{2+}]_i$ responses were unaffected by Joro spider toxin (Blaschke *et al.*, 1993) or Evan's Blue (Keller *et al.*, 1993), selective antagonists for Ca^{2+}-permeable AMPA receptors (Morley *et al.*, 2000). MIN6 cells also showed large rapid increases in $[Ca^{2+}]_i$ when stimulated with 45 mM K^+ (Morley *et al.*, 2000). The AMPA- and K^+-triggered $[Ca^{2+}]_i$ responses were blocked in Na^+-free medium, by the general VDCC antagonist lanthanum (La^{3+}) and by the dihydropyridines, nifedipine, and nimodipine (Morley *et al.*, 2000). Coupled with the high expression of the GluR2 subunit in its edited form in MIN6 cells, the data showing a block of AMPA-triggered $[Ca^{2+}]_i$ responses by Ca^{2+} channel blockers and by incubating the cells in Na^+-free medium, suggests that MIN6 cells possess Ca^{2+}-impermeable, Na^+-permeable AMPA receptors (Morley *et al.*, 2000). Thus, the AMPA-triggered $[Ca^{2+}]_i$ responses would result from the opening of VDCCs which occurred subsequent to the depolarization caused by Na^+ entry through these Ca^{2+}-impermeable AMPA receptors (Morley *et al.*, 2000). GK-P3 cells also responded to glutamate with increases in $[Ca^{2+}]_i$ that were potentiated by cyclothiazide and were blocked by La^{3+}, suggesting that like in MIN6 cells, AMPA-triggered Ca^{2+} entry was due to a secondary activation of VDCCs following membrane depolarization (Weaver *et al.*, 1998).

β-Cell insulin release is stimulated by elevations of $[Ca^{2+}]_i$. Gonoi *et al.* (1994) showed that AMPA-, NMDA-, and KA-stimulated insulin secretion in MIN6 cultures in the presence of 3.3 mM glucose. The NMDA-induced insulin secretion in MIN6 cells was blocked by Mg^{2+} (Gonoi *et al.*, 1994). The actions of mGluRs on insulin secretion have been studied in MIN6 cells using the group I (mGluR1/5), group II (mGluR2/3), and group III (mGluR4/6–8) specific agonists DHPG, L-CCG-1, and L-AP4, respectively (Brice *et al.*, 2002). The effects were studied in cells at glucose concentrations of 3, 10, or 25 mmol/L. The group I and II agonists, DHPG and L-CCG-1 increased insulin secretion at low glucose concentrations and to a lesser degree at higher glucose concentrations. The group III agonist L-AP4 enhanced insulin secretion at 3 mmol/L glucose, but inhibited secretion at the high glucose concentration. These actions of mGluR agonists reflect the distribution of mGluR subtypes demonstrated in the same study by RT-PCR and immunoblot (Brice *et al.*, 2002).

Glutamate, formed from the amination of citric acid cycle-derived α-ketoglutarate in β-cell mitochondria by glutamate dehydrogenase, may be an intracellular messenger in glucose-induced insulin secretion in rat insulinoma INS-1 cells (Maechler and Wollheim, 1999). Another report casts doubt on the concept of an intracellular signaling action of glutamate. MacDonald and Fahien (2000) found no increase in glutamate levels in pancreatic islets following the stimulation of insulin secretion by glucose or several other inducers. In addition, increasing the intracellular glutamate levels 10-fold above basal level by adding exogenous glutamine did not stimulate insulin release (MacDonald and Fahien, 2000). Contrary to the findings of Maechler and Wollheim (1999), experiments with mitochondria from pancreatic islets suggest that glutamate dehydrogenase is quiescent during glucose-induced insulin secretion (MacDonald and Fahien, 2000).

As mentioned previously, elevated glutamate release and stimulation of postsynaptic receptors is believed to be an important contributor to ischemic neuronal death. Glutamate triggers large $[Ca^{2+}]_i$ surges in neurons following ischemia which have been implicated as key components of the mechanism responsible for neuronal death (Lipton and Rosenberg, 1994). The mechanism by which a rapid rise in $[Ca^{2+}]_i$ can effect the death of neurons is unknown. It is known, however, that the magnitude of the excitatory amino acid-triggered $[Ca^{2+}]_i$ response is not the only parameter that determines the neurotoxic effects of Ca^{2+} (Michaels and Rothman, 1990; Dubinsky and Rothman, 1991). These observations have led to the hypothesis that the location or mode of entry of the Ca^{2+} may be responsible, at least in part, for the neurotoxic effects of some agents (Dubinsky and Rothman, 1991; Tymianski et al., 1993). Ca^{2+} from different cellular sites may have access to different intracellular compartments or enzymes and therefore activate different downstream signaling pathways. An early and persistent loss of protein kinase C (PKC) activity is another characteristic of cerebral ischemia (Crumrine et al., 1990; Domanska-Janik and Zalewska, 1992). A persistent loss of membrane-associated PKC activity occurs in the membranes of cultured cortical neurons following stimulation with glutamate that may be an essential step in the death process (Durkin et al., 1997). Some cell lines that express glutamate receptors do not survive in culture (Anegawa et al., 1995). This is not the case with the MIN6 cell line (Goroi et al., 1994; Morley et al., 2000). Unlike cortical neuronal cultures, which show a loss of membrane-associated PKC activity (Black et al., 1995, 1996) and die in response to excitatory amino acid exposure (Choi, 1987; Durkin et al., 1996), glutamate, NMDA, AMPA, or KA were not toxic to MIN6 cells and they did not decrease PKC activity (Morley et al., 2000). Therefore, MIN6 cells may provide a model to further understand the role of Ca^{2+} and PKC in glutamate toxicity by separating PKC-dependent and PKC-independent cellular events in the cell death cascade. These observations also raise the question of whether iGluR in these β-cell lines are regulated differently than in neurons.

2.3. *In Vitro* Studies—Primary β-Cells

As in β-cell lines, various combinations of glutamate receptor subunits have been reported in freshly isolated rat pancreatic islet preparations (Inagaki et al., 1995; Molnar et al., 1995; Weaver et al., 1996). RT-PCR showed mRNA expression of GluR1, GluR2, GluR3, GluR6, GluR7, KA2 NR1, NR2A, NR2C, and NR2D in rat islets (Inagaki et al., 1995). However, GluR1, NR2A, and NR2C mRNA expression levels were very low. The presence of both NR1 and NR2 subunits suggests the presence of functional NMDA receptors. Western analysis showed NR1, KA2, but not GluR1, protein in rat islet of Langerhans membranes, however there was much lower expression than that seen in neuronal membranes (Molnar et al., 1995). Rat islets were also shown by RT-PCR and immunoblot to express mRNA and protein for mGluR3, mGluR4, mGluR5, and mGluR8 (Brice et al., 2002). In contrast, mGluR3 and mGluR5 were detected in human islets (Brice et al., 2002). In another study, mGluR8 expression in rat islets was shown by immunoreactivity to be primarily located on the glucagon-secreting α-cells (Tong et al., 2002). The receptors appear functional since $[Ca^{2+}]_i$ responses to AMPA, KA, and NMDA were seen in Fura-2-loaded single β-cells (Molnar et al., 1995). KA-induced $[Ca^{2+}]_i$ responses were blocked by CNQX and NMDA-induced responses blocked by MK-801 (Molnar et al., 1995). Patch clamp has also been used to look at the effect of glutamate on membrane potential and ionic currents in freshly isolated islet cells. In one study, current clamp was used to show that KA, AMPA, and NMDA rapidly depolarized islet cells (Inagaki et al., 1995). In some cases, the depolarization evoked action potentials (Weaver et al., 1996). Exposure to glutamate in voltage clamp studies produced very small conductance currents unless the recordings were performed in the presence of cyclothiazide (Weaver *et al.*, 1996). In all, about 25% of islet cells showed responses to glutamate receptor activation, however in this study NMDA-induced currents were not present (Weaver *et al.*, 1996). The glutamate-evoked ionic currents were blocked by CNQX and potentiated by cyclothiazide, just like neuronal AMPA receptors (Weaver *et al.*, 1996).

Insulin secretion studies have also been done in freshly isolated islet preparations. The rate of insulin secretion from rat islets at 3.3 mM glucose was not affected by glutamate, NMDA, or KA (Molnar *et al.*, 1995). However, at 8.3 mM glucose, the rate of insulin secretion was increased by 47% by NMDA, and at 16.7 mM glucose, NMDA increased insulin secretion by 37% (Molnar *et al.*, 1995). Neither glutamate nor KA affected insulin secretion (Molnar *et al.*, 1995). In another report, KA and AMPA, but not NMDA, in the presence of 8.3 mM glucose, moderately (1.2–1.8-fold) stimulated insulin secretion from rat islets (Inagaki *et al.*, 1995). The lack of effect of NMDA on insulin secretion in this study may be related to the low level of expression of NMDA receptor subunits and/or the low magnitude $[Ca^{2+}]_i$ responses evoked by NMDA (Inagaki *et al.*, 1995). The differences in glutamate receptor subunit expression seen in different studies may be related in part to the presence of glutamate receptors on other cell types in the freshly isolated islet preparations. Enzymatic digestion used to isolate islets may also adversely affect glutamate receptor expression and function. Overall, the structural, pharmacological, and ionic permeability properties of glutamate receptors in β-cell lines and in islet preparations are very similar to those of neuronal glutamate receptors. Peripheral AMPA receptors show a similar responsiveness for inhibition of desensitization by cyclothiazide, ion permeability, and modulation by excitatory amino acid receptor agonists and antagonists as central AMPA receptors.

2.4. *In Vitro* Studies—Tissue

Double-labeling immunofluorescence studies using antibodies specific for insulin and the glutamate receptor subunits have shown the localization of AMPA receptor GluR1-4 subunits in the newborn guinea pig pancreas (Liu *et al.*, 1997). GluR1 and GluR4 subunits were localized mostly in insulin-secreting cells in the central mass of the islets, while GluR2/3 were localized to the outer rim of the islet, which tend to be non-insulin-secreting cells. GluR2/3 and GluR4 were also localized in intrapancreatic ganglia as visualized using NADPH-diaphorase histochemistry (Liu *et al.*, 1997). There is a complex intrapancreatic nervous system composed of nerves and ganglia dispersed in the parenchymal tissue. In addition, pancreatic islets are richly innervated by parasympathetic and sympathetic nerves that increase and decrease insulin secretion, respectively.

Similar studies have also been done on rat pancreas slices (Weaver *et al.*, 1996). AMPA receptor GluR2/3 subunits were expressed in α-cells, β-cells, and pancreatic polypeptide-producing cells whereas the KA receptor subunit GluR6/7 was found mostly in glucagon-secreting α-cells and somatostatin-secreting δ-cells (Weaver *et al.*, 1996). In the rat pancreas, GluR1 staining was weak and found only in β-cells and the KA2 subunit was seen only in δ-cells (Weaver *et al.*, 1996). Other subunits, including GluR4 and NR1, showed no staining. In support of the immunofluorescence data, GluR2/3 protein was also observed in Western blots of rat pancreas tissue (Weaver *et al.*, 1996). Northern blot analysis in human pancreatic tissue showed NR2C, but not NR1, receptor subunit expression (Lin *et al.*, 1996). The sensitivity of Northern analysis is less than that of RT-PCR, so these subunits may be present, but expressed at very low levels. These findings suggest that pancreatic tissues lack functional NMDA channels or that a yet unidentified subunit can combine with NR2C in the periphery to form a functional channel.

Immunofluorescence studies have also suggested potential sites of glutamate release in the pancreas (Inagaki *et al.*, 1995). Islets stained for phosphate-activated glutaminase, an enzyme involved in glutamate generation, showed intense staining in the glucagon-secreting α-cells and in the intrapancreatic ganglia (Inagaki *et al.*, 1995). These results are compatible with glutamate being released from neurons as well as from α-cells to act on β-cells in a paracrine manner.

The effect of glutamate and glutamate receptor agonists on insulin and glucagon secretion in the perfused rat pancreas has been investigated. Glutamate, AMPA, KA, or quisqualate receptor activation transiently stimulates an immediate and concentration-dependent increase of insulin release at stimulating (8.3 mM), but not resting (2.8 mM), glucose concentrations (Teitelman and Lee, 1987). These observations suggest that glutamate is a potentiator of glucose-induced insulin secretion, but it is not an initiator of secretion. The concentrations of glutamate required to affect insulin secretion (i.e., 30–100 μM), are physiological blood levels in both rats and humans. The magnitude of the insulin response to glutamate and KA was 3-fold higher than that to AMPA or quisqualate (Teitelman and Lee, 1987). NMDA, even in the presence of glycine and absence of Mg^{2+}, did not affect insulin release. Insulin secretory responses were blocked by the AMPA receptor antagonist CNQX, but were unaffected by the NMDA receptor antagonist MK-801 (Teitelman and Lee, 1987). Likewise, Inagaki *et al.* (1995) found KA and AMPA, but not NMDA, increased insulin secretion in response to glucose in rat islets. However, contrary to these findings, Molnar *et al.* (1995) reported NMDA, but not glutamate or KA, to enhance glucose-stimulated insulin secretion in rat islets. In one study, glutamate was reported to not stimulate insulin secretion in mouse pancreatic islets (Sehlin, 1972), however, these 1-h static incubations would not have seen the rapid transient responses detected in the perfusion studies.

The action of glutamate on insulin secretion does not appear to be mediated via the nervous system (Teitelman and Lee, 1987). If atropine is used to block acetylcholine release from pancreatic parasympathetic nerves or tetrodotoxin is used to block action potentials, the insulin response to glutamate is not affected (Teitelman and Lee, 1987). Therefore, glutamate likely mediates its actions through direct effects on β-cells. Generally, the reports to date suggest that insulin secretion from the rat pancreas is stimulated by AMPA receptor activation.

Micromolar concentrations of glutamate, AMPA, KA, or quisqualate also transiently stimulate an immediate and concentration-dependent increase of glucagon in the rat pancreas perfused with 2.8 mM glucose (Bertrand *et al.*, 1993). This is a stimulating concentration of glucose for glucagon secretion. On the other hand, perfusion of the rat pancreas with high millimolar concentrations of glutamate induces sustained glucagon secretion (Assan *et al.*, 1977). The magnitude of the transient glucagon response to KA was 4-fold higher than that to AMPA, quisqualate, or glutamate (Bertrand *et al.*, 1993). NMDA did not affect glucagon release, even in the presence of glycine and absence of Mg^{2+}. Glucagon secretory responses were blocked by the AMPA receptor antagonist CNQX (Bertrand *et al.*, 1993). As for insulin secretion, the concentrations of glutamate required to affect glucagon secretion (i.e., 30–100 μM), are physiological blood levels in both rats and humans. Glutamate appears to stimulate glucagon secretion from the rat pancreas by activating AMPA receptors.

Since both insulin- and glucagon-secreting cells possess AMPA receptors it is not surprising that AMPA stimulates the secretion of both hormones. However, the opposing actions of the two hormones on blood glucose concentration would make it unlikely that glutamate would release both of them. At equivalent doses, glutamate has a stronger stimulatory effect on insulin than it does on glucagon secretion. In addition, the actions of glutamate on insulin and glucagon secretion differ depending on the glucose concentration. At a resting concentration of 2.8 mM glucose, glutamate stimulates glucagon secretion, but does not affect insulin secretion. For now, the relative contribution and role of AMPA in insulin and glucagon secretion under normal physiological conditions is not known, nor is the mechanism by which AMPA interacts with other physiological modulators of insulin and glucagon secretion.

The presence of a high affinity glutamate/aspartate uptake system, similar to that described in the CNS, has also been demonstrated in pancreatic tissue (Manfras *et al.*, 1994; Weaver *et al.*, 1998; Howell *et al.*, 2001). Glutamate transporters in the CNS pump the synaptically-released glutamate back into neurons and glia to prevent excitotoxicity. While early reports suggested that glutamate is very poorly taken up by mouse pancreatic islets (Sehlin, 1972), more recent work has shown the uptake of [^{3}H]glutamate by a temperature- and Na$^+$-dependent mechanism that is blocked by L-*trans*-pyrrolidine-2,4-dicarboxylic acid (L-*trans*-PDC), an inhibitor of neuronal and glial glutamate transporters (Weaver *et al.*, 1998). In perfusion experiments using isolated islets, L-*trans*-PDC potentiated glucose-stimulated insulin secretion by about 20% (Weaver *et al.*, 1998). The transporter seems to be localized to α-cells (Weaver *et al.*, 1998). Using immunoblot analysis, the glutamate transporter proteins EAAC1, GLT-1, and GLAST1 have been reported in homogenates of pancreatic tissue from sheep and cows (Howell *et al.*, 2001). To corroborate protein expression data, GLAST1, EAAC1, and GLT-1 mRNA transcripts were detected in pancreatic tissue by Northern analysis. The expression of GLT-1 mRNA was confirmed by RT-PCR. Sequencing of the resulting partial-length ovine GLT-1 cDNA revealed 100% identity with the rat homolog (Howell *et al.*, 2001). The role of these transporters is still unknown, but they may be an important component of a complete glutamate signaling system in pancreatic islets consisting of glutamate release, specific glutamate receptors and a high affinity uptake system.

2.5. *In Vivo* Studies

As demonstrated in perfused pancreatic tissue, AMPA receptors also mediate insulin and glucagon release in rats. Oral or intravenous glutamate stimulates a transient, dose-dependent and CNQX-sensitive insulin release in fed, but not in fasted, rats (Bertrand *et al.*, 1995). Glutamate was effective at plasma concentrations of 200–300 μM (Bertrand *et al.*, 1995). The effect of glutamate on insulin secretion is therefore dependent upon the fed state of the rat. So, like in the *in vitro* studies, glutamate is not an initiator of secretion, but instead it potentiates glucose-induced insulin secretion (Bertrand *et al.*, 1995).

3. Glutamate Receptors in the Adrenal Gland

The adrenal gland is composed of two major structural parts: An outer cortex and an inner medulla. The adrenal cortex primarily produces three steroid hormones: aldosterone, cortisol (corticosterone in rodents), and dehydroepiandrosterone, from distinct zones of the adrenal cortex known as the zona glomerulosa, zona fasciculata, and zona reticularis, respectively. Of these three steroids, cortisol, a glucocorticoid (GC), is produced most copiously. Its secretion is predominantly mediated by the activity of the hypothalamo–pituitary–adrenal (HPA) axis: corticotropin-releasing hormone (CRH) released by the hypothalamus stimulates secretion of adrenocorticotropin hormone (ACTH) from the anterior pituitary which in turn stimulates GC secretion from the adrenal cortex. While GC are secreted in a circadian rhythm, their secretion is additionally stimulated by a variety of physical and psychological stressors, such as hypoglycemia and hypoxia. Both the circadian rhythm and the stress-induced activity of the HPA axis are inhibited by GC feedback on the pituitary, the hypothalamus and possibly on other brain regions such as the hippocampus (Jacobson and Sapolsky, 1991).

The adrenal medulla, which is coupled to the sympathetic nervous system, plays a pivotal role in the stress response. The adrenal medulla is composed principally of chromaffin cells, with minor populations of small intensely fluorescent cells and noradrenergic or cholinergic ganglionic neurons. The chromaffin cells, which synthesize, store, and secrete the catecholamines epinephrine or norepinephrine, are innervated by preganglionic sympathetic fibers from the splanchnic nerve. Stresses lead to an increase in splanchnic nerve discharge and the release of acetylcholine which binds to nicotinic acetylcholine receptors on the chromaffin cells. Stimulation of these receptors leads to catecholamine secretion and the induction of enzymes involved in catecholamine synthesis. Epinephrine exerts widespread effects on organ systems that are ideally suited for fight or flight responses, such as elevating heart rate, increasing blood glucose levels, and increasing blood flow to muscle (Nankova and Sabban, 1991).

While the secretion of cortisol and epinephrine are regulated mainly by the HPA axis and the sympathetic nervous system, respectively, other regulatory mechanisms are involved. For instance, the adrenal cortex is richly innervated by a dense network of nerve terminals that originate from intrinsic or extrinsic neurons. Intrinsic fibers arise from ganglion cells of the medulla that synthesize norepinephrine, vasoactive intestinal polypeptide (VIP), neuropeptide tyrosine, and nitric oxide-synthase (NOS) (Engeland, 1998). Extrinsic fibers are mainly sympathetic noradrenergic fibers that innervate the subcapsular region where they contact adrenocortical cells and blood vessels. These synthesize norepinephrine (Kleitman and Holzwarth, 1985), acetylcholine (Holgert et al., 1995), neuropeptides (Kondo, 1985), and NOS (Afework et al., 1994). Intra-adrenal interactions between cortical and medullary cells provide an additional regulatory component. Catecholamines, neuropeptides, growth factors, and cytokines secreted from chromaffin cells mediate cell-to-cell communication in an endocrine or paracrine fashion. The secretory activity of the different types of adrenal cells may thus be controlled by their particular receptor pattern. For example, in addition to acetylcholine receptors, chromaffin cells express histamine (Noble et al., 1988), bradykinin (Houchi et al., 1999), and angiotensin (Belloni et al., 1998) receptors which lead to catecholamine release when stimulated. The differential distribution of receptors between epinephrine- and norepinephrine-secreting chromaffin cells, along with their specific innervation patterns, regulate the nature of the secretory response (Aunis and Langley, 1999). While less is known about the role of glutamate receptors in the adrenal gland, a number of expression studies suggest that these receptors are expressed in both adrenal cortical and medullary cells (Kristensen, 1993; Watanabe et al., 1994; Hinoi and Yoneda, 2001; Schwernd and Jezova, 2001; Hinoi et al., 2002). Defining the function and signal transduction pathways of these receptors may therefore be of importance for better understanding the regulation of adrenal hormone secretion under different pathological conditions.

Like the β-cells of the pancreas, adrenal chromaffin cells share common characteristics with neurons and, as such, have served as a model to offer insight on the production and secretion of neurotransmitters in neurons. They too are electrically excitable and they contain tyrosine hydroxylase (Nankova and Sabban, 1999), DOPA decarboxylase (Nankova and Sabban, 1999), dopamine β-hydroxylase (Nankova and Sabban, 1999), and glutamic acid decarboxylase (Ahonen et al., 1989). Furthermore, chromaffin cells, as well as some cells of the adrenal cortex, express the L-glutamate transporter GLAST (Lee *et al.*, 2001), suggestive of a potential role for glutamate signaling in the adrenal gland.

3.1. *In Vitro* Studies—Primary Chromaffin Cell Cultures and Tissue

Studies investigating the expression of glutamate receptors in the adrenal gland have mostly used radioligand binding, *in situ* hybridization, and RT-PCR techniques. Primary chromaffin cultures and the perfused adrenal gland are commonly used for *in vitro* functional studies. *In vivo* studies investigating glutamate signaling in the adrenal are discussed by Jezova and Schwendt in another chapter of this book.

3.1.1. Binding Studies

Binding studies identified sites for [³H]glutamate ([³H]Glu) in both the adrenal cortex and medulla of adult rodent (Yoneda and Ogita, 1986; O'Shea *et al.*, 1992), bovine (O'Shea *et al.*, 1992) and canine (O'Shea *et al.*, 1992) preparations. These binding sites consisted of a single population (Yoneda and Ogita, 1986; O'Shea *et al.*, 1992), were saturable (Yoneda and Ogita, 1986; O'Shea *et al.*, 1992) and stereo-selective (Yoneda and Ogita, 1986; O'Shea *et al.*, 1992), appeared intrinsically different from those in central structures (Yoneda and Ogita, 1986; O'Shea *et al.*, 1992), and also appeared to differ among species (O'Shea *et al.*, 1992). For instance, binding activity in membranous particulate preparations obtained from rat adrenal glands was not affected by inorganic ions known to enhance central binding, such as Cl^- and Cl^-/Ca^{2+} (Yoneda and Ogita, 1986). Furthermore, Na^+ enhanced binding in rat cerebral preparations, but inhibited binding in adrenal preparations (Yoneda and Ogita, 1986). In contrast, in bovine adrenal sections, [³H]Glu binding was stimulated by Ca^{2+} and was Na^+-independent (O'Shea *et al.*, 1992).

Displacement studies using various agonists for subtypes of central glutamate receptors were performed to determine the pharmacological profile of the glutamate binding sites in the adrenal. In bovine adrenal gland sections, 100 μM of the subtype-specific glutamate receptor agonists NMDA, KA, L-AP4, and ACPD failed to displace [³H]Glu binding (O'Shea *et al.*, 1992). Quisqualate was also unable to displace [³H]Glu binding in bovine adrenal sections (O'Shea *et al.*, 1992) but was found to elicit a prominent suppression of [³H]Glu binding in rat adrenal preparations (Yoneda and Ogita, 1986). However, AMPA was not able to displace [³H]Glu binding in rat adrenal preparations (Yoneda and Ogita, 1989). Since quisqualate is both a potent AMPA agonist and a group I mGluR antagonist, this suggests that quisqualate was acting on mGluR and not AMPA receptors in rat preparations. This belief was supported by the observation that AP3, a group I mGluR antagonist, inhibited [³H]Glu binding in both bovine (O'Shea *et al.*, 1992) and rat (Yoneda and Ogita, 1986) adrenal. However, in later studies using rat adrenal preparations, [³H]quisqualate binding sites were not detected (Hinoi *et al.*, 2001) while [³H]AMPA binding sites were (Hinoi and Yoneda, 2001). In a similar fashion, while initial displacement studies showed that KA had no effect on [³H]Glu binding in rat (Yoneda and Ogita, 1986) or bovine (O'Shea *et al.*, 1992) adrenal preparations, significant [³H]KA binding was later detected in the rat adrenal (Hinoi and Yoneda, 2001).

Displacement studies targeting the NMDA receptor further underline the differences between central and adrenal binding sites. NMDA, at a concentration of 100 μM, was unable to displace [³H]Glu binding in either rat (Yoneda and Ogita, 1986) or bovine (O'Shea *et al.*, 1992) adrenal preparations. Interestingly, at a higher dose (1 mM), NMDA exerted a significant augmentation of [^{3}H]Glu binding in rat adrenal (Yoneda and Ogita, 1986, 1987), a result that cannot be explained in terms of classical excitatory amino acid receptor pharmacology. This NMDA-potentiation was not affected by the addition of the specific NMDA antagonists, AP5 or AP7. In contrast, the tryptophan metabolites, kynurenine, xanthurenic acid, and quinolinic acid, diminished the NMDA-induced enhancement of the binding and attenuated basal binding in a concentration-dependent manner (10^{-6}–10^{-3} M) (Yoneda and Ogita, 1987). Kynurenic acid (1 mM), a noncompetitive antagonist of NMDA receptors that acts on the strychnine-insensitive glycine site, selectively diminished the NMDA-induced potentiation without affecting basal binding, suggesting that NMDA potentiates glutamate binding by interacting through a kynurenic acid sensitive site (Yoneda and Ogita, 1987). It should be noted that, in the millimolar range, kynurenic acid is also able to interact as a competitive antagonist to AMPA and KA receptors (Stone, 2000). These binding studies show that the adrenal gland does possess binding sites for glutamate that appear to have a different pharmacological profile than central receptors. However, many of the antagonists used in these studies were nonselective making interpretation of the results difficult.

The localization of the [^{3}H]Glu binding sites within the adrenal gland may help evaluate the functional significance, if any, of these receptors. Autoradiography used to identify [^{3}H]Glu binding sites in adrenal sections from different adult species showed differential localization (O'Shea *et al.*, 1992). In bovine sections, binding sites were mainly confined to the cortex, with lower amounts in the medulla. In canine sections, binding was seen preferentially in the medulla, but significant levels were also present in the cortex (O'Shea *et al.*, 1992). In rat and guinea pig adrenal sections, the density of binding sites was greater in the cortex than the medulla, with appreciable level of binding present in both regions (O'Shea *et al.*, 1992). Another study found that rat adrenal medulla preparations possessed more than 5-fold [^{3}H]Glu binding activity than that found in adrenal cortical preparations (Yoneda and Ogita, 1986), a discrepancy that may be due to differences in experimental techniques. While these studies are useful in that they determine whether [^{3}H]Glu binding is found in the cortex or medulla of the adrenal, they do not provide information as to which types of cells within these regions express the receptors.

3.1.2. Expression Studies

Binding studies alone do not provide conclusive evidence that glutamate receptors are expressed in the adrenal gland. It is possible that [^{3}H]Glu was binding to an unidentified protein that has an affinity for glutamate comparable to the receptor. Therefore *in situ* hybridization, RT-PCR, and immunoblotting techniques have been used to assay the presence of mRNA or protein encoding glutamate receptors in the adrenal (Table 8.2). Of the NMDA receptor subunits, NR1 mRNA was detected in the medulla of P21 mice (Watanabe *et al.*, 1994) and in both the cortex and medulla of adult rats (Schwendt and Jezova, 2001; Hinoi *et al.*, 2002). mRNA and protein was detected for NR2C and NR2D in rat cortex and medulla (Hinoi *et al.*, 2002), but not for the NR2A and NR2B subunits in either mouse (Watanabe *et al.*, 1994) or rat (Schwendt and Jezova, 2001; Hinoi *et al*, 2002). AMPA subunits displayed a differential pattern of expression in the rat (Kristensen, 1993). *In situ* hybridization detected mRNA for GluR1 and GluR4 in the zona glomerulosa, for GluR3 in the remaining parts of the cortex, and for GluR2 in medullary cells. The sympathetic ganglion cells of the medulla expressed all four AMPA subunit mRNAs. In this same study, RT-PCR using mRNA isolated from the whole adrenal gland of rat was positive for GluR2 and GluR3 mRNA, but not for

Table 8.2.   Glutamate Receptor Subtypes in the Adrenal Gland

Tissue type	Study type	NMDA	AMPA	Kainate	References
Mouse medulla	*In situ* hybridization	NR1			Watanabe *et al.* (1994)
Rat adrenal	RT-PCR		GluR2		Kristensen *et al.* (1993)
			GluR3		
Rat adrenal	RT-PCR ± restriction digest		GluR1 GluR3	GluR5 KA1 KA2	Hinoi *et al.* (2001)
Rat adrenal	RT-PCR ± restriction digest	NR1 NR2C NR2D			Hinoi *et al.* (2002)
Rat medulla	RT-PCR	NR1			Schwendt and Jezova (2001)
Rat medulla	*In situ* hybridization		GluR1 GluR2 GluR3 GluR4		Kristensen *et al.* (1993)
Rat medulla	Northern	NR1			Hinoi *et al.* (2001)
Rat medulla	Western	NR1			Hinoi *et al.* (2001)
Rat medulla	Immuno	NR1 NR2C NR2D			Hinoi *et al.* (2001)
Rat cortex	RT-PCR	NR1			Schwendt and Jezova (2001)
Rat cortex	*In situ* hybridization		GluR1 GluR3 GluR4		Kristensen *et al.* (1993)

GluR1 or GluR4 mRNA, presumably because levels were too low in total adrenal gland mRNA. In the brain, all four AMPA receptor subunits occur in two alternatively spliced versions, flip and flop. The flip forms of most subunits desensitize more slowly and less profoundly that the flop forms (Dingledine *et al.*, 1999). In the adrenal, the flip form of GluR2 and GluR3 dominated and the GluR2 mRNA was present in its edited version at the Q/R site (Kristensen, 1993). After E14 in rats, more that 99% of central GluR2 mRNA is edited (Dingledine *et al.*, 1999). Receptors with edited GluR2 subunits have low permeability to Ca^{2+} (Dingledine *et al.*, 1999). Another study that examined AMPA receptor subunit expression using mRNA isolated from the whole adrenal gland of rat, found mRNA for GluR1 and GluR3, but not for GluR2 or GluR4 (Hinoi and Yoneda, 2001). Nonetheless, Western blotting assays using antibodies against GluR1, GluR2-3, and GluR4 subunits failed to confirm the expression of receptor proteins in adrenal cortical or medullary preparation (Hinoi and Yoneda, 2001). RT-PCR analysis for KA receptor subunits revealed expression of mRNA for the GluR5, KA1, and KA2 subunits but not for the GluR6 or GluR7 subunits (Hinoi and Yoneda, 2001). Neither RT-PCR nor Western blot analysis demonstrated the expression of the metabotropic subunits, mGluR1 or mGluR5, in the rat adrenal (Hinoi *et al.*, 2001).

While most studies that examined the expression of protein for different glutamate receptor subunits failed to detect it, the binding studies, together with functional studies, suggest that glutamate receptors are present in the adrenal gland. Additional work is needed to localize the expression of the different subtypes of receptors to the different types of cells within the adrenal cortex and medulla. This will help better evaluate the potential function of glutamate signaling within the adrenal gland. The expression of several cell surface receptors are known to vary between epinephrine- and norepinephrine-secreting chromaffin cells. For example, epinephrine-secreting chromaffin cells preferentially express histamine, angiotensin II, and muscarinic acetylcholine receptors (Aunis and Langley, 1999). Perhaps glutamate receptors are also differentially expressed among the different cells of the adrenal.

3.1.3. Functional Studies

Cultured bovine chromaffin cells have been used to study the effects of glutamate on catecholamine secretion (Boksa, 1990; Gonzalez et al., 1998). In one study, glutamate (100 μM to 1 mM) failed to stimulate significant release of either norepinephrine or epineprhine in such cultures (Boksa, 1990). In a more recent study, however, Gonzalez et al. (1998) found that glutamate (1 nM–1 mM) and the glutamate receptor agonists, NMDA, quisqualate, KA, and t-ACPD, stimulated basal catecholamine secretion in a dose-dependent manner. In this study, the percentage of catecholamine secretion produced by 100 μM glutamate or by glutamate receptor agonists was between 20% and 60% of that obtained by 10 μM nicotine, an activator of the nicotinic acetylcholine receptor (Gonzalez et al., 1998). With the exception of KA, the agonists were also able to potentiate nicotine-evoked catecholamine secretion. Their effect on basal secretion was prevented by treatment with the appropriate glutamate receptor antagonists and, with the exception of KA, catecholamine release induced by the glutamate agonists was not inhibited by hexamethonium, an antagonist of the nicotinic receptor, suggesting that glutamate was indeed acting via glutamate receptors (Gonzalez et al., 1998).

The release of catecholamines necessitates a rise in $[Ca^{2+}]_i$ to trigger translocation of secretory granules and their fusion with the plasma membrane. The rise in $[Ca^{2+}]_i$ results from the opening of VDCCs after membrane depolarization (Aunis and Langley, 1999). The stimulatory action on catecholamine secretion produced by glutamate, NMDA, quisqualate, KA, and t-ACPD was accompanied by membrane depolarization, as measured using a bisoxonol fluorescence assay, and small but significant increases in $[Ca^{2+}]_i$, as monitored by Fura-2 fluorescence, providing further evidence for the presence of functional glutamate receptors in chromaffin cells (Gonzalez et al., 1998).

No studies have yet investigated the effects of glutamate on corticosterone or aldosterone secretion in adrenocorticol cultures. Furthermore, electrophysiological studies have not been done to determine the effects of glutamate on membrane potential in either adrenocortical or medullary cultures.

Unlike the findings of Gonzalez et al. (1998), which strongly support a role for glutamate in modulating catecholamine release in primary bovine chromaffin cells, O'Shea et al. (1992) found that glutamate, at concentrations of 1 and 10 mM, had no effect on either basal or nicotine-evoked catecholamine secretion from perfused isolated bovine adrenal glands. Differences in the findings of these two studies may reflect differences in preparations (primary cultures vs whole gland preparations) and in methods of measuring catecholamine secretion. In canine isolated perfused adrenal glands, glutamate (50 μM–50 mM) produced a concentration-dependent increase in the rate of catecholamine release (Nishikawa et al., 1982). However, high concentrations of other amino acids, such as serine, cysteine, and arginine, also had this effect, albeit to a lesser extent. Since the effect was dependent on Ca^{2+}, it was proposed that glutamate and the other amino acids induced catecholamine release by depolarizing chromaffin cells through an increased permeability to Na^+ followed by an enhancement of Ca^{2+} entry. While the authors fail to speculate on the mechanisms by which these amino acids depolarize cells, they did not appear to be acting on nicotinic or muscarinic acetylcholine receptors since perfusion with atropine and hexamethonium did not affect the amino acid induced catecholamine release (Nishikawa *et al.*, 1982). Again, eletrophysiological data is needed to determine whether glutamate and glutamate receptor agonists are able to induce membrane depolarization in chromaffin cells. Such data would also help establish whether glutamate receptors in the adrenal gland are similar to central receptors.

The findings that: (a) GluR subunits are expressed in the adrenal medulla and (b) NMDA and other glutamate receptor agonists induce catecholamine secretion from chromaffin cells, suggest that glutamate signaling is involved in mediating catecholamine secretion in the adrenal gland. This is further supported by *in vivo* studies, which are discussed by Jezova and Schwendt. Glutamate receptor subunits are expressed in the adrenal cortex as well, but studies examining the direct effect of glutamate on adrenocortical cells are lacking. Overall, the exact physiological significance of glutamate signaling in the adrenal gland remains to be evaluated and the source of glutamate responsible for activating these peripheral receptors will have to be determined. The concentration of circulating glutamate is relatively high. In bovine serum, the glutamate concentration is 3.4 ± 0.2 mM (Gonzalez *et al.*, 1998). It could be argued that such concentrations of glutamate would be sufficient to saturate peripheral glutamate receptors, arguing against a physiological role. However, the true local concentration of glutamate within the adrenal gland is not known.

Both the adrenal cortex and medulla are richly innervated. Perhaps these innervations release glutamate along with other neurotransmitters. Sympathetic preganglionic neurons that send axons to the adrenal medulla receive synapses that are immunoreactive for glutamate (Llewellyn-Smith *et al.*, 1992). Other possible sources of glutamate include chromaffin cells, and cells of the immune system and of the vascular wall, which secrete diverse corticotropic factors involved in the paracrine control of adrenocortical cell activity. Medullary ganglion cells or cells from the adrenal cortex may supply the glutamate acting in the medulla. In the rat, 92.6% of the total blood flow reaches the medulla after having irrigated the cortex (Pignatelli *et al.*, 1998). In addition to glutamate, other substances may be acting on adrenal glutamate receptors to mediate adrenal hormone secretion. Histogranin is abundantly expressed and released in bovine adrenal medulla upon cholinergic stimulation and has been shown to possess NMDA receptor antagonist activity (Lemaire *et al.*, 1993). It displaces NMDA receptor ligand binding in rat brain homogenates (Lemaire *et al.*, 1993; Shukla *et al.*, 1995), blocks NMDA-induced convulsions in mice (Lemaire *et al.*, 1993; Shukla *et al.*, 1995) and prevents NMDA-mediated spinal hyperexcitability (Hama *et al.*, 1999). The release of histogranin from perfused bovine adrenal glands after bolus stimulation with carbamylcholine is observed only after the release of other granule constituents such as catecholamines and enkephalins (Lemaire *et al.*, 1995). It may thus be part of a mechanism that protects adrenal cells against the overstimulation of NMDA receptors.

4. Conclusion

The physiological roles of peripheral glutamate receptors in the endocrine pancreas and the adrenal gland have not yet been elucidated. Future studies investigating the function of glutamate signaling in these tissues will have to focus on the localization and pharmacology of the glutamate receptors being expressed, the source of glutamate responsible for activating these receptors, the expression of other proteins known to be involved in glutamate signaling, and the physiological and pathophysiological conditions under which these receptors are activated. If peripheral glutamate receptors have similar structural and pharmacological properties as those in the CNS, a better understanding of their function is required if glutamate receptor antagonists are to be administered for the treatment of neurodegenerative diseases such as stroke (Herman, 2002). It may be possible, or indeed necessary, to develop glutamate receptor antagonists which target only brain-specific glutamate receptors to avoid side-effects in endocrine tissues. On the other hand, glutamate receptor antagonists targeted to peripheral sites may have a therapeutic role. Abnormal expression of adrenal receptors for neuropeptides and neurotransmitters appear to be involved in adrenal disorders such as Cushing's syndrome and hyperaldosteronism (Lacroix *et al.*, 2001). If such is the case with glutamate receptors, then the use of glutamate receptor antagonists may be an effective pharmacological therapy for such disorders. Furthermore, type II diabetics, who show tendencies towards hyperglycemia, may have increased pancreatic sensitivity to glutamate and/or glutamate receptor antagonists. Also of interest is the observation that Rasmussen's encephalitis may be an autoimmune disorder in which autoantibodies to GluR3 activate neuronal AMPA receptors causing brain damage (Twyman *et al.*, 1995). It is possible that autoantibodies to glutamate receptors could also be involved in some forms of diabetes. Answers to these many questions should be forthcoming with the increased interest in peripheral glutamate receptors.

References

Afework, M., A. Tomlinson, and G. Burnstock (1994). Distribution and colocalization of nitric oxide synthase and NADPH-diaphorase in adrenal gland of developing, adult and aging Sprague-Dawley rats. *Cell Tissue Res.* **276**(1), 133–142.

Ahonen, M., T.H. Joh, J.-Y. Wu, and O. Happola (1989). Immunocytochemical localization of L-glutamate decarboxylase and catecholamine-synthesizing enzymes in the retroperitoneal sympathetic tissue of the newborn rat. *J. Auton. Nerv. Syst.* **26**(2), 89–96.

Anegawa, N.J., D.R. Lynch, T.A. Verdoorn, and D.B. Pritchett (1995). Transfection of N-methyl-D-aspartate receptors in a non-neuronal cell line leads to cell death. *J. Neurochem.* **64**(5) 2004–2012.

Assan, R., J.R. Attali, G. Ballerio, J. Boillot, and J.R. Girard (1977). Glucagon secretion induced by natural and artificial amino acids in the perfused rat pancreas. *Diabetes* **26**(4), 300–307.

Aunis, D. and K. Langley (1999). Physiological aspects of exocytosis in chromaffin cells of the adrenal medulla. *Acta. Phsyiol. Scand.* **167**(2), 89–97.

Belloni, A.S., P.G. Andreis, V. Macchi, G. Gottardo, L.K. Malendowicz, and G.G. Nussdorfer (1998). Distribution and functional significance of angiotensin-II AT1- and AT2-receptor subtypes in the rat adrenal gland. *Endrocr. Res.* **24**(1), 1–15.

Bertrand, G., R. Gross, R. Puech, M.M. Loubatieres-Mariani, and J. Bockaert (1992). Evidence for a glutamate receptor of the AMPA subtype which mediates insulin release from rat perfused pancreas. *Br. J. Pharmacol.* **106**(2), 354–359.

Bertrand, G., R. Gross, R. Puech, M.M. Loubatieres-Mariani, and J. Bockaert (1993). Glutamate stimulates glucagon secretion via an excitatory amino acid receptor of the AMPA subtype in rat pancreas. *Eur. J. Pharmacol.* **237**(1), 45–50.

Bertrand, G., R. Puech, M.M. Loubatieres-Mariani, and J. Bockaert (1995). Glutamate stimulates insulin secretion and improves glucose tolerance in rats. *Am. J. Physiol.* **269**(3 Pt 1), E551–E556.

Black, M.A., R. Tremblay, G. Mealing, R. Ray, J.P. Durkin, J.F. Whitfield *et al.* (1995). N-Methyl-D-aspartate- or glutamate-mediated toxicity in cultured rat cortical neurons is antagonized by FPL 15896AR. *J. Neurochem.* **65**(5), 2170–2177.

Black, M.A., R. Tremblay, G.A.R. Mealing, J.P. Durkin, J.F. Whitfield, and P. Morley (1996). The desglycinyl metabolite of remacemide hydrochloride is neuroprotective in cultured rat cortical neurons. *J. Neurochem.* **66**(3), 989–995.

Blaschke, M., B.U. Keller, R. Rivosecchi, M. Hollmann, S. Heinemann, and A. Konnerth (1993). A single amino acid determines the subunit-specific spider toxin block of α-amino-3-hydroxy-5-methylisoxazole-4-propionate/kainate receptor channels. *Proc. Natl. Acad. Sci. USA* **90**(14), 6528–6532.

Boksa, P. (1990). Dopamine release from bovine adrenal medullary cells in culture. *J. Auton. Nerv. Syst.* **30**(1), 63–74.

Brice, N.L., A. Varadi, S.J.H. Ashcroft, and E. Molnar (2002). Metabotropic glutamate and GABAB receptors contribute to the modulation of glucose-stimualted insulin secretion in pancreatic beta cells. *Diabetologia* **45**(2), 242–252.

Chenu, C., C.M. Serre, C. Raynal, B. Burt-Pichat, and P.D. Delmas (1998). Glutamate receptors are expressed by bone cells and are involved in bone resorption. *Bone* **22**(4), 295–299.

Choi, D.W. (1987). Ionic dependence of glutamate neurotoxicity. *J. Neurosci.* **7**(2), 369–379.

Choi, D.W. (1988). Glutamate neurotoxicity and diseases of the nervous system. *Neuron* **1**(8), 623–634.

Conn, P.J. and J.-P. Pin (1997). Pharmacology and functions of metabotropic glutamate receptors. *Annu. Rev. Pharmacol. Toxicol.* **37**, 205–237.

Crumrine, R.C., G. Dubyak, and J.C. LaManna (1990). Decreased protein kinase C activity during cerebral ischemia and after reperfusion in the adult rat. *J. Neurochem.* **55**(6), 2001–2007.

Dingledine, R., K. Borges, D. Bowie, and S.F. Traynelis (1999). The glutamate receptor ion channels. *Pharmacol. Rev.* **51**(1), 7–61.

Dingledine, R. and P.J. Conn (2000). Peripheral glutamate receptors: Molecular biology and role in taste sensation. *J. Nutr.* **130**(4S Suppl.), 1039S–1042S.

Domanska-Janik, K. and T. Zalewska (1992). Effect of brain ischemia on protein kinase C. *J. Neurochem.* **58**(4), 1432–1439.

Dubinsky, J.M. and S.M. Rothman (1991). Intracellular calcium concentrations during "chemical hypoxia" and excitotoxic neuronal injury. *J. Neurosci.* **11**(8), 2545–2551.

Durkin, J.P., R. Tremblay, A. Buchan, J. Blosser, B. Chakravarthy, G. Mealing *et al.* (1996). An early loss in membrane protein kinase C activity precedes the excitatory amino acid-induced death of primary cortical neurons. *J. Neurochem.* **66**(3), 951–962.

Durkin, J.P., R. Tremblay, B. Chakravarthy, G. Mealing, P. Morley, D. Small *et al.* (1997). Evidence that the early loss of membrane protein kinase C is a necessary step in the excitatory amino acid-induced death of primary cortical neurons. *J. Neurochem.* **68**(4), 1400–1412.

Engeland, W.C. (1998). Functional innervation of the adrenal cortex by the splanchnic nerve. *Horm. Metab. Res.* **30**(6–7), 311–314.

Erdo, S.L. (1991). Excitatory amino acid receptors in the mammalian periphery, *Trends Pharmacol. Sci.* **12**(11), 426–429.

Genever, P.G., S.J. Maxfield, G.D. Kennovin, J. Maltman, C.J. Bowgen, M.J. Raxworthy *et al.* (1999). Evidence for a novel glutamate-mediated signaling pathway in keratinocytes. *J. Invest. Dermatol.* **112**(3), 337–342.

Gill, S.S., R.W. Mueller, P.F. McGuire, and O.M. Pulido (2000). Potential target sites in peripheral tissues for excitatory neurotransmission and excitotoxicity. *Toxicol. Pathol.* **28**(2), 277–284.

Gill, S.S. and O.M. Pulido (2001). Glutamate receptors in peripheral tissues: Current knowledge, future research, and implications for toxicology. *Toxicol. Pathol.* **29**(2), 208–223.

Gill, S.S., O.M. Pulido, R.W. Mueller, and P.F. McGuire (1998). Potential target/effector sites for excitatory neurotransmission and excitotoxicity, *Soc. Neurosci.* **24**, 341.

Gill, S.S., O.M. Pulido, R.W. Mueller, and P.F. McGuire (1999). Immunochemical localization of the metabotropic glutamate receptors in the rat heart. *Brain Res. Bull.* **48**(2), 143–146.

Gonoi, T., N. Mizuno, N. Inagaki, H. Kuromi, Y. Seino, J.I. Miyazaki *et al.* (1994). Functional neuronal ionotropic glutamate receptors are expressed in the non-neuronal cell line MIN6. *J. Biol. Chem.* **269**(25), 16989–16992.

Gonzalez, M.P., M.T. Herrero, S. Vicente and M.J. Oset-Gasque (1998). Effect of glutamate receptor agonists on catecholamine secretion in bovine chromaffin cells. *Neuroendocrinology* **67**(3), 181–189.

Gonzalez-Cadavid, N.F., I. Ryndin, D. Vernet, T.R. Magee, and J. Rajfer (2000). Presence of NMDA receptor subunits in the male lower urogenital tract. *J. Androl.* **21**(4), 566–78.

Gu, Y. and S.J. Publicover (2000). Expression of functional metabotropic glutamate receptors in primary cultured rat osteoblasts. Cross-talk with N-methyl-D-aspartate receptors. *J. Biol. Chem.* **275**(44), 34252–34259.

Hama, A.T., J.B. Siegan, U. Herzberg, and J. Sagen (1999). NMDA-induced spinal hypersensitivity is reduced by naturally derived peptide analog [Ser1]histogranin. *Pharmacol. Biochem. Behav.* **62**(1), 67–74.

Herman B. (ed.) (2002). *Glutamate and Addiction.* (Humana Press, New Jersey).

Hinoi, E., S. Fujimori, Y. Nakamura, V. Balcar, K. Kubo, K. Ogita *et al.* (2002). Constitutive expression of heterologous N-methyl-D-aspartate receptor subunits in rat adrenal medulla. *J. Neurosci. Res.* **68**(1), 36–45.

Hinoi, E., K. Ogita, Y. Takeuchi, H. Ohashi, T. Maruyama, and Y. Yoneda (2001). Characterization with [³H]quisqualate of group I metabotropic glutamate receptor subtype in rat central and peripheral excitable tissues. *Neurochem. Int.* **38**(3), 277–285.

Hinoi, E. and Y. Yoneda (2001). Expression of GluR6-7 subunits of kainate receptors in rat adenohypophysis. *Neurochem. Int.* **38**(6), 539–547.

Holgert, H., K. Aman, C. Cozzari, B.K. Hartman, S. Brimijoin, P. Emson *et al.* (1995). The cholinergic innervation of the adrenal gland and its relation to enkephalin and nitric oxide synthase. *Neuroreport* **6**(18), 2576–2580.

Houchi, H., M. Azuma, M. Yoshizumi, T. Tamaki, and K. Minakuchi (1999). Possible role of bradykinin on stimulus-secretion coupling in adrenal chromaffin cells. *J. Med. Invest.* **46**(1–2), 1–9.

Howell, J.A., A.D. Matthews, K.C. Swanson, D.L. Harmon, and J.C. Matthews (2001). Molecular identification of high-affinity glutamate transporters in sheep and cattle forestomach, intestine, liver, kidney, and pancreas. *J. Anim. Sci.* **79**(5), 1329–1336.

Inagaki, N., H. Kuromi, T. Gonoi, Y. Okamoto, H. Ishida, Y. Seino *et al.* (1995). Expression and role of ionotropic glutamate receptors in pancreatic islet cells. *FASEB J.* **9**(8), 686–691.

Isa, T., M. Iino, and S. Ozawa (1996). Spermine blocks synaptic transmission mediated by Ca²⁺-permeable AMPA receptors. *NeuroReport* **7**(3), 689–692.

Ishii, T., K. Moriyoshi, H. Sugihara, K. Sakurada, H. Captain, M. Yokoi *et al.* (1993). Molecular characterization of the family of N-methyl-D-aspartate receptor subunits. *J. Biol. Chem.* **268**(4), 2836–2843.

Jacobson, L. and R. Sapolsky (1991). The role of the hippocampus in feedback regulation of the hypothalamic-pituitary-adrenocortical axis. *Endocr. Rev.* **12**(2), 118–134.

Keller, B.U., M. Blaschke, R. Rivosecchi, M. Hollmann, S.F. Heinemann, and A. Konnerth (1993). Identification of a subunit-specific antagonist of α-amino-3-hydroxy-5-methyl-4-isoxazolepropionate/kainate receptor channels. *Proc. Natl. Acad. Sci. USA* **90**(2), 605–609.

Kleitman, N. and M.A. Holzwarth (1985). Catecholaminergic innervation of the rat adrenal cortex. *Cell Tissue Res.* **241**(1), 139–147.

Kondo, H. (1985). Immunohistochemical analysis of the localization of neuropeptides in the adrenal gland. *Arch. Histo. Jap.* **48**(5), 453–481.

Kristensen, P. (1993). Differential expression of AMPA glutamate receptor mRNAs in the rat adrenal gland. *FEBS Lett.* **322**(1–2), 14–18.

Lacroix, A., N. N'Diaye, J. Tremblay, and P. Hamet (2001). Ectopic and abnormal hormone receptors in adrenal Cushing's syndrome. *Endroc. Rev.* **22**(1), 75–110.

Laketic-Ljubojevic, I., L.J. Suva, F.J.M. Maathuis, D. Sanders, and T.M. Skerry (1999). Functional characterization of N-methyl-D-aspartic acid-gated channels in bone cells. *Bone* **25**(6), 631–637.

Lee, J.-A., Z. Long, N. Nimura, T. Iwatsubo, K. Imai, and H. Homma (2001). Localization, transport, and uptake of D-aspartate in the rat adrenal and pituitary glands. *Arch. Biochem. Biophys.* **385**(2), 242–249.

Lemaire, S., C. Rogers, M. Dumont, V.K. Shukla, C. Lapierre, J. Prasad *et al.* (1995). Histogranin, a modified histone H4 fragment endowed with N-methyl-D-aspartate antagonist and immunostimulatory activities. *Life Sci.* **56**(15), 1233–1241.

Lemaire, S., V.K. Shukla, C. Rogers, I.H., Ibrahim, C. Lapierre, P. Parent *et al.* (1993). Isolation and characterization of histogranin, a natural peptide with NMDA receptor antagonist activity. *Eur. J. Pharmacol.* **245**(3), 247–256.

Lin, Y.J., S. Bovetto, J.M. Carver, and T. Giordano (1996). Cloning of the cDNA for the human NMDA receptor NR2C subunit and its expression in the central nervous system and periphery. *Brain Res. Mol. Brain Res.* **43**(1–2), 57–64.

Lin, W. and S.C. Kinnamon (1999). Physiological evidence for ionotropic and metabotropic glutamate receptors in rat taste cells. *J. Neurophysiol.* **82**(5), 2061–2069.

Lipton, S.A. and P.A. Rosenberg (1994). Excitatory amino acids as a final common pathway for neurologic disorders. *New Engl. J. Med.* **330**(9), 613–622.

Liu, H.-P., S.S.-W. Tay, and S.K. Leong (1997). Localization of glutamate receptor subunits of the α-amino-3-hydroxy-5-methyl-4-isoxazolepropionate (AMPA) type in the pancreas of newborn guinea pigs. *Pancreas* **14**(4), 360–368.

Llewellyn-Smith, I.J., K.D. Phend, J.B. Minson, P.M. Pilowsky, and J.P Chalmers (1992). Glutamate-immunoreactive synapses on retrogradely-labelled sympathetic preganglionic neurons in rat thoracic spinal cord. *Brain Res.* **581**(1), 67–80.

MacDonald, M.J. and L.A. Fahien (2000). Glutamate is not a messenger in insulin secretion. *J. Biol. Chem.* **275**(44), 34025–34027.

Maechler, P. and C.B. Wollheim (1999). Mitochondrial glutamate acts as a messenger in glucose-induced insulin exocytosis. *Nature* **402**(6762), 685–689.

Manfras, B.J., W.A. Rudert, M. Trucco, and B.O. Boehm (1994). Cloning and characterization of a glutamate transporter cDNA from human brain and pancreas. *Biochem. Biophys. Acta* **1195**(1), 185–188.

Meldrum, B. and J. Garthwaite (1990). Excitatory amino acid neurotoxicity and neurodegenerative disease. *Trends Pharmacol. Sci.* **11**(9), 379–387.

Michaels, R.L. and S.M. Rothman (1990). Glutamate neurotoxicity in vitro: Antagonist pharmacology and intracellular calcium concentrations. *J. Neurosci.* **10**(1), 283–292.

Miyazaki, J., K. Araki, E. Yamato, H. Ikegami, T. Asano, T. Shibasaki *et al.* (1990). Establishment of a pancreatic β-cell line: Special reference to expression of glucose transporter isoforms. *Endocrinology* **127**(1), 126–132.

Molnar, E., A. Varadi, A.R.J. McIlhinney, and S.J.H. Ashcroft (1995). Identification of functional ionotropic glutamate receptor proteins in pancreatic β-cells and in islets of Langerhans. *FEBS Lett.* **371**(3), 253–257.

Morhenn, V.B., N.S. Waleh, J.N. Mansbridge, D. Unson, A. Zolotorev, P. Cline *et al.* (1994). Evidence for an NMDA receptor subunit in human keratinocytes and rat cardiocytes. *Eur. J. Pharmacol.* **268**(3), 409–414.

Morley, P., S. MacLean, T.F. Gendron, D.L. Small, R. Tremblay, J.P. Durkin *et al.* (2000). Pharmacological and molecular characterization of glutamate receptors in the MIN6 pancreatic β-cell line. *Neurol. Res.* **22**(4), 379–385.

Moroni, F., S. Luzzi, S. Franchi-Micheli, and L. Zilletti (1986). The presence of N-methyl-D-aspartate-type receptors for glutamic acid in the guinea pig myenteric plexus. *Neurosci. Lett.* **68**(1), 57–62.

Munger, B.L. (1981). Morphological characterization of islet cell diversity. In S.J. Cooperstein and D. Watkins, (eds), *The Islets of Langerhans*. Academic Press, New York, pp. 3–34.

Nankova, B.B. and E.L. Sabban (1999). Multiple signaling pathways exist in the stress-triggered regulation of gene expression for catecholamine biosynthetic enzymes and several neuropeptides in the rat adrenal medulla. *Acta. Physiol. Scand.* **167**(1), 1–9.

Nishikawa, T., K. Morita, K. Kinjo, and A. Tsujimoto (1982). Stimulation of catecholamine release from isolated adrenal glands by some amino acids. *Japan. J. Pharmacol.* **32**(2), 291–297.

Noble, E.P., M. Bommer, D. Liebisch, and A. Herz (1988). H1-histaminergic activation of catecholamine release by chromaffin cells. *Biochem. Pharmacol.* **37**(2), 221–228.

Okada, Y. (1986). In S.L. Erdo and N.G. Bowery (eds), Localization and function of GABA in the pancreatic islets. *GABAergic Mechanisms in the Mammalian Periphery*. Raven Press, New York, pp. 223–240.

O'Shea, R.D., P.D. Marley, L.D. Mercer, and P.M. Bert (1992). Biochemical, autoradiographic and functional studies on a unique glutamate binding site in adrenal gland. *J. Auton. Nerv. Syst.* **40**(1), 71–86.

Ozawa, S., H. Kamiya, and K. Tsuzuki (1998). Glutamate receptors in the mammalian central nervous system. *Prog. Neurobiol.* **54**(5), 581–618.

Patton, A.J., P.G. Genever, M.A. Birch, L.J. Suva, and T.M. Skerry (1998). Expression of an N-methyl-D-aspartate-type receptor by human and rat osteoblasts and osteoclasts suggests a novel glutamate signaling pathway in bone. *Bone* **22**(6), 645–649.

Pignatelli, D., M.M. Magalhaes, and M.C. Magalhaes (1998). Direct effects of stress on adrenocortical function. *Horm. Metab. Res.* **30**(6–7), 464–474.

Pipeleers, D., R. Kiekens and P. In't Veld (1981). In F.M. Ashcroft and S.J.H. Ashcroft (eds), Morphology of the pancreatic B-cells. *Insulin, Molecular Biology to Pathology*. IRL Press, Oxford, pp. 5–31.

Polack, J.M., S.R. Bloom, and P.J. Marangos (1984). In S. Falkner, R. Hakanson, and F. Sundler (eds), Neuron-specific enolase, a marker of neuroendocrine cells. *Evolution and Tumor Pathology of the Neuroendocrine System*, Elsevier, Amesterdam, pp. 433–542.

Rajan, A.S., L. Aguilar-Bryan, D.A. Nelson, G.C. Yaney, W.H. Hsu, D.L. Kunze *et al.* (1990). Ion channels and insulin secretion. *Diabetes Care* **13**(3), 340–363.

Ren, J., H.-Z. Hu, S. Liu, Y. Xia, and J.D. Wood (2000). Glutamate receptors in the enteric nervous system: Ionotropic or metabotropic? *Neurogastroenterol. Mot.* **12**(3), 257–264.

Robertson, B.S., B.E. Satterfield, S.I. Said, and R.D. Dey (1998). N-methyl-D-aspartate receptors are expressed by intrinsic neurons of rat larynx and esophagus. *Neurosci. Lett.* **244**(2), 77–80.

Rothman, S.M. and J.W. Olney (1986). Glutamate and the pathophysiology of hypoxic-ischemic brain damage. *Ann. Neurol.* **19**(2), 105–111.

Rorsman, P., P.-O. Berggren, K. Bokvist, H. Ericson, H. Mohler, C.-G. Ostenson *et al.* (1989). Glucose-inhibition of glucagons secretion involves activation of GABA$_A$-receptor chloride channels. *Nature* **341**(6239), 233–236.

Rorsman, P. and G. Trube (1986). Calcium and delayed potassium currents in mouse pancreatic β cells under voltage clamp conditions. *J. Physiol.* **374**, 531–550.

Said, S.I. (1999). Glutamate receptors and asthmatic airway disease. *Trends Pharmacol. Sci.* **20**(4), 132–134.

Said, S.I., H.I. Berisha, and H. Pkbaz (1996). Excitotoxicity in the lung: N-methyl-D-aspartate-induced nitric oxide-dependent, pulmonary edema is attenuated by vasoactive intestinal peptide and by inhibitors of poly(ADP-ribose) polymerase. *Proc. Natl. Acad. Sci. USA* **93**(10), 4688–4692.

Said, S.I., R.D. Dey, and K. Dickman (2001). Glutamate signaling in the lung. *Trends Pharmacol. Sci.* **22**(7), 344–345.

Satin, L.S. and T.A. Kinard (1998). Neurotransmitters and their receptors in the islets of Langerhans of the pancreas: What messages do acetylcholine, glutamate and GABA transmit? *Endocrine* **8**(3), 213–223.

Schwendt, M. and D. Jezova (2001). Gene expression of NMDA receptor subunits in rat adrenals under basal and stress conditions. *J. Physiol. Pharmacol.* **52**(4), 719–727.

Sehlin, J. (1972). Uptake and oxidation of glutamic acid in mammalian pancreatic islets. *Hormones* **3**(3), 156–166.

Shannon, H.E. and B.D. Sawyer (1989). Glutamate receptors of the N-methyl-D-aspartate subtype in the myenteric plexus of the guinea pig ileum. *J. Pharmacol. Exp. Ther.* **251**(2), 518–523.

Shukla, V.K., S. Lemaire, M. Dumont, and Z. Merali (1995). N-methyl-D-aspartate receptor antagonist activity and phencyclidine-like behavioral effects of the pentadecapeptide, [Ser1]histogranin. *Pharmacol. Biochem. Behav.* **50**(1), 49–54.

Skerry, T.M. and P.G. Genever (2001). Glutamate signaling in non-neuronal tissues. *Trends Pharmacol. Sci.* **22**(4), 174–181.

Small, D.L., P. Morley, and A.M. Buchan (1999). In A. Shuaib (ed.), Glutamate receptor antagonists for the treatment of acute cerebral ischemia. *Management of Acute Stroke*. Martin Dekker, New York, pp. 341–361.

Sommer, B., M. Kohler, R. Sprengel, and P.H. Seeburg (1991). RNA editing in brain controls a determinant of ion flow in glutamate-gated channels. *Cell* **67**(1), 11–19.

Stone, T.W. (2000). Development and therapeutic potential of kynurenic acid and kynurenine derivatives for neuroprotection. *Trends Pharmacol. Sci.* **21**(4),149–154.

Storto, M., U. de Grazia, G. Battaglia, M.P. Felli, M. Maroder, A. Gulino *et al.* (2000). Expression of metabotropic glutamate receptors in murine thymocytes and thymic stromal cells. *J. Neuroimmunol.* **109**(2), 112–120.

Sugiyama, H., I. Ito, and C. Hirono (1987). A new type of glutamate receptor linked to inositol phospholipid metabolism. *Nature* **325**(6104), 531–533.

Sureda, F., A. Copani, V. Bruno, T. Knopfel, G. Meltzger, and F. Nicoletti (1997). Metabotropic glutamate receptor agonists stimulate polyphosphoinositide hydrolysis in primary cultures of rat hepatocytes. *Eur. J. Pharmacol.* **338**(2), R1–R2.

Taverna, F.A., B.-R. Cameron, D.L. Hampson, L.-Y. Wang, and J.F. MacDonald (1994). Sensitivity of AMPA receptors to pentobarbital. *Eur. J. Pharm.* **267**(3), R3–R5.

Teitelman, G. and J.K. Lee (1987). Cell lineage analysis of pancreatic islet cell development: Glucagons and insulin cells arise from catecholaminergic precursors present in the pancreatic duct. *Dev. Biol.* **121**(2), 454–466.

Tong, Q., R. Ouedraogo and A.L. Kirchgessner (2002). Localization and function of group III metabotropic glutamate receptors in rat pancreatic islets. *Am. J. Physio. Endocrinol. Metab.* **282**(6), E1324–E1333.

Twyman, R.E., L.C. Gahring, J. Spiess, and S.W. Rogers (1995). Glutamate receptor antibodies activate a subset of receptors and reveal an agonist binding site. *Neuron* **14**(4), 755–762.

Tymianski, M., M.P. Charlton, P.L. Carlen, and C.H. Tator (1993). Source specificity of early calcium neurotoxicity in cultured embryonic spinal neurons. *J. Neurosci.* **13**(5), 2085–2104.

Wang, Y., D.L. Small, D.B. Stanimirovic, P. Morley, and J.P. Durkin (1997). AMPA receptor-mediated regulation of a G_i-protein in cortical neurons. *Nature* **389**(6650), 502–504.

Watanabe, M., M. Mishina, and Y. Inoue (1994). Distinct gene expression of the N-methyl-D-aspartate receptor channel subunit in peripheral neurons of the mouse sensory ganglia and adrenal gland. *Neurosci. Lett.* **165**(1–2), 183–186.

Weaver, C.D., V. Gundersen, and T.A. Verdoorn (1998). A high affinity glutamate/aspartate transport system in pancreatic islets of Langerhans modulates glucose-stimulated insulin secretion. *J. Biol. Chem.* **273**(3), 1647–1653.

Weaver, C.D., J.G. Partridge, T.L. Yao, J.M. Moates, W.A. Magnuson, and T.A. Verdoorn (1998). Activation of glycine and glutamate receptors increases intracellular calcium in cells derived from the endocrine pancreas. *Mol. Pharmacol.* **54**(4), 639–646.

Weaver, C.D., T.L. Yao, A.C. Powers, and T.A. Verdoorn (1996). Differential expression of glutamate receptor subtypes in rat pancreatic islets. *J. Biol. Chem.* **271**(22), 12977–12984.

Yamada, K.A. and C.-M. Tang (1993). Benzothiadiazides inhibit rapid glutamate receptor desensitisation and enhance glutamatergic synaptic currents. *J. Neurosci.* **13**(9), 3904–3915.

Yoneda, Y. and K. Ogita (1986). Localization of [^{3}H]glutamate binding sites in rat adrenal medulla. *Brain Res.* **383**(1–2), 387–391.

Yoneda, Y. and K. Ogita (1987). Enhancement of [^{3}H]glutamate binding by N-methyl-D-aspartic acid in rat adrenal. *Brain Res.* **406**(1–2), 24–31.

Yoneda, Y. and K. Ogita (1989). Characterization of quisqualate-sensitive [^{3}H]Glutamate binding activity solubilized from rat adrenal. *Neurochem. Int.* **15**(2), 137–143.

9

Adrenal Glutamate Receptors: A Role in Stress and Drug Addiction?

Daniela Jezova and Marek Schwendt

The relationship between excitatory amino acids and the endocrine system is evident and has been intensively investigated. The main attention has been given to the role of central glutamate neurotransmission in neuroendocrine regulation (Brann, 1995). Brain glutamate receptors are involved in the control of hypothalamic and pituitary hormone release with subsequent activation of peripheral endocrine glands (Oliver *et al.*, 1996). Moreover, endocrine organs are vulnerable to neurotoxic action of excitatory amino acids (EAAs) and administration of glutamate to neonatal rodents results in many endocrine disturbances (Skultetyova *et al.*, 1998). This applies also to the hypothalamic–pituitary–adrenocortical (HPA) axis, which is affected by both exogenous and endogenous EAAs. Much less information is available on possible contribution of peripheral glutamate receptors, which were found to be located also in some endocrine cells (Erdö, 1991; Gill and Pulido, 2001). Of particular interest is the adrenal gland, the "executive part" of the HPA axis, which is involved in several physiological functions.

1. Physiology of the Adrenal Gland

Just to provide basic information, the adrenal gland is a paired organ located in the retroperitoneum at the superior poles of the kidneys. It is composed of two developmentally and functionally distinct parts, namely the outer cortex and the inner medulla.

As it is well known, the adrenal cortex consists of steroidogenic cells arranged into three zones. The superficial zona glomerulosa is producing the mineralocorticoid aldosterone, which acts mainly on the kidney to regulate electrolyte and fluid balance. The largest layer is the zona fasciculata secreting glucocorticoids. In humans, the dominant glucocorticoid is cortisol, while in rodents and some other species the dominant hormone is corticosterone. Glucocorticoids are essential for life-affecting carbohydrate, protein, and lipid metabolism.

Daniela Jezova and Marek Schwendt • Laboratory of Pharmacological Neuroendocrinology, Institute of Experimental Endocrinology, Slovak Academy of Sciences, Bratislava, Slovakia.

Glutamate Receptors in Peripheral Tissue, edited by Santokh Gill and Olga Pulido.
Kluwer Academic / Plenum Publishers, New York, 2005.

They are inevitable in the stress response and also exert immunomodullatory effects. The deepest layer is the zona reticularis that secretes adrenal androgens. Secretion of glucocorticoids and androgens is regulated by pituitary adrenocorticotropic hormone (ACTH); aldosterone is controlled mainly by angiotensin II. Further details can be found in books related to the pathophysiology of the adrenal gland (Margioris and Chrousos, 2001).

The adrenal medulla contains cells of neuroectoderm origin, which include catecholamine-secreting chromaffin cells. These cells are actually postganglionic sympathetic neurons (without axons) with storage granules containing catecholamines, mainly epinephrine and norepinephrine. Release of catecholamines is stimulated by sympathetic terminals from splanchnic nerves. Chromaffin cells also synthesize several peptides, such as neurotensin, substance P, and opioids. Catecholamines exert many cardiovascular and metabolic effects, which are particularly needed in acute stress situations (Jezova et al., 1987).

A very important physiological role of the adrenal gland is its involvement in the stress response. Glucocorticoids and catecholamines represent peripheral limbs of the stress system (Stratakis and Chrousos, 1995; Jezova et al., 1996). As mentioned above, glucocorticoids are "executive" hormones of the HPA axis (Figure 9.1). The activation of the HPA axis is initiated

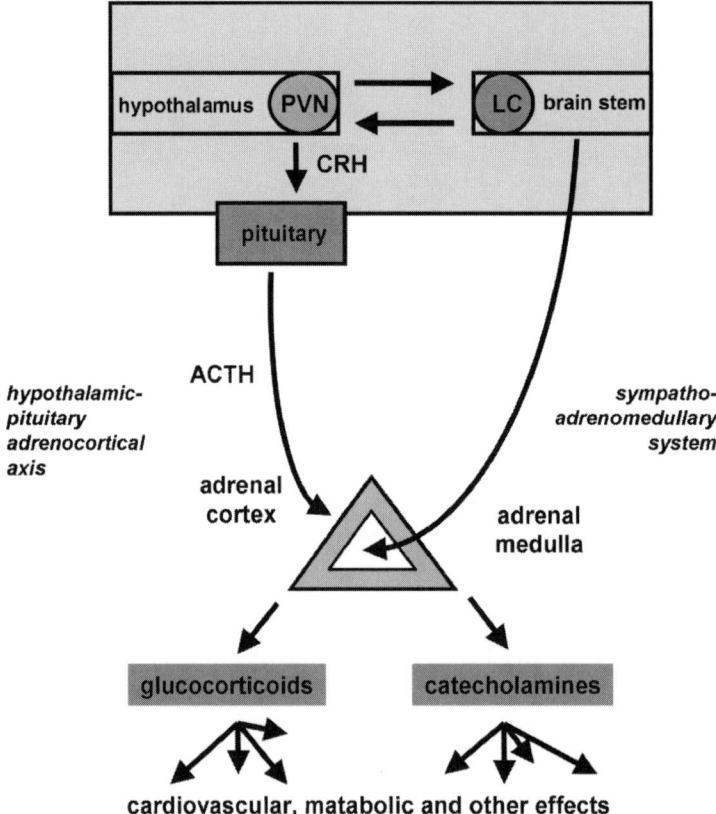

Figure 9.1.   Adrenal hormones involved in stress and their major regulatory factors. ACTH—adrenocorticotropic hormone, PVN—paraventricular nucleus, LC—locus coeruleus.

by enhanced release of hypothalamic regulatory factors, particularly that of corticotropin-releasing hormone (CRH). The main source of CRH in the brain is hypothalamic paraventricular nucleus (PVN) and it is believed to be an integrative center of the stress response. CRH release into the median eminence is followed by pituitary ACTH release and finally secretion of glucocorticoids from the adrenal cortex. Glucocorticoids are thought to play a crucial role in stress and its negative consequences.

Catecholamines released from the adrenal medulla form a part of the sympatho-adrenomedullary system, which is regulated by brain stem nuclei (Figure 9.1). Increased sympathoadrenal activity during stress seems to participate on the development of cardiovascular diseases mainly by catecholamine actions on the heart, vessels, and platelets (Black and Garbutt, 2002).

2. Adrenal Hormones and Excitatory Amino Acids

Peripheral administration of ionotropic glutamate receptor agonists evokes an activation of HPA axis activity (Farah *et al.*, 1991; Jezova *et al.*, 1991) accompanied by increased release of glucocorticoids from the adrenal cortex. Although the glutamate agonists used do not readily cross the blood–brain barrier, their action appears to be centrally mediated. Indeed, peripheral injection of *N*-methyl-D-aspartic acid (NMDA) and other glutamate agonists induce pituitary ACTH release. These data were obtained in experimental animals. In human beings, only ingestion of high doses of glutamate results in a mild activation of cortisol release (Jezova *et al.*, 1995a). So far, there is no evidence on a direct action of glutamate on glucocorticoid secretion at the level of the adrenal cortex.

Catecholamine release is also activated by peripheral administration of glutamate receptor agonists, though higher doses of the drugs are required (Jezova *et al.*, 1991; Yousef *et al.*, 1994). Similarly, administration of a high dose of glutamate to healthy humans resulted in a rise in plasma norepinephrine but not epinephrine levels (Jezova *et al.*, 1995a). In contrast to the adrenal cortex, some evidence is available indicating that glutamate or its agonists may act directly at the level of the adrenal gland. Thus, an old paper by Nishikawa *et al.* (1982) described a stimulation of catecholamine release from isolated dog adrenal glands. On the other hand, glutamate had no effect on epinephrine and norepinephrine secretion from perfused bovine adrenal glands (O'Shea *et al.*, 1992). However, Gonzales *et al.* (1998) reported in a complex study, a significant action of various glutamate receptor agonists on catecholamine secretion in isolated bovine chromaffin cells. Even though the effects observed *in vitro* need not to reflect the action of EAAs in the integrated body *in vivo*, modulation of catecholamine release by glutamate at the level of the adrenal gland is possible.

3. Identification of Glutamate Receptors in the Adrenals

It is clear that the requirement for the action of glutamate within the adrenal gland is the presence of appropriate receptors. First evidence of glutamate-binding sites in the adrenal gland of rats was provided by Yoneda and Ogita (1986, 1987). Glutamate binding was also described in sections of bovine adrenal glands (O'Shea *et al.*, 1992).

The presence of glutamate receptors in adrenal tissue is supported by some molecular biological analyses. Using *in situ* hybridization technique, Watanabe *et al.* (1994) identified gene expression of zeta1 receptor subunit in the adrenal medulla of mice. Levels of mRNA coding for several subunits of 2-amino-3-hydroxy-5-methyl-4-isoxazole propionic acid (AMPA) receptor were localized in the rat adrenal gland (Kristensen, 1993). On the other hand, group 1 metabotropic receptor subunits failed to be demonstrated in the adrenal gland (Hinoi et al., 2001).

First description of gene expression of NMDA receptor subunits in rat adrenals has been provided only recently (Schwendt and Jezova, 2001; Pirnik et al., 2001). The results of a RT-PCR analysis showed that NR1 mRNA, the main NMDA glutamate receptor subunit, was present in the cortex and medulla of the rat adrenal gland, while specific mRNAs coding for NR2A and NR2B failed to be detected in the adrenal tissue. This indicates that the heteromeric NR1/NR2A-B receptors, which are frequently present in the brain, are absent in the adrenals. The presence of NR1 subunit protein in the adrenal gland was confirmed by Western blot analysis (unpublished results). As to the distribution in the adrenal gland, NR1 mRNA was found to be predominantly localized in the adrenal medulla (Schwendt and Jezova, 2001), which is in agreement with the results of previously published results of binding studies (Yoneda and Ogita, 1986). In agreement with the results described by Kristensen et al. (1993), genetic message of the AMPA receptor subunit GluR1 is localized both in the adrenal cortex and medulla (Figure 9.2).

Our data on NMDA receptor subunit gene expression in the adrenal medulla have been independently confirmed by Hinoi et al. (2002a). In a detailed study involving investigation of the expression of NMDA receptor subunit proteins, these authors provided further important information, such as the presence of NR2C and NR2D subunits in adrenal medullary tissue. Thus, heterogenic NMDA receptor channels appear to be constitutively expressed in rat adrenal medulla. Moreover, their data suggest that NR1 subunit in the adrenal medulla may be glycosylated. In contrast to our observation (Schwendt and Jezova, 2001), NR1 mRNA was not detected in the adrenal cortex, which might be due to methodological differences. In the experiments of Hinoi et al. (2002b), RT-PCR technique appears to be used for analysis of whole adrenal tissue, while the distribution of NR1 mRNA levels in the medulla and the cortex was determined by Northern blotting.

To our knowledge, the presence of glutamate receptor message in human adrenal gland has not been reported yet. In a pilot study, we have succeeded to identify NR1 mRNA levels in RNA extract from human adrenal gland (Figure 9.3).

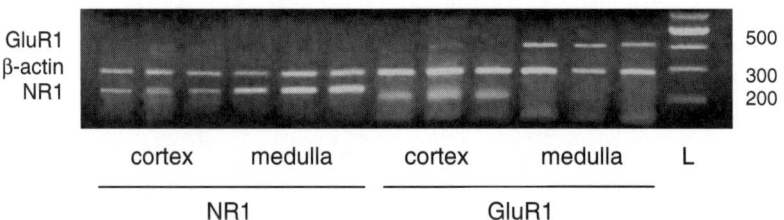

Figure 9.2. Profile of gene expression of the main NMDA (NR1) and AMPA (GluR1) receptor subunits in rat adrenal gland as detected by RT-PCR. Total RNA was extracted from adrenal cortex or medulla and subjected to duplex RT-PCR as described previously (Schwendt and Jezova, 2001). PCR reactions were performed in the presence of two pairs of primers, one for NR1 or GluR1 and another for control gene b-actin. Amplified products were separated on 2% agarose gel, stained with ethidium bromide. Intensity of each band was measured under UV illumination. A 100 bp marker (lane L) was used as a size standard. Numbers on the right indicate size in base pairs.

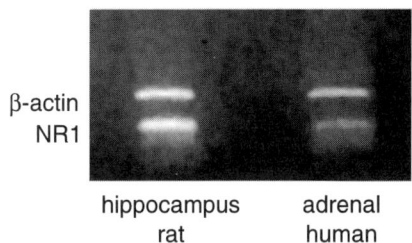

β-actin
NR1

hippocampus   adrenal
rat   human

Figure 9.3. Identification of NR1 mRNA in human adrenal gland by RT-PCR. Total RNA was extracted from whole human adrenal gland and rat hippocampus (used as a control tissue) and subjected to RT-PCR as described in Figure 9.1.

4. Functional Aspects of Adrenal Glutamate Receptors

Although central glutamate receptors are known to be involved in many physiological functions and to contribute to the development of several pathological states, very little information is available on the function of glutamate receptors in peripheral tissues. As the adrenal gland is a part of the HPA axis and thus it is under the control of higher brain centers, we have suggested that adrenal glutamate receptors may play a role in situations, in which central glutamate receptors were shown to have regulatory influence. These situations involve stress response and drug addiction.

4.1. Glutamate Receptors in Stress

Pharmacological blockade of central glutamate receptors results in an attenuation of stress-induced responses of several hormones and mediators, such as ACTH, prolactin, and catecholamines (Jezova et al., 1995b; Tokarev and Jezova, 1997; Zelena et al., 1999). The contribution of glutamatergic neurotransmission to the neuroendocrine response during stress seems to vary with regard to the stress stimulus applied and the stress hormone studied. To reveal the role of central glutamate receptors, a simultaneous blockade of NMDA and AMPA receptors may be required (Zelena et al., 1999). As to the glutamate receptors themselves, we and others have described selective changes in NMDA and AMPA receptor subunits in stress-related regions of the brain after acute and repeated stress exposure (Bartanusz et al., 1995; Fitzgerald et al., 1996; Schwendt and Jezova, 2000).

Our hypothesis that adrenal glutamate receptors participate in the control of hormone release has also been based on pharmacological data. Namely, blockade of NMDA receptors by antagonists, which do not cross the blood–brain barrier, led to an inhibition of stress-induced catecholamine release (Jezova et al., 1995b). These data suggest the involvement of peripheral glutamate receptors.

To evaluate possible changes in adrenal glutamate receptors during stress, we have investigated gene expression of NMDA receptor subunits in adrenal cortex and medulla in rats subjected to an intensive stressor, immobilization for 2 hr (Schwendt and Jezova, 2001). No changes were noticed 3 hr following the stress exposure. However, a significant rise in NR1 mRNA levels in both the adrenal medulla and cortex was observed at a delayed time interval (24 hr).

Possible functional consequences of increased gene expression of the main subunit of NMDA receptor in the adrenal gland remain to be elucidated. No data are available to suggest

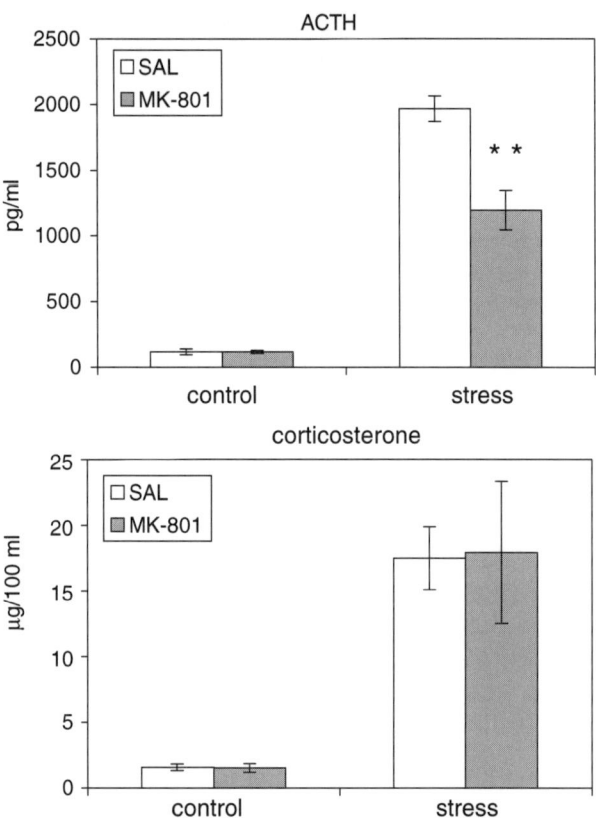

Figure 9.4. Effects of NMDA receptor blockade by MK-801 on stress-induced ACTH and corticosterone levels in plasma. Adult male Sprague–Dawley rats ($n = 7$–8) were cannulated into the tail artery and peritoneum (Jezova *et al.*, 1995b). 24 hr later rats were administered either with saline or MK-801 (racemic form, 1 mg/kg, i.p.). Animals were sacrificed by decapitation under resting conditions (control) and following 20 min of immobilization stress. Plasma ACTH and corticosterone levels were measured by radioimmunoassays (Jezova *et al.*, 1987, 1994). Values represents means $\pm$ SEM. Statistical significance between between SAL and MK-801 animals of the stressed group—$**p < 0.01$.

a role of adrenal glutamate receptors in the adrenal cortex. A very weakly substantiated speculation is that adrenal glutamate receptors may be involved in ACTH-independent modulation of glucocorticoid release. Under several experimental conditions including those using glutamate receptor antagonists (Figure 9.4), changes in corticosterone level do not exactly follow those observed in ACTH. As to the adrenal medulla, glutamate receptors may be associated with catecholamine release. This suggestion is supported by modulation of catecholamine release by glutamate receptor agonists or antagonists both *in vitro* (Nishikawa *et al.*, 1982; Gonzalez *et al.*, 1998) and *in vivo* (Jezova *et al.*, 1995b). However, the onset of stress-induced alteration of NR1 subunit gene expression was delayed and it cannot be related to rapid changes in catecholamine release. We suggest that increase in NR1 expression may participate in modulation of the sensitivity of adrenal tissue to subsequent stress stimuli or to circulating EAAs. In support to this suggestion, stress-induced activation of several transcription factors and *de novo* protein synthesis in rat adrenals were supposed to be mediated by NMDA receptors (Hinoi *et al.*, 2002). Thus, changes in glutamate-receptor gene-expression may belong to molecular mechanisms through which stress exerts long-term effects on adrenal function.

4.2. Glutamate Receptors in Drug Addiction

An increasing body of evidence support the idea that central glutamatergic neurotransmission is involved in the action of drugs of abuse and the development of addictive behavior (Herman *et al.*, 2002). Concerning the dependence to morphine-like drugs, ionotropic receptors of NMDA type seem to be of particular importance. NMDA receptors are thought to be involved in plastic changes in the brain occurring during long-term treatment with opioids and their pharmacological blockade inhibited the development of tolerance, dependence, and withdrawal signs (Mao, 1999). Chronic morphine treatment was found to be associated with changes in gene expression of NMDA receptor subunits in several brain regions (Zhu *et al.*, 1999). Changes in central NMDA receptor gene expression were reported even following an acute administration of morphine (LeGreves *et al.*, 1998). To evaluate possible modulation of adrenal glutamate receptors, we have investigated the effect of a single treatment with morphine on adrenal NR1 mRNA levels in female rats (Pirnik *et al.*, 2001). Interestingly, single injection of morphine resulted in a decrease in NR1 mRNA concentrations in the adrenals as observed 24 hr following the treatment (Figure 9.5). While opioid receptors of μ and κ type were identified in rat adrenal tissue (Wittert *et al.*, 1996), the mechanisms of this effect are completely unknown. Observed down-regulation of NR1 mRNA levels is in agreement with

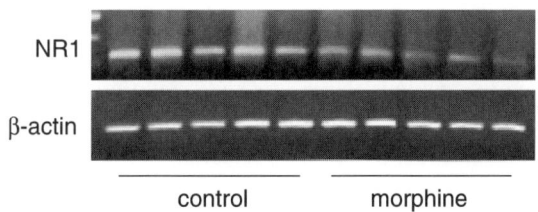

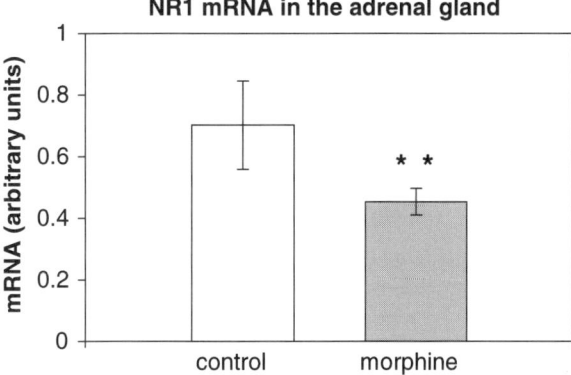

Figure 9.5. mRNA levels of the NR1 subunit of NMDA receptor in the adrenal gland of adult female Sprague–Dawley rats 24 hr after single dose of morphine (10 mg/kg, s.c.). Rats were subjected to handling and s.c. saline (0.9% NaCl) injections for several days before acute treatment with morphine (Pirnik *et al.*, 2001; with permission of *Endocrine Regul.*). NR1 expression was detected by RT-PCR (see Figure 9.1). Values represent means $\pm$ SEM ($n = 6$). Statistical significance compared to control—**$p < 0.01$.

the hypothesis that transient facilitation of NMDA receptor function by morphine may be followed by decreased expression of NMDA receptor subunits, and probably by changes in the number of NMDA receptors in some susceptible regions (LeGreves *et al.*, 1998; Bespalov *et al.*, 2001). Changes in glutamate-receptor gene-expression induced by a single dose of morphine may result in a modulation of adrenal function in response to subsequent exposure to opioids and may contribute to some alterations occurring during morphine dependence. Thus, the presented findings are in favor of an interaction between glutamatergic and opioid systems at the level of the adrenal gland. This possibility is worth further investigation.

5. General Aspects

The most difficult question in relation to the function of glutamate receptors in peripheral tissues is the origin of their endogenous agonists. Glutamatergic innervation of adrenal medulla is uncertain (Llevellyn-Smith *et al.*, 1995). Circulating concentrations of glutamate are high enough to saturate adrenal glutamate receptors. There is no evidence to support the speculation that circulating glutamate may be unable to gain access to adrenal glutamate receptors (Erdö, 1991).

An alternative explanation has been provided by Hinoi *et al.* (2002), who made an attempt to investigate functional characteristics of adrenal NMDA receptors. In the adrenal gland, DNA binding activity of the nuclear transcription factor activator protein-1 failed to be influenced by *in vivo* administration of NMDA, but increased significantly following peripheral injection of the NMDA receptor antagonist (+) -5-methyl-10,11-dihydro-5H-dibenzo [a,d]cyclohepten-5,10-imine (MK-801) (Hinoi *et al.*, 2002). The action of the same treatment with the agonist and/or antagonist was exactly opposite in the brain tissue. The authors have suggested that a sudden interruption and/or blockade could be a signal for peripheral glutamate receptors to be released from a tonic control by circulating glutamate. This hypothesis seems conceivable, but needs further verification. For example, both drugs were used in very high doses and additional mechanisms involved in their action cannot be excluded. Nevertheless, the suggestion that adrenal glutamate receptors are under the influence of tonic stimulation by circulating glutamate is certainly worth consideration.

Acknowledgments

The performance of Western blot analysis of NR1 protein in the adrenal gland by Silke Seeber, Erlangen, Germany, is gratefully acknowledged. These studies were performed within the project of the Centre of Excellence supported by EC (ICA1-CT-2000-70008) and a grant of VEGA 2/2007.

References

Bartanusz, V., J.-M. Aubry, S. Pagliusi, D. Jezova, J. Baffi, and J.Z. Kiss (1995). Stress-induced changes in messenger RNA levels of N-methyl-D-aspartate and AMPA receptor subunits in selected regions of the rat hippocampus and hypothalamus. *Neuroscience* **66**, 247–252.

Bespalov, A.Y., E.E. Zvartau, and P.M. Beardsley (2001). Opioid-NMDA receptor interactions may clarify conditioned (associative) components of opioid analgesic tolerance. *Neurosci. Biobehav. Rev.* **25**, 343–353.

Black, P.H. and L.D. Garbutt (2002). Stress, inflammation and cardiovascular disease. *J. Psychosom. Res.* **52**, 1–23.

Brann, D.W. (1995). Glutamate: A major excitatory transmitter in neuroendocrine regulation. *Neuroendocrinology* **61**, 213–225.

Erdö, E.L. (1991). Excitatory amino acid receptors in the mammalian periphery. *Trends Pharmacol. Sci.* **12**, 426–429.

Farah, J.M.Jr., T.S. Rao, S.J. Mick, K.E. Coyne, and S. Iyengar (1991). N-methyl-D-aspartate treatment increases circulating adrenocorticotropin and luteinizing hormone in the rat. *Endocrinology* **128**, 1875–1880.

Fitzgerald, L.W, J. Ortiz, A.G. Hamedani, and E.J. Nestler (1996). Drugs of abuse and stress increase the expression of GluR1 and NMDAR1 glutamate receptor subunits in the rat ventral tegmental area: Common adaptations among cross-senzting agents. *J. Neurosci.* **16**, 274–282.

Gill, S.S. and O.M. Pulido (2001). Glutamate receptors in peripheral tissues: Current knowledge, future research, and implications for toxicology. *Toxicol. Pathol.* **29**, 208–223.

Gonzalez, M.P., M.T. Herrero, S. Vicente, and M. Oset-Gasque (1998). Effect of glutamate receptor agonists on catecholamine secretion in bovine chromaffin cells. *Neuroendocrinology* **67**, 181–189.

Herman, B.H., J. Frankenheim, R.Y. Litten, P.H. Sheridan, F.F. Weight, and S.R. Yukin (2002). *Glutamate and Addiction*. Humana Press, Totowa, NJ.

Hinoi, E., K. Ogita, Y. Takeuchi, H. Ohashi, T. Maruyama, and Y. Yoneda (2001). Characterization with [^{3}H]quisqualate of group I metabotropic glutamate receptor subtype in rat central and peripheral excitable tissues. *Neurochem. Int.* **38**, 277–285.

Hinoi, E., S. Fujimori, Y. Nakamura, V.J. Balcar, K. Kubo, K. Ogita *et al.* (2002a). Constitutive expression of heterologous N-methyl-D-aspartate receptor subunits in rat adrenal medulla. *J. Neurosci. Res.* **68**, 36–45.

Hinoi, E., S. Fujimori, M. Yoneyama, and Y. Yoneda (2002b). Blockade by N-methyl-D-aspartate of elevation of activator protein-1 binding after stress in rat adrenal gland. *J. Neurosci. Res.* **70**, 161–171.

Jezova, D., R. Kvetnansky, K. Kovacs, Z. Oprsalova, M. Vigas, and G.B. Makara (1987). Insulin-induced hypoglycemia activates the release of adrenocorticotropin predominantly via central and propranolol insensitive mechanisms. *Endocrinology* **120**, 409–415.

Jezova, D., C. Oliver, and J. Jurcovicova (1991). Stimulation of adrenocorticotropin but not prolactin and catecholamine release by N-methyl-aspartic acid. *Neuroendocrinology* **54**, 488–492.

Jezova, D., V. Guillaume, E. Jurankova, P. Carayon, and C. Oliver (1994). Studies of the physiological role of ANF in ACTH regulation. *Endocr. Regul.* **28**, 163–169.

Jezova, D., E. Jurankova, and M. Vigas (1995a). Glutamate neurotransmission, stress and hormone secretion. *Bratisl. Lek. Listy* **96**, 588–596.

Jezova, D., D. Tokarev, and M. Rusnak (1995b). Endogenous excitatory amino acids are involved in stress-induced adrenocorticotropin and catecholamine release. *Neuroendocrinology* **62**, 326–332.

Jezova, D., E. Jurankova, A. Mosnarova, M. Kriska, and I. Skultetyova (1996). Neuroendocrine response during stress with relation to gender differences. *Acta Neurobiol. Exp.* **56**, 779–785.

Kristensen, P. (1993). Different expression of AMPA glutamate receptor mRNA in the rat adrenal gland. *FEBS Lett.* **332**, 183–186.

LeGreves, P., W. Huang, Q. Zhou, M. Thornwall, and F. Nyberg (1998). Acute effects of morphine on the expression of mRNAs for NMDA receptor subunits in the rat hippocampus, hypothalamus and spinal cord. *Eur. J. Pharmacol.* **341**, 161–164.

Llevellyn-Smith T.J., J.B. Minson, P.M. Pilowski, L.F. Arnolda, and J.P. Chalmers (1995). The one hundred percent hypothesis: Glutamate or GABA in synapses on sympathetic preganglion neurons. *Clin. Exp. Hypertens.* **17**, 323–333.

Mao, J.(1999). NMDA and opioid receptors: Their interactions in antinociception, tolerance and neuroplasticity. *Brain Res. Rev.* **30**, 289–304.

Margioris A.N. and G.P. Chrousos (eds.) (2001). *Adrenal Disorders*. Humana Press, Totowa, NJ.

Nishikawa, T., K. Morita, K. Kinjo, and A. Tsujimoto (1982). A stimulation of catecholamine release from isolated adrenal glands by some amino acids. *Jpn. J. Pharmacol*, **32**, 291–297.

Oliver, C., D. Ježová, M. Grino, O. Paulmyer-Lacroix, F. Boudouresque, and P. Joanny (1996). Excitatory amino acids and the hypothalamic-pituitary-adrenal axis. In D.W. Brann and V.B. Mahesh (eds.), *Excitatory Amino Acids. Their Role in Neuroendocrine Function*. CRC Press, New York, pp.167–185.

O'Shea, R.D., P.D. Marley, L.D. Mercer, and P.M. Beart (1992). Biochemical, autoradiographic and functional studies on a unique glutamate binding site in adrenal gland. *J. Autonom. Nerv. Sys.* **40**, 71–86.

Pirnik, Z., M. Schwendt, and D. Jezova (2001). Single dose of morphine influences plasma corticosterone and gene expression of main NMDA receptor subunit in the adrenal gland but not in the hippocampus. *Endocr. Regul.* **35**, 187–193.

Schwendt, M. and D. Jezova (2000). Gene expression of two glutamate receptor subunits in response to repeated stress exposure in rat hippocampus. *Cell. Mol. Neurobiol.* **20**, 319–329.

Schwendt, M. and D. Jezova (2001). Gene expression of NMDA receptor subunits in rat adrenals under basal and stress conditions. *J. Physiol. Pharmacol.* **52**, 719–727.

Skultetyova, I., A. Kiss, and D. Jezova (1998). Neurotoxic lesions induced by monosodium glutamate result in increased adenopituitary proopiomelanocortin gene expression and decreased corticosterone clearance in rats. *Neuroendocrinology* **67**, 412–420.

Stratakis, A.S. and G.P. Chrousos (1995). Neuroendocrinology and pathophysiology of the stress system. *Ann. N. Y. Acad. Sci.* **771**, 1–18.

Tokarev, D. and D. Jezova (1997). Effect of central administration of the non-NMDA receptor antagonist DNQX on ACTH and corticosterone release before and during immobilization stress. *Meth. Find. Exp. Clin. Pharmacol.* **19**, 323–328.

Watanabe, M., M. Mishina, and Y. Imoue (1994). Distinct gene expression of the N-methyl-D-aspartate receptor channel subunit in peripheral neurons of the mouse sensory ganglia and adrenal gland. *Neurosci. Lett.* **165**, 183–186.

Wittert, G., P. Hope, and D. Pyle (1996). Tissue distribution of opioid receptor gene expression in the rat. *Biochem. Biophys. Res. Commun.* **218**, 877–881.

Yoneda, Y. and K. Ogita (1986). Localization of glutamate binding sites in rat adrenal medulla. *Brain Res.* **383**, 387–391.

Yoneda, Y. and K. Ogita (1987). Enhancement of [3H] glutamate binding by N-methyl-D-aspartic acid in rat adrenal. *Brain Res.* **406**, 24–31.

Yousef, K.A., P.G. Tepper, P.E. Molina, N.N. Abumrad, and C.H. Lang (1994). Differential control of glucoregulatory hormone response and glucose metabolism by NMDA and kainite. *Brain. Res.* **634**, 131–140.

Zelena, D., G.B. Makara, and D. Jezova (1999). Simultaneous blockade of two glutamate receptor subtypes (NMDA and AMPA) results in stressor-specific inhibition of prolactin and corticotropin release. *Neuroendocrinology* **69**, 316–323.

Zhu, H., C.G. Jang, T. Ma, S. Oh, R.W. Rockhold, and I.K. Ho (1999). Region specific expression of NMDA receptor NR1 subunit mRNA in hypothalamus and pons following chronic morphine treatment. *Eur. J. Pharmacol.* **365**, 47–54.

10

Glutamate Receptors in the Stomach and their Implications

Li Hsueh Tsai and Jang-yen Wu

1. Introduction

Excitatory amino acids (EAAs), for example, L-glutamate (L-Glu) and L-aspartate (L-Asp), are present in various parts of the vertebrate central nervous system (CNS) and serve as major excitatory neurotransmitters (Bowman and Kimelberg, 1984; Marmo, 1988; Monaghan *et al.*, 1989; Gasic, 1995; Aschner *et al.*, 2001). EAAs play important roles in many nervous systems including the gastrointestinal tract (Tsai *et al.*, 1994, 1999; Kirchgessner *et al.*, 1997; Liu *et al.*, 1997). The glutamate receptors have been shown to be present in the submucous and myenteric plexuses (Liu *et al.*, 1997; Liu and Kirchgessner, 2000; Reis *et al.*, 2000; Ren *et al.*, 2000). Several lines of evidence indicate a role of EAAs in the regulation of gastric motility, secretion, and gastric reflexes. However, the receptor subtypes and mechanisms that mediate the effects of EAAs in the stomach are still poorly understood. In the last few years, investigators have demonstrated that in addition to ionotropic glutamate receptors (iGluR), the enteric nervous system (ENS) also contains functional group I metabotropic glutamate (mGluR) receptors that appear to participate in enteric reflexes. These findings open up an entirely new area to study the roles of EAAs in gastric function and present potential new target sites for drug development.

2. Glutamatergic and Aspartatergic Neurons in the Stomach

Immunohistochemical studies support the notion that glutamatergic and aspartatergic neurons are present in the stomach (Figure 10.1). The distribution of glutamatergic and aspartatergic neurons and their processes in both myenteric ganglia and circular muscle is heterogeneous within the stomach (Tsai *et al.*, 1994, 1999).

Li Hsueh Tsai • Department of Physiology, School of Medicine, Taipei Medical University, Taipei 11014, Taiwan.
Jang-yen Wu • Department of Biomedical Science, Florida Atlantic University, Boca Raton, FL.

Glutamate Receptors in Peripheral Tissue, edited by Santokh Gill and Olga Pulido.
Kluwer Academic / Plenum Publishers, New York, 2005.

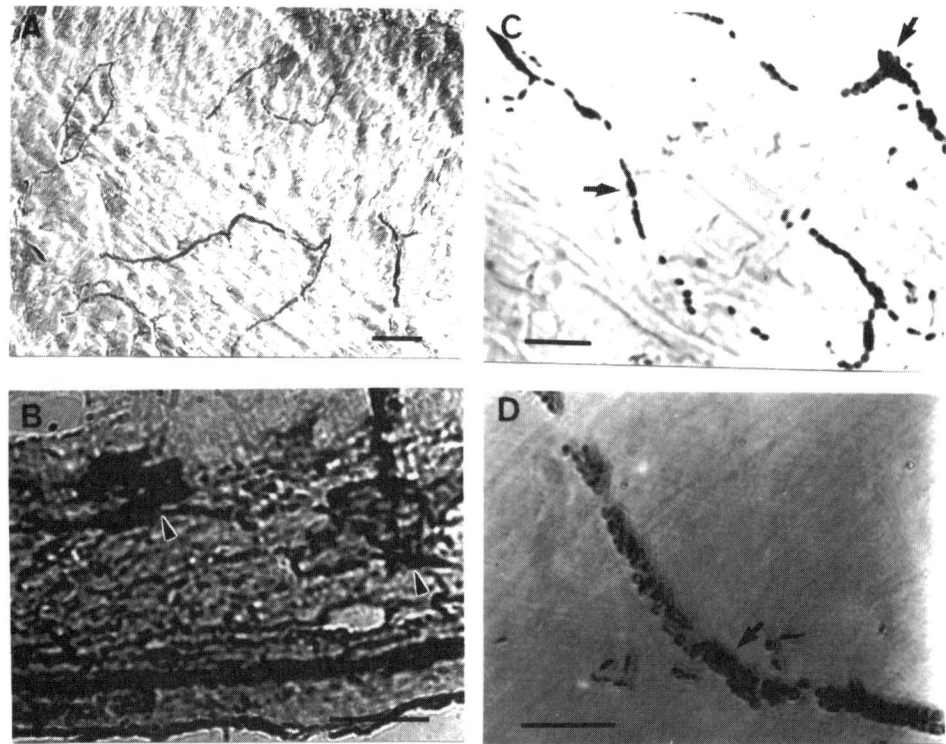

Figure 10.1. Immunohistochemical localization of glutamatergic and aspartatergic neurons in the rat stomach. Glutamate positive processes in the muscle layers (A, B). Aspartate positive processes in the muscle layers (C, D). Bar = 50 μm. The illustrations are modified from Tsai *et al*. (1994, 1999).

Different experimental models clearly suggest that most of the glutamate-containing axons in the intestinal and stomach wall originate from cell bodies within the myenteric and submucous plexus (Stefani *et al*., 1994; Kirchgessner and Liu, 1998). Furthermore, there is evidence suggesting that the gastrointestinal tract is also innervated by extrinsic glutamate-immunoreactive axons. Axon terminals contain large numbers of small and clear synaptic vesicles (Burns *et al*., 1994; Tsai *et al*., 1994; Burns and Stephens, 1995; Cleland and Selverston, 1995, 1997; Panico *et al*., 1995; Johnson and Harris-Warrick, 1997; Krenz *et al*., 2000; Gutovitz *et al*., 2001; Tong *et al*., 2001). Glutamatergic varicosities, which often contain choline acetyltransferase (ChAT) and the vesicular acetylcholine transporter, are apposed to a subset of neuronal cell bodies in the submucous and myenteric plexus. In the stomach, glutamatergic neurons coexpress ChAT (Tong *et al*., 2001). The submucosal neurons that contain ChAT are thought to be the primary afferent neurons that project to the mucosa and respond to the sensory stimuli from the gut (Kirchgessner *et al*., 1992). Thus, some of the enteric glutamatergic neurons are likely to be sensory neurons and glutamate may be involved in the transfer of sensory information from the mucosa to the enteric plexuses.

The synapse in the ENS appears to possess a high-affinity glutamate uptake system. This is compatible with the observation that proteins expressed by epithelia isolated from the rumen showed high-affinity glutamate transporter activity. Excitatory amino acid carrier 1 (proteins EAAC1) is widely distributed in the stomach as shown by immunohistochemical studies (Howell *et al.*, 2001). However, the function of EAAC1 in the stomach still remains poorly understood.

3. Types of Receptor

There are two main classes of receptors for glutamate, namely, iGluR (ligand-gated ion channels) and mGluR (coupled to G proteins) (Ozawa *et al.*, 1998). The iGluR, which can be further divided into NMDA (*N*-methyl-D-aspartate) and non-NMDA receptors, are responsible for fast excitatory synaptic transmission (Bochet and Rossier, 1993; Michaelis, 1998). Both iGluR and mGluR have been shown to localize in enteric ganglia using immunocytochemistry and *in situ* hybridization (Kirchgessner *et al.*, 1997; Larzabal *et al.*, 1999; Ren *et al.*, 2000; Kirchgessner, 2001).

3.1. Ionotropic Glutamate Receptors

Ionotropic glutamate receptors are ligand-gated ion channels comprising of three subtypes, NMDA, α-amino-3-hydroxy-5-methylisoxazole-4-propionic acid (AMPA), and kainate (KA) receptors (Table 10.1).

3.1.1. NMDA Receptors

NMDA receptors are heteromeric pentamers. The subunits are the products of two gene families, one NR1 gene and four NR2 (NR2A-D) genes. Although NR1 provides a functional receptor, it is thought that NR2 increases the activity of the channel. Results from *in situ* hybridization studies have demonstrated the presence of mRNA for NR1 (Burns *et al.*, 1994; Burns and Ritter, 1997) and NMDA (Burns and Stephens, 1995) in both myenteric and submucosal ganglia. These findings are also consistent with the idea that the enteric glutamate receptors are involved in neurogenic motility (Bongianni *et al.*, 1998; Jankovic *et al.*, 1999) or secretion of the stomach (Tsai *et al.*, 1994, 1999). Recently, the NR1 subunits have been localized immunohistochemically in peripheral terminals of primary afferent nerves of dorsal root ganglia (McRoberts *et al.*, 2001). In addition, the peripheral NMDA receptors are also

Table 10.1.   Ionotropic Glutamate Receptor Subtypes

Receptor type	Subunits	Related protein complexes	Agonists	Antagonists
NMDA	NR1	AMPA receptor	NMDA	D-AP5
	NR2A-D	Kainate receptor		MK-801
AMPA	GluR1-4	NMDA receptor	AMPA	CNQX
		Kainate receptor		DNQX
Kainate	GluR5-7	MPA receptor	Kainic acid	CNQX
	KA1	MDA receptor		DNQX
	KA2			

CNQX, 6-cyano-7-nitroquinosaline-2,3-dione, D-AP5, D-2-amino-5-phosphonopentanoic acid, DNQX, 6,7-dinitroquinoxaline-2,3-dione, MK-801, Dizoclipine.

found in both vagal and spinal primary afferent in the stomach, where the function of NMDA receptors is believed to be involved in gastric motility and gastric secretion.

3.1.2. Non-NMDA Receptors

AMPA receptors are mainly involved in mediating fast glutamatergic neurotransmission. There are four subunits, known as GluR1 to GluR4 (see Table 10.1) (also called GluRA to GluRD) (Hollmann and Heinemann, 1994). These AMPA receptor subunits are widely, but differentially, distributed throughout the CNS (Martin et al., 1993a, b). The functional AMPA receptors could be tetrameric or pentameric subunit assemblies.

AMPA receptors mediate the fast component of excitatory postsynaptic currents, whereas the slow component is contributed by NMDA receptors (Stern et al., 1992). The latter can be viewed as coincidence detectors of pre- and postsynaptic activity, since the gating of the integral ion channel requires two close and simultaneous events, namely, presynaptic release of glutamate and depolarization of the postsynaptic membrane. Depolarization is induced by the activation of AMPA receptors. Coincidence detection by the NMDA receptors is rest on their voltage-dependent channels which are blocked by extracellular Mg^{2+}. NMDA receptors are designed for high Ca^{2+} permeability, and Ca^{2+} influx through the NMDA receptor channel is thought to be essential for activity-dependent synaptic modulation (Bliss and Collingridge, 1993).

Other non-NMDA ionotropic receptor subunits are known as GluR5, GluR6, Kal, and KA2. They normally form receptor assemblies which have high affinity for kainate and hence are designated as KA receptors, although some researchers classify GluR5 as an AMPA receptor subunit. KA receptors were previously thought to be mostly presynaptic. In addition, they are expressed in the dorsal root ganglia (Huettner, 1990; Wong and Mayer, 1993; Wong et al., 1994). In contrast to AMPA, which facilitates glutamate release via presynaptic action, kainate acting on presynaptic autoreceptors decreases glutamate release from rat hippocampal synaptosomes and also depresses glutamatergic synaptic transmission (Chittajallu et al., 1996). Recent evidence indicates that non-NMDA receptors are also involved in mediating neurotransmission in the spiny lobster stomatogastric ganglion (STG) (Cleland and Selverston, 1997; Krenz et al., 2000). It is conceivable that different glutamate receptor subtypes in the dorsal vagal complex may activate gastric secretion and motility in a different way. This provides a framework for glutamate receptor diversity in regulating gastric function.

3.2. Metabotropic Glutamate Receptors

The ENS in the stomach contains mGluR as the CNS does (Lang and Ajmal, 1995; Riccardi, 1999; Krenz et al., 2000), which are members of the G-protein-coupled receptor family. At present, eight different mGluR have been cloned, termed from mGluR1 to mGluR8 (Pin and Duvoisin, 1995). Based on their sequence similarities, pharmacology, and signal transduction mechanism, mGluR are classified into three groups. Group I receptors (mGluR1 and mGluR5), which are coupled to phospholipase C, exert their effects by activating protein kinase C and releasing Ca^{2+} from intracellular stores. Group II (mGluR2 and mGluR3) and group III (mGluR4 and mGluR6–mGluR8) receptors are negatively coupled to adenylyl cyclase. Group I receptors generally increase cell excitability by inhibiting K^+ channels and are mostly postsynaptic, although presynaptic effects have also been reported. Group II and III receptors are mostly present in glutamatergic presynaptic terminals and are believed to

Table 10.2. Metabotropic Glutamate Receptor Subtypes

Receptor type	Subtypes	Transduction mechanism	Agonists	Antagonists
Group I mGluR	mGluR 1 mGluR 5	PLC↑	DHPG, Quisqualate CHPG, DHPG, Quisqualate	4-CPG, AIDA, MPEP
Group II mGluR	mGluR 2 mGluR 3	AC↑	L-CCG-I, ACPD, LY354740 NAAG	(RS)-APICA; MCCG; S-MCPG
Group III mGluR	mGluR 4 mGluR 6 mGluR7 mGluR8	AC↓	L-AP4 L-AP4 L-AP4, L-SOP L-AP4	MPPG MCPA MAP4

PLC, phospholipase C; AC, adenylyl cyclase; DHPG, 3,5-dihydroxyphenylglycine; CHPG, (RS)-2-chloro-5-hydroxy-phenylglycine; L-CCG-I, (2S, 1'S, 2'S)-2-(carboxycyclopropyl)glycine; L-AP4, L-(+)-2-amino-4-phosphonobutyrate; L-SOP, L-serine-O-phosphate; AIDA,(RS)-1-aminoindan-1,5-dicarboxylic acid; MPPG, (RS)—methyl-4-phosphono-phemethylserine; DHPMP, 3,5-dihydroxybenzylmethylphosphinate; 4-CPG, (S)-4-carboxy-phenylglycine; MPEP, 2-methyl-6-(phenylethynyl)-pyridine; (RS)APICA, (RS)-1-amino-5-phosphonoindan-1-carboxylic acid; MCCG, (2S, 3S, 4'S)-2-methyl-2(carboxycyclopropyl)glycine.

exert their action by inhibiting neurotransmitter release (Table 10.2). However, the function of mGluR in the stomach is still poorly understood. The subcellular localization of mGluR in enteric neurons might have functional implications in physiology and pathology of the gut.

4. Function and Implications of Glutamate Receptors in the Stomach

4.1. Gastric Smooth Muscle Contraction

Anatomical, physiological, and pharmacological evidence suggests that glutamate is an excitatory neurotransmitter in the peripheral nervous system (Liu *et al.*, 1997). The glutamate receptors in rat bronchi and gut are located on cholinergic nerves prejunctionally to enhance nerve-mediated responses (Aas *et al.*, 1989; Koyuncuoglu *et al.*, 1992). Glutamate immunore-activity has been detected in cholinergic enteric neurons. The immunoreactivity of both NMDA and non-NMDA receptors are also detected in neurons in submucosal and myenteric plexuses (Liu *et al.*, 1997). *In situ* hybridization studies, mRNA coding for the glutamate NMDA receptor is found to be expressed in rat enteric neurons in the stomach. Enteric neurons expressing mRNA for both NMDA receptors and vasoactive intestinal peptide (VIP) are found in the myenteric and submucosal ganglia, suggesting that the mechanism for gluta-matergic excitation is present in VIP-containing enteric neurons (Burns *et al.*, 1994; Burns and Stephens, 1995).

There is a difference in functional roles of different types of EAA receptors between the isolated rat fundus and intestine. EAAs show a powerful stimulating effect on most smooth muscle layers in the stomach. NMDA and kainic acid stimulate contraction of isolated rat gastric fundus with almost identical strength of action, whereas the metabotropic receptor agonist ACPD has no effect (Jankovic *et al.*, 1999). The gastric excitatory motor response are elicited through NMDA and non-NMDA receptors that activate intrinsic excitatory neurons within the wall of rat gastric fundus, while in intestine, it is due to the release of acetylcholine. (Jankovic et al., 1999).

4.2. Modulation of Gastric Acid Secretion

The effect of glutamate on gastric acid secretion has been investigated on an everted preparation of isolated rat stomach. Glutamate or aspartate alone had no effect on acid secretion. The oxotremorine-, histamine-, or gastrin-stimulated acid secretion is markedly reduced by glutamate. Among glutamate receptor agonists, quisqualic acid (QA) is the most potent, followed by kainic acid (KA) and NMDA in inhibiting oxotremorine-stimulated acid secretion. Aspartate and NMDA inhibit the oxotremorine-stimulated acid secretion, which is antagonized by two specific antagonists of NMDA receptors, namely, 2-amino-5-phosphono-valeric acid (AP-5) and ($\pm$)3-(2-carboxypiperazin-4-yl) propyl-1-phosphonic acid (CPP). These findings suggest that glutamate receptors are involved in the modulation of gastric acid secretion via ionotropic QA/KA receptors and aspartate regulates acid secretion in the stomach by inhibiting histamine release through the NMDA receptors (Tsai et al., 1994, 1999). There is also evidence showing a role for enteric iGluR and mGluR in the regulation of acid secretion (Tsai et al., 1994, 1999). Our recent studies demonstrated that EAAs and their analogues such as glutamate, NMDA, QA, and KA were able to protect against cold-restraint stress (CRS)-induced gastric ulcer formation through their interaction with NMDA and non-NMDA receptors (Chen et al., 2001). These results suggest that glutamate receptor agonists can modulate gastric secretion and protect against CRS-induced gastric ulcer.

4.3. Food Intake

Several groups have reported the increase of food intake after treatment with NMDA or non-NMDA receptor agonists or antagonists (Wirtshafter and Krebs, 1990; Stanley et al., 1993; Burns and Ritter, 1997). NMDA receptor-mediated modulation of gastric motor function may be of great importance for the regulation of acceleration in gastric emptying and daily food intake (Covasa et al., 2000).

MK-801 (dizoclipine), an antagonist of NMDA receptors, increased meal size and duration in rats. MK-801 did not increase sham feeding or attenuate reduction of sham feeding by intra-intestinal nutrient infusions. These results suggested that the MK-801-induced increase in meal size was not due to the antagonism of postgastric satiety signals. Consequently, NMDA antagonists might increase food intake by directly antagonizing gastric mechanosensory signals or by accelerating gastric emptying, thereby reducing gastric mechanoreceptive feedback (Covasa et al., 2000).

Although these results are consistent with NMDA receptor-mediated glutamatergic transmission of vagal satiety signals in general, MK-801 lends limited support for such a role in the transmission of specific gastric distension signals (Zheng et al., 1999). NMDA receptors are expressed in intrinsic gastric neurons (Burns et al., 1994), as well as in CNS (Aicher et al., 1999).

Systemically administered MK-801 could enhance gastric emptying through actions via central and/or peripheral mechanism. Nonetheless, several observations of our own, as well as those from other laboratories, strongly support the hypothesis that acceleration of gastric emptying and increased meal size are mediated by NMDA actions in the dorsal vagal complex.

The source(s) of EAA to the dorsal vagal complex and the precise sites where such afferents might increase gastric emptying and meal sizes are unknown. However, the presence of glutamate in primary vagal sensory neurons (Shinozaki *et al.*, 1990) and the existence of NMDA receptors in both axon terminals of vagal sensory neurons and postsynaptic site of these terminals (Aicher *et al.*, 1999) suggest that modulation of vagal motor output could be mediated by adjusting the responsiveness of primary and higher order viscerosensory neurons within the dorsal vagal complex. Such NMDA receptor-mediated modulation of gastric motor function could play an important role in controlling meal size and, hence, controlling daily food intake and ingestive behavior.

4.4. Role of Hyperalgesic Pain and Antinociception

Over the past two decades, there has been increasing evidence supporting a role of EAAs and their receptors in the transmission and production of nociception (pain) (Dickenson *et al.*, 1997). Glutamate as well as glutamate receptors, including iGluR and mGluR were subsequently found to localize in the spinal cord, including the central terminals of primary afferent nociceptors (Miller *et al.*, 1988). Each of these receptors has been implicated in spinal nociceptive transmission because iontophoretic application of NMDA or AMPA receptor agonists enhances spinal dorsal horn neuron responses to noxious stimulation (Aanonsen *et al.*, 1990). Furthermore, spinal administration of NMDA, AMPA, or mGluR agonists produces spontaneous pain behavior and hyperalgesia (Brambilla *et al.*, 1996; Boxall *et al.*, 1998b; Mills and Hulsebosch, 2002).

A role for peripheral glutamate receptors in nociceptive modulation was further demonstrated by the finding that peripheral injection of selective NMDA or AMPA receptor antagonists into an inflamed hindpaw attenuated the hyperalgesia produced by the inflammation (Jackson *et al.*, 1995). These results support a role for peripheral NMDA and non-NMDA receptors in the activation/sensitization of primary afferent nociceptors and development of hyperalgesia following peripheral tissue injury.

The mGluR are particularly important in the transmission of pain information in the nervous system. A specific role for group I mGluR (mGluR1 and mGluR5) in nociceptive processing has been demonstrated by pharmacological, immunohistochemical and *in situ* hybridization studies (Fisher and Coderre, 1998; Boxall *et al.*, 1998b; Jia *et al.*, 1999). In dorsal horn, group I mGluR particularly mGluR1 are also involved in acute nociception as demonstrated by behavioral (Fisher and Coderre, 1996a, b; Fisher *et al.*, 1998) and electrophysiological studies *in vitro* (Boxall *et al.*, 1998a, b) as well as *in vivo* (Neugebauer *et al.*, 1999; Walker *et al.*, 2001a, b). A potent, subtype-selective antagonist of the metabotropic glutamate-5 (mGlu5) receptor, 2-methyl-6-(phenylethynyl)-pyridine (MPEP), has recently been shown to have antihyperalgesic effects in inflammatory pain (Walker *et al.*, 2001a, b). Although progress has been made in localizing these receptors, lack of selective agonists and antagonists has hampered the attempts to elucidate their physiological functions as reviewed by Pin *et al.* (1999). Several lines of evidence indicate the presence of both iGluR and mGluR in myenteric as well as submucosal ganglia of the rat stomach (Burns *et al.*, 1994). Glutamatergic nerve fibers are abundant in ganglionated plexuses and mucosa. Therefore, at least some of the enteric glutamatergic neurons are likely to be sensory neurons and glutamate may be involved in the transfer of sensory information from mucosa to enteric plexuses. Both C fibers and Aδ fibres are parts of the sensory-glutamate branch of the peripheral nervous system. Their axons pass through the dorsal root ganglion, where the cell bodies are located, and then go into the gray matter of the spinal cord where they make synaptic contact with interneurons. Glutamate receptor agonists may stimulate ENS, spinal, and brain components of the nociceptive pathway through peripheral primary afferent neurons (Crawford *et al.*, 2000; Walker *et al.*, 2001a, b). Recently, it has been demonstrated that the intracellular calcium concentration in cultured rat dorsal root ganglion neurons increased by stimulating with glutamate receptor agonists (Coggeshall *et al.*, 1997).

Traditional therapies for chronic pain are based on two well-established classes of analgesics, namely, opioids and non-steroidal anti-inflammatory drugs (NSAID). Both classes of pain-relieving drugs have significant shortcomings, including some undesirable side effects. The use of upload analgesics is limited by their sedative, respiratory and gastrointestinal side effects as well as potential abuse, whereas chronic NSAID use is restricted by serious gastrointestinal or renal side effects. These problems have led to new approaches focusing on the discovery of novel molecular and pharmacological mechanisms of nociceptive modulation that could allow the development of new classes of pain-relieving drugs without such undesirable side effects as described. Glutamate receptors which play an important role in the development and maintenance of hyperalgesia are potential therapeutic targets in pain management and are warranted for further investigation.

5. Conclusions

Half a century of researches on EAAs, substantially accelerated during the last 10 years, has brought forward a detailed knowledge of the distribution and properties of glutamate receptors. With advances in molecular and cellular biology and improvement in immunocytochemical and *in situ* immunohistochemical techniques, a detailed biochemical and morphological characterization of the neuronal systems containing glutamate receptors has been achieved. In spite of the considerable amount of data on the distribution and actions of glutamate receptors in many parts of the nervous system, the function of glutamate receptors in ENS of the stomach is still poorly understood

It is suggested that the iGluR receptors are involved in the modulation of stomach motility. This conclusion is based on the following observations: first, glutamatergic nerve cells and fibers are present in the ENS, second, release of EAAs has been demonstrated from these neurons, and third, EAAs have been shown to have depolarizing and muscle stimulating effects in the smooth muscle of the stomach.

There is also plenty of evidence suggesting a role for enteric iGluR and mGluR in controlling acid secretion and protecting against CRS-induced gastric ulcer formation. Glutamate receptors are involved in the modulation of acid secretion. The recent finding about NMDA receptor-mediated modulation of gastric motor function can be considerably important for the control of accelerating gastric emptying and daily food intake. NMDA receptors are present in the stomach of ENS. Pharmacological manipulation of these receptors would allow for modulation of enteric and vago-vagal reflexes.

The role of glutamate receptors in transmission of nociceptive information has been extensively studied, but the mechanism remains elusive. Recent observation that highly specific glutamate receptor antagonists are potent antinociceptive agents, however, strongly suggests that glutamate is involved in this mechanism. The discoveries of specific glutamate receptor antagonists and agonists have proven to be invaluable in the study of the physiological significance of glutamate in nociceptive system and also provide molecular basis for developing new types of antinociceptive drugs.

Metabotropic glutamate receptors also play a role in enteric reflexes. It is believed that internalization of receptor molecules might be a major mechanism for regulation of mGluR activity. Enteric mGluR may present potential new target sites for the development of gastrointestinal drugs. Further development of more potent and selective mGluR antagonists and agonists is needed in order to understand fully the role of glutamate in the enteric system.

References

Aanonsen, L.M., S. Lei, and G.L. Wilcox (1990). Excitatory amino acid receptors and nociceptive neurotransmission in rat spinal cord. *Pain* **41**, 309–321.

Aas, P., R. Tanso, and F. Fonnum (1989). Stimulation of peripheral cholinergic nerves by glutamate indicates a new peripheral glutamate receptor. *Eur. J. Pharmacol.* **164**, 93–102.

Aicher, S.A., S. Sharma, and V.M. Pickel (1999). N-methyl-D-aspartate receptors are present in vagal afferents and their dendritic targets in the nucleus tractus solitarius. *Neuroscience* **91**, 119–132.

Aschner, M., L. Mutkus, and J.W. Allen (2001). Aspartate and glutamate transport in acutely and chronically ethanol exposed neonatal rat primary astrocyte cultures. *Neurotoxicology* **22**, 601–605.

Bliss, T.V. and G.L. Collingridge (1993). A synaptic model of memory: Long-term potentiation in the hippocampus. *Nature* **361**, 31–39.

Bochet, P. and J. Rossier (1993). Molecular biology of excitatory amino acid receptors: Subtypes and subunits. *Exs* **63**, 224–233.

Bongianni, F., D. Mutolo, M. Carfi, and T. Pantaleo (1998). Area postrema glutamate receptors mediate respiratory and gastric responses in the rabbit. *Neuroreport* **9**, 2057–2062.

Bowman, C.L. and H.K. Kimelberg (1984). Excitatory amino acids directly depolarize rat brain astrocytes in primary culture. *Nature* **311**, 656–659.

Boxall, S.J., A. Berthele, D.J. Laurie, B. Sommer, W. Zieglgansberger, L. Urban *et al.* (1998a). Enhanced expression of metabotropic glutamate receptor 3 messenger RNA in the rat spinal cord during ultraviolet irradiation induced peripheral inflammation. *Neuroscience* **82**, 591–602.

Boxall, S.J., A. Berthele, T.R. Tolle, W. Zieglgansberger, and L. Urban (1998b). mGluR activation reveals a tonic NMDA component in inflammatory hyperalgesia. *Neuroreport* **9**, 1201–1203.

Brambilla, A., A. Prudentino, N. Grippa, and F. Borsini (1996). Pharmacological characterization of AMPA-induced biting behaviour in mice. *Eur. J. Pharmacol.* **305**, 115–117.

Burns, G.A. and R.C. Ritter (1997). The non-competitive NMDA antagonist MK-801 increases food intake in rats. *Pharmacol. Biochem. Behav.* **56**, 145–149.

Burns, G.A. and K.E. Stephens (1995). Expression of mRNA for the N-methyl-D-aspartate (NMDAR1) receptor and vasoactive intestinal polypeptide (VIP) co-exist in enteric neurons of the rat. *J. Auton. Nerv. Syst.* **55**, 207–210.

Burns, G.A., K.E. Stephens, and J.A. Benson (1994). Expression of mRNA for the N-methyl-D-aspartate (NMDAR1) receptor by the enteric neurons of the rat. *Neurosci. Lett.* **170**, 87–90.

Chen, S.H., H.L. Lei, L.R. Huang, and L.H. Tsai (2001). Protective effect of excitatory amino acids on cold-restraint stress-induced gastric ulcers in mice: Role of cyclic nucleotides. *Dig. Dis. Sci.* **46**, 2285–2291.

Chittajallu, R., M. Vignes, K.K. Dev, J.M. Barnes, G.L. Collingridge, and J.M. Henley (1996). Regulation of glutamate release by presynaptic kainate receptors in the hippocampus. *Nature* **379**, 78–81.

Cleland, T.A. and A.I. Selverston (1995). Glutamate-gated inhibitory currents of central pattern generator neurons in the lobster stomatogastric ganglion. *J. Neurosci.* **15**, 6631–6639.

Cleland, T.A. and A.I. Selverston (1997). Dopaminergic modulation of inhibitory glutamate receptors in the lobster stomatogastric ganglion. *J. Neurophysiol.* **78**, 3450–3452.

Coggeshall, R.E., S. Zhou, and S.M. Carlton (1997). Opioid receptors on peripheral sensory axons. *Brain Res.* **764**, 126–132.

Covasa, M., R.C. Ritter, and G.A. Burns (2000). NMDA receptor participation in control of food intake by the stomach. *Am. J. Physiol. Regul. Integr. Comp. Physiol.* **278**, R1362–1368.

Crawford, J.H., A. Wainwright, R. Heavens, J. Pollock, D.J. Martin, R.H. Scott *et al.* (2000). Mobilisation of intra-cellular Ca2+ by mGluR5 metabotropic glutamate receptor activation in neonatal rat cultured dorsal root gan-glia neurones. *Neuropharmacology* **39**, 621–630.

Dickenson, A.H., V. Chapman, and G.M. Green (1997). The pharmacology of excitatory and inhibitory amino acid-mediated events in the transmission and modulation of pain in the spinal cord. *Gen. Pharmacol.* **28**, 633–638.

Fisher, K. and T.J. Coderre (1996a). Comparison of nociceptive effects produced by intrathecal administration of mGluR agonists. *Neuroreport* **7**, 2743–2747.

Fisher, K. and T.J. Coderre (1996b). The contribution of metabotropic glutamate receptors (mGluRs) to formalin-induced nociception. *Pain* **68**, 255–263.

Fisher, K. and T.J. Coderre (1998). Hyperalgesia and allodynia induced by intrathecal (RS)- dihydroxyphenylglycine in rats. *Neuroreport* **9**, 1169–1172.

Fisher, K., M.E. Fundytus, C.M. Cahill, and T.J. Coderre (1998). Intrathecal administration of the mGluR compound, (S)-4CPG, attenuates hyperalgesia and allodynia associated with sciatic nerve constriction injury in rats. *Pain* **77**, 59–66.

Gasic, G. (1995). Systems and molecular genetic approaches converge to tackle learning and memory. *Neuron* **15**, 507–512.

Gutovitz, S., J.T. Birmingham, J.A. Luther, D.J. Simon, and E. Marder (2001). GABA enhances transmission at an excitatory glutamatergic synapse. *J. Neurosci.* **21**, 5935–5943.

Hollmann, M. and S. Heinemann (1994). Cloned glutamate receptors. *Annu. Rev. Neurosci.* **17**, 31–108.

Howell, J.A., A.D. Matthews, K.C. Swanson, D.L. Harmon, and J.C. Matthews (2001). Molecular identification of high-affinity glutamate transporters in sheep and cattle forestomach, intestine, liver, kidney, and pancreas. *J. Anim. Sci.* **79**, 1329–1336.

Huettner, J.E. (1990). Glutamate receptor channels in rat DRG neurons: Activation by kainate and quisqualate and blockade of desensitization by Con A. *Neuron* **5**, 255–266.

Jackson, D.L., C.B. Graff, J.D. Richardson, and K.M. Hargreaves (1995). Glutamate participates in the peripheral modulation of thermal hyperalgesia in rats. *Eur. J. Pharmacol.* **284**, 321–325.

Jankovic, S.M., D. Milovanovic, M. Matovic, and V. Iric-Cupic (1999). The effects of excitatory amino acids on iso-lated gut segments of the rat. *Pharmacol. Res.* **39**, 143–148.

Jia, H., A. Rustioni, and J.G. Valtschanoff (1999). Metabotropic glutamate receptors in superficial laminae of the rat dorsal horn. *J. Comp. Neurol.* **410**, 627–642.

Johnson, B.R. and R.M. Harris-Warrick (1997). Amine modulation of glutamate responses from pyloric motor neu-rons in lobster stomatogastric ganglion. *J. Neurophysiol.* **78**, 3210–3221.

Kirchgessner, A.L. (2001). Glutamate in the enteric nervous system. *Curr. Opin. Pharmacol.* **1**, 591–596.

Kirchgessner, A.L. and M.T. Liu (1998). Immunohistochemical localization of nicotinic acetylcholine receptors in the guinea pig bowel and pancreas. *J. Comp. Neurol.* **390**, 497–514.

Kirchgessner, A.L., M.T. Liu, and F. Alcantara (1997). Excitotoxicity in the enteric nervous system. *J. Neurosci.* **17**, 8804–8816.

Kirchgessner, A.L., H. Tamir, and M.D. Gershon (1992). Identification and stimulation by serotonin of intrinsic sen-sory neurons of the submucosal plexus of the guinea pig gut: Activity-induced expression of Fos immunoreac-tivity. *J Neurosci.* **12**, 235–248.

Koyuncuoglu, H., Y. Uresin, Y. Esin, and F. Aricioglu (1992). Morphine and naloxone act similarly on glutamate-caused guinea pig ileum contraction. *Pharmacol. Biochem. Behav.* **43**, 479–482.

Krenz, W.D., D. Nguyen, N.L. Perez-Acevedo, and A.I. Selverston (2000). Group I, II, and III mGluR compounds affect rhythm generation in the gastric circuit of the crustacean stomatogastric ganglion. *J. Neurophysiol.* **83**, 1188–1201.

Lang, C.H. and M. Ajmal (1995). Metabolic, hormonal, and hemodynamic changes induced by metabotropic exci-tatory amino acid agonist (1S,3R)-ACPD. *Am. J. Physiol.* **268**, R1026–1033.

Larzabal, A., J. Losada, J.M. Mateos, R. Benitez, I.J. Garmilla, R. Kuhn *et al.* (1999). Distribution of the group II metabotropic glutamate receptors (mGluR2/3) in the enteric nervous system of the rat. *Neurosci. Lett.* **276**, 91–94.

Liu, M. and A.L. Kirchgessner (2000). Agonist- and reflex-evoked internalization of metabotropic glutamate recep-tor 5 in enteric neurons. *J. Neurosci.* **20**, 3200–3205.

Liu, M.T., J.D. Rothstein, M.D. Gershon, and A.L. Kirchgessner (1997). Glutamatergic enteric neurons. *J. Neurosci.* **17**, 4764–4784.

Marmo, E. (1988). L-glutamic acid as a neurotransmitter in the CNS. *Med. Res. Rev.* **8**, 441–458.

Martin, L.J., C.D. Blackstone, A.I. Levey, R.L. Huganir, and D.L. Price (1993a). AMPA glutamate receptor subunits are differentially distributed in rat brain. *Neuroscience* **53**, 327–358.

Martin, L.J., C.D. Blackstone, A.I. Levey, R.L. Huganir, and D.L. Price (1993b). Cellular localizations of AMPA glutamate receptors within the basal forebrain magnocellular complex of rat and monkey. *J. Neurosci.* **13**, 2249–2263.

McRoberts, J.A., S.V. Coutinho, J.C. Marvizon, E.F. Grady, M. Tognetto, J.N. Sengupta *et al.* (2001). Role of peripheral N-methyl-D-aspartate (NMDA) receptors in visceral nociception in rats. *Gastroenterology* **120**, 1737–1748.

Michaelis, E.K. (1998). Molecular biology of glutamate receptors in the central nervous system and their role in excitotoxicity, oxidative stress and aging. *Prog. Neurobiol.* **54**, 369–415.

Miller, K.E., J.R. Clements, A.A. Larson, and A.J. Beitz (1988). Organization of glutamate-like immunoreactivity in the rat superficial dorsal horn: Light and electron microscopic observations. *Synapse* **2**, 28–36.

Mills, C.D. and C.E. Hulsebosch (2002). Increased expression of metabotropic glutamate receptor subtype 1 on spinothalamic tract neurons following spinal cord injury in the rat. *Neurosci. Lett.* **319**, 59–62.

Monaghan, D.T., R.J. Bridges, and C.W. Cotman (1989). The excitatory amino acid receptors: their classes, pharmacology, and distinct properties in the function of the central nervous system. *Annu. Rev. Pharmacol. Toxicol.* **29**, 365–402.

Neugebauer, V., P.S. Chen, and W.D. Willis (1999). Role of metabotropic glutamate receptor subtype mGluR1 in brief nociception and central sensitization of primate STT cells. *J. Neurophysiol.* **82**, 272–282.

Ozawa, S., H. Kamiya, and K. Tsuzuki (1998). Glutamate receptors in the mammalian central nervous system. *Prog. Neurobiol.* **54**, 581–618.

Panico, W.H., N.J. Cavuto, G. Kallimanis, C. Nguyen, D.M. Armstrong, S.B. Benjamin *et al.* (1995). Functional evidence for the presence of nitric oxide synthase in the dorsal motor nucleus of the vagus. *Gastroenterology* **109**, 1484–1491.

Pin, J.P., C. De Colle, A.S. Bessis, and F. Acher (1999). New perspectives for the development of selective metabotropic glutamate receptor ligands. *Eur. J. Pharmacol.* **375**, 277–294.

Pin, J.P. and R. Duvoisin (1995). The metabotropic glutamate receptors: Structure and functions. *Neuropharmacology* **34**, 1–26.

Reis, H.J., A.R. Massensini, M.A. Prado, R.S. Gomez, M.V. Gomez, and M.A. Romano-Silva (2000). Calcium channels coupled to depolarization-evoked glutamate release in the myenteric plexus of guinea-pig ileum. *Neuroscience* **101**, 237–242.

Ren, J., H.Z. Hu, S. Liu, Y. Xia, and J.D. Wood (2000). Glutamate receptors in the enteric nervous system: Ionotropic or metabotropic? *Neurogastroenterol Motil.* **12**, 257–264.

Riccardi, D. (1999). Cell surface, Ca2+(cation)-sensing receptor(s): One or many? *Cell Calcium* **26**, 77–83.

Shinozaki, H., Y. Gotoh, and M. Ishida (1990). Selective N-methyl-D-aspartate (NMDA) antagonists increase gastric motility in the rat. *Neurosci. Lett.* **113**, 56–61.

Stanley, B.G., L.H. Ha, L.C. Spears, and M.G. Dee, 2nd. (1993). Lateral hypothalamic injections of glutamate, kainic acid, D,L-alpha-amino-3-hydroxy-5-methyl-isoxazole propionic acid or N-methyl-D-aspartic acid rapidly elicit intense transient eating in rats. *Brain Res.* **613**, 88–95.

Stanley, B.G., V.L. Willett, 3rd, H.W. Donias, M.G. Dee, 2nd, and M.A. Duva (1996). Lateral hypothalamic NMDA receptors and glutamate as physiological mediators of eating and weight control. *Am. J. Physiol.* **270**, R443–449.

Stefani, A., A. Pisani, N.B. Mercuri, G. Bernardi, and P. Calabresi (1994). Activation of metabotropic glutamate receptors inhibits calcium currents and GABA-mediated synaptic potentials in striatal neurons. *J. Neurosci.* **14**, 6734–6743.

Stern, P., P. Behe, R. Schoepfer, and D. Colquhoun (1992). Single-channel conductances of NMDA receptors expressed from cloned cDNAs: Comparison with native receptors. *Proc. R. Soc. Lond. B. Biol. Sci.* **250**, 271–277.

Tong, Q., J. Ma, and A.L. Kirchgessner (2001). Vesicular glutamate transporter 2 in the brain-gut axis. *Neuroreport* **12**, 3929–3934.

Tsai, L.H., Y.J. Lee, and J. Wu (1999). Effect of excitatory amino acid neurotransmitters on acid secretion in the rat stomach. *J. Biomed. Sci.* **6**, 36–44.

Tsai, L.H., W. Tsai, and J.Y. Wu (1994). Effect of L-glutamic acid on acid secretion and immunohistochemical localization of glutamatergic neurons in the rat stomach. *J. Neurosci. Res.* **38**, 188–195.

Walker, K., M. Bowes, M. Panesar, A. Davis, C. Gentry, A. Kesingland *et al.* (2001a). Metabotropic glutamate receptor subtype 5 (mGlu5) and nociceptive function. I. Selective blockade of mGlu5 receptors in models of acute, persistent and chronic pain. *Neuropharmacology* **40**, 1–9.

Walker, K., A. Reeve, M. Bowes, J. Winter, G. Wotherspoon, A. Davis *et al.* (2001b). mGlu5 receptors and nociceptive function II. mGlu5 receptors functionally expressed on peripheral sensory neurones mediate inflammatory hyperalgesia. *Neuropharmacology* **40**, 10–19.

Wirtshafter, D. and J.C. Krebs (1990). Control of food intake by kainate/quisqualate receptors in the median raphe nucleus. *Psychopharmacology* **101**, 137–141.

Wong, L.A. and M.L. Mayer (1993). Differential modulation by cyclothiazide and concanavalin A of desensitization at native alpha-amino-3-hydroxy-5-methyl-4-isoxazolepropionic acid- and kainate-preferring glutamate receptors. *Mol. Pharmacol.* **44**, 504–510.

Wong, L.A., M.L. Mayer, D.E. Jane, and J.C. Watkins (1994). Willardiines differentiate agonist binding sites for kainate- versus AMPA-preferring glutamate receptors in DRG and hippocampal neurons. *J. Neurosci.* **14**, 3881–3897.

Zheng, H., L., Kelly, L.M. Patterson, and H.R. Berthoud (1999). Effect of brain stem NMDA-receptor blockade by MK-801 on behavioral and fos responses to vagal satiety signals. *Am. J. Physiol.* **277**, R1104–1111.

11

Glutamate Toxicity in Lung and Airway Disease

Sami I. Said

Introduction: Glutamate as a Physiological Transmitter and Toxic Agent

The amino acids, glutamate and aspartate, abundantly present in the mammalian central nervous system (CNS), are the principal excitatory neurotransmitters. Acting on glutamate receptors, these excitatory amino acids play an important physiological role in learning, memory, development, and other functions (Mayer and Westbrook, 1987; Malenka and Nicoll, 1993; Hollmann and Heinemann, 1994). But overactivation of glutamate receptors has been recognized as an important mechanism of neuronal toxicity and neuronal cell death in a variety of acute and chronic neurological disorders, ranging from stroke to Alzheimer's disease (Olney, 1990).

Among the glutamate receptors involved in these processes are a subtype of *ionotropic receptors*, known as *N-methyl-D-aspartate* (*NMDA, a glutamate agonist*) *receptors*. Other known ionotropic receptors, named for their respective specific agonists, are AMPA (α-amino-3-hydroxy-5-methyl-4-isoxazolepropionic acid and kainate receptors (Madden, 2002). Toxicity mediated by excessive activation of these receptors is known as *excitotoxicity* (Olney, 1990; Choi, 1992).

Another class of glutamate receptors, the *metabotropic receptors*, is the family of heterogeneous, G-protein-coupled receptors that regulate the activity of membrane enzymes and ion channels, and act through different second-messenger systems (Schoepp *et al.*, 1999).

Glutamate may also cause cell injury and death through non-receptor-mediated mechanisms. High concentrations of extracellular glutamate interfere with the uptake of cystine into the cell, leading to decreased intracellular levels of cystine and its reduction product cysteine, with resultant decrease in glutathione synthesis and eventual cell death (Murphy *et al.*, 1989).

Although predominantly located in the CNS, glutamate receptors have now been described in peripheral sites, both neuronal and non-neuronal (Erdö, 1991; Skerry and Genever, 2001). This chapter summarizes evidence that glutamate signaling may be a previously unsuspected mechanism in the pathogenesis of lung and airway disease, including acute

---

Sami I. Said • Pulmonary and Critical Care Medicine, T-17 025 Health Sciences Center, SUNY at Stony Brook, Stony Brook, NY.

Glutamate Receptors in Peripheral Tissue, edited by Santokh Gill and Olga Pulido. Kluwer Academic / Plenum Publishers, New York, 2005.

lung injury and bronchial asthma. Glutamate receptors may also serve novel, yet to-be-characterized regulatory functions in the peripheral respiratory system. The potential for such physiological and pathophysiological roles has recently received impetus from the demonstration of several NMDA receptor subtypes in the lungs and airways.

1. Glutamate as a Trigger of Acute Lung Injury

Our interest in the potential occurrence of glutamate toxicity in the lung, and the relationship of such toxicity to human disease, began with the observation that the excitotoxin NMDA could induce acute lung injury in the rat (Said *et al.*, 1996). The injury was in the form of acute high-permeability edema, evidenced by increased lung weight, leakage of protein into the bronchoalveolar lavage fluid, and elevated airway and pulmonary artery pressures. The lung injury was prevented by NMDA receptor blockers or antagonists, and, as is the case with neuronal excitotoxicity, it was nitric oxide (NO)-dependent (Said *et al.*, 1996). The injury could also be prevented by the neuropeptide vasoactive intestinal peptide (VIP), acting downstream from the stimulation of NO production.

2. Are Glutamate Receptors Involved in Mediating Oxidant Lung Injury?

These observations raised the possibility that endogenous glutamate toxicity might be a novel mechanism in the pathogenesis of acute lung injury, as seen in the acute respiratory distress syndrome (ARDS; Ware and Matthay, 2000). Two complementary lines of evidence would validate this hypothesis: (a) if glutamate toxicity could be shown to be involved in mediating the lung injury induced by some of the more common insults, and (b) if the existence of NMDA receptors could be demonstrated in the lung. The first line of evidence was verified with the demonstration that the NMDA receptor antagonist MK-801 attenuated not only the pulmonary injury caused by perfusion with NMDA, but also that induced by acute oxidant stress due to paraquat or xanthine + xanthine oxidase (Said *et al.*, 2000). These findings suggested that *endogenous* activation of NMDA receptors, probably by glutamate released from damaged cells, could play an important role in the pathogenesis of oxidant lung injury. It has been shown that intracellular glutamate levels are a 1,000-fold greater than extracellular levels, and that glutamate released from injured neurons and activated neutraphils acts as a source of excitotoxic injury of other cells (Collard *et al.*, 2002). This mechanism of positive feedback, by which glutamate toxicity can be perpetuated and amplified, has not yet been documented in the lung.

Evidence for the expression of NMDA receptors in the lungs and airways is discussed below.

3. Glutamate and Airway Reactivity—A Role in Bronchial Asthma?

In the course of our studies on NMDA-induced pulmonary edema, we observed that the onset of edema was associated with a pronounced increase in airway resistance. We therefore wondered whether NMDA receptor activation might also enhance the airway responses to injury and inflammation. We tested this possibility in an experimental model in which the vanilloid agonist capsaicin was administered intra-tracheally into isolated guinea pig lungs perfused via the airway, rather than via the circulation. In this preparation, capsaicin causes profound airway constriction and increased airway vascular permeability, two of the principal features of bronchial asthma. In our experiments, capsaicin (10^{-8} to 2×10^{-7} M) elicited a sharp increase in airway perfusion pressure, which was greatly attenuated in magnitude and duration by the selective NMDA channel blocker dizocilpine (MK-801) (Figure 11.1; Said, 1999). This observation suggested that glutamate receptor activation plays a dominant role in mediating capsaicin-induced airway smooth muscle contraction and increased vascular permeability. In this model, the glutamate receptors might be directly activated by capsaicin itself, as well as indirectly by endogenously released tachykinins (Liu *et al.*, 1997) and glutamate (Ueda *et al.*, 1994). It is worth noting here that 90% of tachykinin-containing sensory neurons, which are stimulated by capsaicin, are also rich in glutamate (Battaglia and Rustioni, 1988).

Additional, more direct, evidence supports a link between NMDA receptor activation and airway smooth muscle responsiveness. Applied to isolated, perfused tracheal segments of guinea pig, NMDA (10^{-8}–10^{-4} M) increased resting muscle tone and enhanced the contractile response to acetylcholine (10^{-5}–10^{-4} M) or methacholine (10^{-7}–10^{-5}). This means that NMDA *induced airway hyper-reactivity* (Figure 11.2). In guinea pig lungs perfused via the trachea, NMDA (1–2 mM) increased airway perfusion pressure, and this increase was totally abolished by dizoclipine (10 mM; Said, 1999).

With the understanding that these data rest on the premise that dizocilpine is a specific blocker of NMDA receptors (Galligan and North, 1990), the findings provide an array of pharmacological evidence that NMDA glutamate receptors are present in the airways, participate in mediating the acute airway response to capsaicin and, when activated, enhance airway smooth muscle responsiveness. The unavoidable implication here is that peripheral NMDA receptor activation might be an important, previously unrecognized, mechanism of

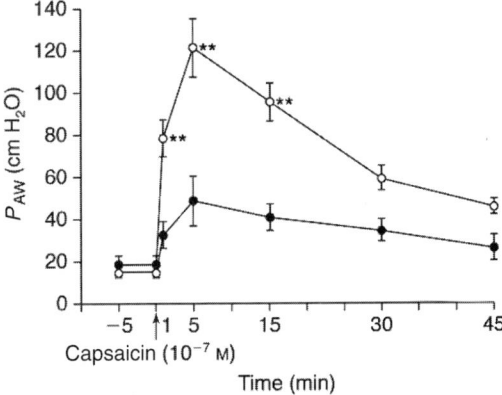

Figure 11.1. Capsaicin (10^{-7} M) infused into guinea-pig lung via the trachea, caused a dramatic increase in airway perfusion pressure (P_{AW}) (open circles; $n = 15$). This pressure increase was greatly attenuated in the presence of the NMDA receptor channel blocker dizoclipine (closed circles; $n = 4$). **Significantly higher than corresponding values in lower tracing ($p < 0.01$). (Reproduced with permission from Said, 1999.)

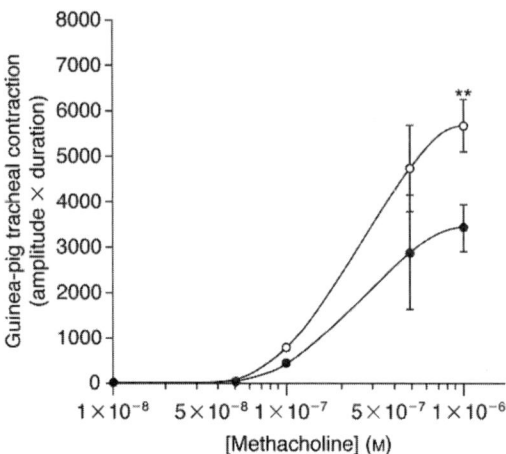

Figure 11.2. Concentration-dependent increase in smooth muscle tone of guinea-pig tracheal strips with methacholine (open circles), and its enhancement in the presence of 10^{-3} M NMDA (closed circles). **Significantly higher than corresponding value in lower tracing ($p < 0.01$). (Reproduced with permission from McDowell, 2002.)

the airway inflammation and hyper-reactivity that are the hallmarks of bronchial asthma (McDowell, 2002). In particular, the results might also help explain the triggering or worsening of acute asthmatic symptoms, in some patients, following a meal of glutamate-rich food (Allen *et al.*, 1987).

4. Glutamate (NMDA) Receptors in the Lung and Airways

The first indications of the presence of glutamate receptors in the respiratory system were: (a) the immunocytochemical demonstration of NMDA receptors in neurons in the rat lung and esophagus, from which nerves supplying airway smooth muscle originate (Robertson *et al.*, 1998) and (b) the report of autoradiographic binding sites for radio-labeled dizoclipine in rat peripheral lung, including alveolar walls, bronchial smooth muscle, and bronchial epithelium (Robertson *et al.*, 1997). The latter report is of special significance because it implies the presence of glutamate receptors in areas of the lung that are devoid of neurons.

More recently, we explored the distribution of NMDA receptor subunits in rat lungs by gene expression and Western blot immunofluorescence (Dickman *et al.*, 2003). Receptor subunit 1 was expressed in all segments of the lungs and airways, while receptor subunits 2C and 2D were localized in the peripheral (gas-exchanging) regions and the mid-segments of the lung containing both parenchymal (alveolar–microvascular) and smaller-airway components (Table 11.1).

These observations provide a molecular–biological basis for the toxic effects of glutamate in rat lungs and airways, and raise the question of a possible regulatory role for these receptors in lung and airway function.

Table 11.1. NMDA Receptor (NMDA)
Subunits Demonstrated in Rat Lung

Site	Receptor subunit
Large airways	1
Mid-lung	1, 2C, 2D
Peripheral lung	1, 2C, 2D

Source: Data from Said and Dickman (2000).

5. Counteracting Glutamate Toxicity: Novel Therapeutic Strategies for Lung and Airway Disease

Insights into the possible role of glutamate toxicity in the pathogenesis of lung and airway diseases provide a rationale for the introduction of novel therapeutic interventions. Channel blockers such as dizoclipine, which target the first step in the excitotoxic cascade (Said, 1999), have already been tested, but their usefulness has been curtailed by undesirable side effects (Olney, 1994).

Other agents potentially capable of blocking or attenuating glutamate toxicity comprise inhibitors or antagonists of selected processes or pathways that contribute significantly to the genesis of glutamate toxicity. These processes include: (a) NO synthesis (*NO synthase inhibitors*; Said *et al.*, 1996); (b) overactivation of DNA repair enzyme poly(ADP) ribose polymerase (PARP) and consequent depletion of cell energy sources ADP and ATP (*PARP inhibitors*; Said *et al.*, 1996); (c) apoptotic cell death (*caspase inhibitors and bcl$_2$ upregulators*); and (d) oxidant stress (*antioxidant measures*; Murphy *et al.*, 1989). An example of one compound that exerts several of these actions is the neuropeptide VIP, which has been shown to protect against excitotoxicity in the lung and neuronal cells, as well as against oxidant stress in PC-12 cells (Said *et al.*, 1998). VIP appears to counteract glutamate toxicity by mechanisms that include anti-apoptotic activities (inhibition of caspase activity, upregulation of Bcl2, and suppression of cytochrome *c* translocation); inhibition of PARP activation; and antioxidant defenses (Said, 2000; Said and Dickman, 2000; Filippatos *et al.*, 2001; Antonawich and Said, 2002).

6. Conclusions

1. Experimental glutamate toxicity can occur in the respiratory system. Acute administration of high concentrations of glutamate or glutamate agonist NMDA can elicit high-permeability pulmonary edema and airway constriction.

2. Endogenous glutamate receptor activation may play a key role in mediating oxidant lung injury, and airway hyper-responsiveness, a major feature of bronchial asthma.

3. NMDA receptor subunits have now been identified and their distribution characterized in rat lungs and airways. Cellular localization and the definition of possible physiological regulatory functions are subjects of future investigation.

4. Understanding glutamate signaling should lead to the introduction of novel strategies for the management of lung and airway diseases which have glutamate toxicity as a pathogenetic mechanism.

References

Allen, D.H., J. Delohery, and G. Baker (1987). Monosodium L-glutamate-induced asthma. *J. Allergy Clin. Immunol.* **80**, 530–537.

Antonawich, F.J. and S.I. Said (2002). VIP attenuates the apoptotic cascade in hippocampal stem cells. *Neurosci. Lett.* **325**, 151–154.

Battaglia, G. and A. Rustioni (1988). Coexistence of glutamate and substance P in dorsal root ganglion neurons of the rat and monkey. *J. Comp. Neurol.* **277**, 302–312.

Choi, D.W. (1992). Excitotoxic cell death. *J. Neurobiol.* **23**, 1261–1276.

Collard, C.D., K.A. Park, N.C. Montalto, S. Alapapti, J.A. Buras, G.L. Stahl *et al.* (2002). Neutrophil-derived glutamate regulates vascular endothelial barrier function. *J. Biol. Chem.* **277**, 14801–14811.

Dickman, K.G., J.G. Youssef, S.M. Mathew, and S.I. Said (2003). Ionotropic glutamate receptors in lungs and airways: Molecular basis for glutamate toxicity. *Am. J. Respir. Cell Mol. Biol.* **30**, 139–144.

Erdö, S.L. (1991). Excitatory amino acid receptors in the mammalian periphery. *Trends Pharmacol. Sci.* **12**, 426–429.

Filippatos, G.S., O. Lalude, N. Parameswaran, S.I. Said, W. Spielman, and B.D. Uhal (2001). Regulation of apoptosis by vasoactive peptides. *Am. J. Physiol.* **281**, L749–L761.

Galligan, J.J. and R.A. North (1990). MK-801 blocks nicotinic depolarizations of guinea pig myenteric neurons. *Neurosci. Lett.* **108**, 105–109.

Hollmann, M. and S. Heinemann (1994). Cloned glutamate receptors. *Annu. Rev. Neurosci.* **17**, 31–108.

Liu, H., P.W. Mantyh, and A.I. Basbaum (1997). NMDA-receptor regulation of substance P release from primary afferent nociceptors. *Nature* **386**, 721–724.

Madden, D.R. (2002). The structure and function of glutamate receptor ion channels. *Nat. Rev. Neurosci.* **3**, 91–101.

Mayer, M.L. and G.L. Westbrook (1987). The physiology of excitatory amino acids in the vertebrate central nervous system. *Prog. Neurobiol.* **28**, 197–276.

Malenka, R.C. and R.A. Nicoll (1993). NMDA-receptor-dependent synaptic plasticity: Multiple forms and mechanisms. *Trends Neurosci.* **16**, 521–527.

McDowell, K.M. (2002). Pathophysiology of asthma. *Respir. Care Clin. N. Am.* **6**, 15–26.

Murphy, T.H., M. Miyamoto, A. Sastre, R.L. Schnaar, and J.T. Coyle (1989). Glutamate toxicity in a neuronal cell line involves inhibition of cystine transport leading to oxidative stress. *Neuron* **2**, 1547–1558.

Olney, J.W. (1990). Excitotoxic amino acids and neuropsychiatric disorders. *Annu. Rev. of Pharmacol. Toxicol.* **30**, 47–71.

Olney, J.W. (1994). Neurotoxicity of NMDA receptor antagonists: An overview. *Psychopharmacol. Bull.*, **30**, 533–540.

Robertson, B., R.D. Dey, L.J. Huffman, M.J. Polak, and S.I. Said (1997). Autoradiographic localization of NMDA receptors in the rat lung and implications in lung injury. *Soc. Neurosci. Abstr.* **23**, 931.

Robertson, B.S., B.E. Satterfield, S.I. Said, and R.D. Dey (1998). *N*-methyl-D-aspartate receptors are expressed by intrinsic neurons of rat larynx and esophagus. *Neurosci. Lett.*, **244**, 77–80.

Said, S.I., H.I. Berisha, and H. Pakbaz (1996). Excitoxicity in the lung: *N*-methyl-D-aspartate-induced, nitric oxide-dependent, pulmonary edema is attenuated by vasoactive intestinal peptide and by inhibitors of poly (ADP-ribose) polymerase. *Proc. Natl. Acad. Sci. USA* **93**, 4688–4692.

Said, S.I., K. Dickman, R.D. Dey, A. Bandyopadhyay, P. DeStefanis, S. Raza *et al.* (1998). Glutamate toxicity in the lung and neuronal cells: Prevention or attenuation by VIP and PACAP. *Ann. NY Acad. Sci.* **865**, 226–237.

Said, S.I., H. Pakbaz, H.I. Berisha, and S. Raza, (2000). NMDA receptor activation: Critical role in oxidant tissue injury. *Free Radic. Biol. Med.* **28**, 1300–1302.

Said, S.I. and K.G. Dickman (2000). Pathways of inflammation and cell death in the lung: Modulation by vasoactive intestinal peptide. *Regul. Pept.* **93**, 21–29.

Said, S.I. (1999). Glutamate receptors and asthmatic airway disease. *Trends Pharmacol. Sci.* **20**, 132–134.

Said, S.I. (2000). The Viktor Mutt Memorial Lecture: Protection by VIP and related peptides against cell death and tissue injury. *Ann. NY Acad. Sci.* **921**, 264–274.

Schoepp, D.D., D.E. Jane, and J.A. Monn (1999). Pharmacological agents acting at subtypes of metabotropic glutamate receptors. *Neuropharmacology* **38**, 1431–1476.

Skerry, T.M. and P.G. Genever (2001). Glutamate signaling in non-neuronal tissues. *Trends Pharmacol. Sci.* **22**, 174–181.

Ueda, M., Y. Kuraishi, K. Sugimoto, and M. Satoh (1994). Evidence that glutamate is released from capsaicin-sensitive primary afferent fibers in rats: Study with on-line continuous monitoring of glutamate. *Neurosci. Res.* **20**, 231–237.

Ware, L.B. and M.A. Matthay (2000). The acute respiratory distress syndrome. *N. Engl. J. of Med.* **342**, 1334–1349.

12

Glutamate: Teaching Old Bones New Tricks—Implications for Skeletal Biology

Gary J. Spencer, Ian S. Hitchcock, and Paul G. Genever

1. Bone Remodeling

Bone is a dynamic tissue. Vertebrate skeletons are continuously remodeled through successive cycles of destruction and reformation at specific microscopic sites that are organized as functional basic multicellular units (BMUs). At these sites, aged or damaged bone is removed by the resorptive activity of osteoclasts. These are large multinucleated cells originating from hematopoietic precursors of the monocyte/macrophage lineage that reside in the bone marrow. Osteoclasts generate peripheral actin ring assemblies and form tight adhesive interactions with the bone surface, creating a sealing zone that delineates a resorbing compartment beneath each osteoclast. Matrix degrading enzymes such as cathepsin K, B, and L, and matrix metalloproteinases are synthesized and secreted by the osteoclast into the underlying resorbing compartment, which is acidified by H^+ ions driven across the basolateral membrane of the osteoclast by an electrogenic ATPase proton-pump. Bone resorption is followed by an intermediate reversal phase that preludes the recruitment and differentiation of preosteoblasts from mesenchymal stem cells located in the bone marrow compartment. The mature differentiated osteoblasts synthesize a new extracellular matrix, composed predominantly of type I collagen with non-collagenous proteins, such as osteopontin, osteocalcin, and bone sialoprotein, which is later mineralized to replace the resorbed bone. During bone formation, some osteoblasts are encased within the developing matrix and persist as viable osteocytes, which are believed to act as cellular sentinels to monitor bone microarchitecture. Osteocytes communicate with cells on the bone surface via membranous extensions through channels (canaliculi) in the bone matrix and these surface cells in turn interact with bone marrow cells to form a continuous and multicellular network throughout the tissue. The entire remodeling cycle through resorption to formation lasts between 6 and 9 months and on completion, the remaining osteoblasts either become quiescent bone lining cells or die through apoptosis. Bone remodeling allows the skeleton to adapt to alterations in an individual's physical activity; mechanical strain is a potent stimulator of bone

Gary J. Spencer, Ian S. Hitchcock, and Paul G. Genever • Biomedical Tissue Research, Department of Biology, University of York, York, UK.

Glutamate Receptors in Peripheral Tissue, edited by Santokh Gill and Olga Pulido.
Kluwer Academic / Plenum Publishers, New York, 2005.

formation, whereas disuse or inactivity causes bone loss. It is estimated that the remodeling activity within bone is sufficient to regenerate an entire adult skeleton every 10 years (Manolagas, 2000), however a number of regulatory safeguards exist to maintain bone homeostasis. Systemic hormones (parathyroid hormone, 1,25 dihydroxyvitamin D_3, sex hormones) and locally released growth factors and cytokines (bone morphogenetic proteins, transforming growth factor β, fibroblast growth factor, interleukins) help couple the activities of osteoclasts and osteoblasts to ensure that resorption equals formation to avert a net gain or net loss of bone mass. Disturbances of the remodeling cycle are clinically manifested as osteopetrotic and osteoporotic conditions, which are characterized by increased and decreased bone mass respectively.

2. Osteoporosis

Osteoporosis is a prevalent age-related disease caused by excessive bone loss, which results in bone fragility and increased susceptibility to fracture. In adulthood, we begin to lose bone in our 40s, however in the years following the menopause, women experience an accelerated reduction in bone mass caused by the loss of estrogen. It is estimated that 1 in 3 women and 1 in 12 men will get osteoporosis in their lifetime, with an annual healthcare cost of over $14 billion in the United States alone. Current therapies used for the treatment of osteoporosis largely rely on antiresorptive drugs, such as estrogen-based agents (hormone replacement therapy, selective estrogen receptor modulators (SERMs)), which can reduce osteoclastogenesis or bisphosphonates (alendronate, etidronate, residronate) that act by inhibiting osteoclast function (see Rodan and Martin, 2000). However, more effective osteogenic therapies are required to promote bone formation and replace lost bone mass to a level above the fracture risk threshold. This demands a thorough comprehension of the regulatory inputs received by osteoclasts, osteoblasts, osteocytes, and their primitive marrow precursor cells, to identify possible novel therapeutic targets for the treatment of osteoporosis and related skeletal disorders. Over the last few years, evidence has emerged to suggest that glutamate acts as an intercellular signaling molecule in bone. This chapter will review these advances in bone biology and discuss the relevance of non-neuronal glutamate signaling in skeletal tissues.

3. Glutamate Transporters in Bone Cells

Studies on glutamate signaling in bone, originated from experiments designed to identify novel genes involved in the regulated control of bone mass, with the aim of developing new therapeutic strategies for the treatment of bone disorders characterized by aberrant bone mass homeostasis. Differential RNA display was used to quantitate changes in osteocyte gene expression *in vivo* following brief periods of mechanical loading, known to physiologically induce new bone formation (Mason *et al.*, 1997). One gene downregulated in response to loading was the glutamate/aspartate transporter GLAST, whose expression had previously only been described in the CNS (Storck *et al.*, 1992). Immunohistochemistry using neonatal rat tibiae confirmed expression of GLAST and localized it to osteoblasts and newly embedded osteocytes *in vivo*. Following mechanical loading, GLAST expression was upregulated by osteoblasts on surfaces where loading induced new bone formation, while on quiescent surfaces, expression remained at low levels. Northern blot analysis confirmed expression of a 2.5 kb GLAST mRNA species in cortical bone, which corresponded in size to one of two expressed transcripts in rat brain, while *in situ* hybridization confirmed the localization of the transporter to osteoblasts and osteocytes. The expression of a smaller GLAST transcript in the CNS was inconsistent with previously published data (Storck *et al.*, 1992), which described the expression of a single 4.5 kb mRNA species in rat brain, suggesting the possible existence of a novel GLAST splice variant. A recent report has described the cloning of this variant, GLAST-1a from whole rat bone, which is identical to neuronal full-length GLAST but lacks exon 3 (Huggett *et al.*, 2000). The size of this mRNA species, the origin of expression within bone, or whether this splice variant would give rise to a functional transporter remains unclear. The authors hypothesize that deletion of exon 3 may alter the membrane topology of the transporter reversing the direction of transport, causing the cellular efflux of glutamate rather than its accumulation. Using identical methods, we have been unable to confirm the expression of the 2.5 kb GLAST transcript in osteoblast enriched primary cultures, which represent the primary site of GLAST expression *in vivo*, although we routinely detect the 4.5 kb species (unpublished observations). However, we cannot discount the possibility that other cells within the bone microenvironment express this splice variant. In addition to GLAST, expression of related neuronal glutamate transporters have also been investigated by immunohistochemistry in bone cells (Mason *et al.*, 1997). GLT-1 specifically localizes to a distinct subset of mononuclear marrow cells, while expression of EAAC1 was not detected in any cell type in neonatal rat tibiae *in vivo*. Significantly, the expression and localization of multiple subtypes of glutamate transporters in bone cells was consistent with considerable variation in their sites of expression in the CNS, raising the possibility that each transporter may have distinct functions related to their site of expression in the bone microenvironment.

4. Identification of Glutamate Receptors in Bone

Considering the fundamental importance of glutamate transporters in neurotransmitter clearance at central synapses, their identification in bone suggested that bone cells might use a system analogous to glutamate signaling in the CNS as means of cellular communication. Over the last few years, a small but growing number of researchers have provided an increasing body of evidence to support this hypothesis, implicating a role for glutamate as an important mediator of bone cell function. At central excitatory synapses, transporters constitute only a small proportion of the complex signaling milieu required for discrete glutamatergic signaling events between juxtaposed cells. Membrane receptors play a crucial role in transducing glutamate signals into an intracellular response. Therefore, if bone cells were to use glutamate as a means of cellular communication one would expect to identify the expression of functional glutamate receptors on cells within the bone microenvironment. Subsequent investigations have revealed the expression of a diverse range of glutamate receptors in bone cells, of which the NMDA-type glutamate receptor has received the most attention.

Expression of the primary subunit of the receptor, NMDAR1, has been demonstrated by RT-PCR, immunohistochemistry, and *in situ* hybridization in osteoblasts, osteocytes, and osteoclasts *in vitro* and *in vivo* (Chenu *et al.*, 1998; Patton *et al.*, 1998: see Figure 12.1). Furthermore, concurrent expression of NMDAR2 subunits has been demonstrated in rat bone marrow and osteoclasts, indicating that like neurons, NMDA receptors expressed by bone cells probably exist as heteromers. Studies have revealed that cells within the marrow cavity express NMDAR2D, but not NMDAR2A, B, and C, while more recent data based on RT-PCR and *in situ* hybridizations have revealed expression of NMDAR2B and NMDAR2D subunits

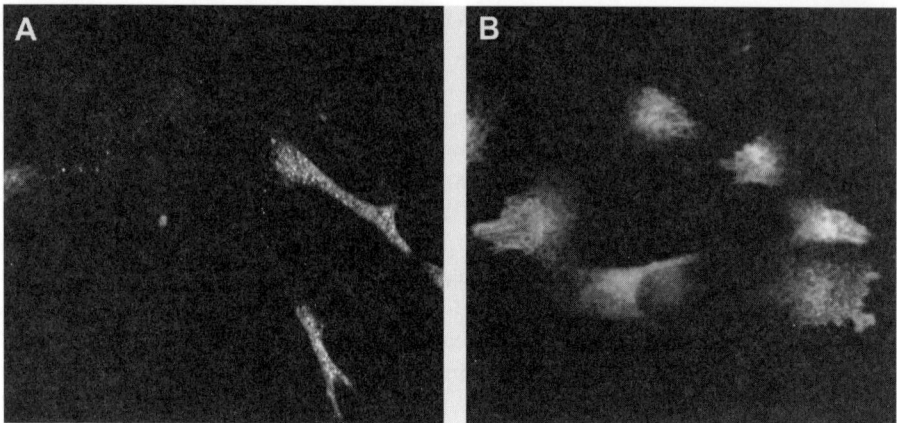

Figure 12.1.   Focal immunoreactivity for the NMDAR1 glutamate receptor subunit (A, primary human osteoblast) and the t-SNARE SNAP-25 (B, SaOS-2 cells) in osteoblastic cells.

by osteoclasts (Itzstein *et al.*, 2001). The identification of both NMDAR1 and NMDAR2 subunits by osteoblasts and osteoclasts provides compelling evidence for the presence of cells within the bone microenvironment that express the minimum molecular machinery required to form functional NMDA receptors and thus receive glutamatergic signals.

Although expression of NMDAR1 has been consistently demonstrated, significant disparity exists between data obtained from studies of NMDAR2 subunit expression. Hinoi *et al.*, describe the expression of NMDAR2D, but not NMDAR2A, B, or C in primary rat calvarial osteoblasts (Hinoi *et al.*, 2001), which is in agreement with our data (Patton *et al.*, 1998). While others have described the expression of NMDAR1 and NMDAR2C, but not NMDAR2A, B, or D (Gu and Publicover, 2000) or NMDAR1, NMDAR2A, B, and D, but not NMDAR2C (Itzstein *et al.*, 2001). In support of Itzstein's data, recent studies have demonstrated that bone marrow stromal cells express NMDAR2A, B, and D but not NMDAR2C compared to total bone marrow where expression of NMDAR2B and C was detected but not NMDAR2A or D (Dobson and Skerry, 2000). Variations in NMDAR2 subunit expression may be a reflection of differences in the sensitivity of detection methods used (combinations of RT-PCR, immunohistochemistry, and *in situ* hybridizations) or differences in osteoblastic models, which include clonal osteoblastic cells, primary rat femoral osteoblasts, bone marrow stromal cells, and primary rat calvarial osteoblasts. Interpretation of these studies is likely to be further complicated by differentiation-dependent expression of receptor subunits in bone, which may occur in manner analogous to the developmental regulation of NMDA receptor subunits in the CNS.

Chenu and coworkers provided some of the first functional evidence for a potential role of glutamate signaling in bone (Chenu *et al.*, 1998). Using freshly isolated disaggregated rabbit osteoclasts they demonstrated that the NMDA receptor antagonists MK-801 and APV, inhibited bone resorption *in vitro* through an action on mature osteoclasts, in a manner that did not involve changes in osteoclast apoptosis. Further characterizations revealed that MK-801 specifically bound to mature osteoclasts and that the antagonists DEP (1-(1,2-diphenylethyl) piperidine) and L-689-560 had similar effects to MK-801 on disaggregated osteoclast resorption activity (Itzstein *et al.*, 2000). MK-801 was demonstrated to have no effect on osteoclast adhesion or viability but was reported to significantly disrupt actin ring formation. These data implicated a functional role for NMDA receptors in the control of osteoclastic resorption, through regulation of actin organization and sealing zone formation essential for the resorptive activity of osteoclasts.

These findings are in contrast to other reported observations. Using identical methodology Peet *et al.* were unable to demonstrate any effect of MK-801 on disaggregated mature osteoclast activity and did not observe any changes in actin ring formation (Peet *et al.*, 1999). However, using a coculture technique in which osteoblasts regulate the formation of osteoclasts from bone marrow mononuclear cells, it was demonstrated that MK-801 significantly reduced bone resorption *in vitro* by inhibiting osteoclast differentiation. These effects were only apparent if the antagonist was included in the coculture medium during the first 7 days of the assay or during the entire culture period but not when MK-801 was added in later stages of culture (>7 days). The temporal nature of this inhibition was indicative of a function for NMDA receptors in the control of osteoclastogenesis rather than regulation of mature osteoclast activity. Naturally, it is possible that some or all of the effect of MK-801 in cocultures is mediated by NMDA receptors expressed by osteoblasts, which may act to disrupt signaling between osteoblasts and osteoclast lineage progenitors required to support osteoclast formation. One possibility is that glutamate regulates the expression and/or function of receptor activator of NFκB ligand (RANKL), an osteoblast membrane-associated protein responsible for driving osteoclastogenesis through a direct interaction with its receptor RANK on osteoclast progenitors. Similarly, NMDA receptor modulation may affect expression of the soluble decoy protein osteoprotogerin (OPG), which acts to negatively regulate osteoclast formation through the sequestration of osteoblastic RANKL.

The molecular identification of glutamate receptor subunits provides compelling evidence for the existence of glutamate-mediated signaling mechanisms in bone cells although their expression is not necessarily indicative of the activity of functional channels. Electrophysiological studies using the patch clamp technique have been performed to study the effects of glutamate and glutamate receptor antagonists on membrane currents in both osteoblasts and osteoclasts. Laketic-Ljubojevic *et al.*, studied the effects of glutamate and NMDA on human clonal osteoblastic cells and demonstrated increases in whole cell membrane currents that were blocked by Mg^{2+}, the competitive NMDA receptor antagonist R-CPP (3-((R)-2-carboxypiperazin-4-yl)-propyl-1-phosphonic acid), and the NMDA receptor channel blocker MK-801 (Laketic-Ljubojevic *et al.*, 1999). The responses of these cells to antagonists resembled those observed for NMDAR1/NMDAR2D heteromers in the CNS, supporting molecular data describing expression of NMDAR1 and NMDAR2D but not other NMDAR2 receptor subunits by osteoblasts (Patton *et al.*, 1998). In these studies the authors also demonstrated the ability of glutamate to bind with high affinity to osteoblastic cells, and reported that NMDA receptor activation increased calcium influx, an effect antagonized by Mg^{2+} and MK-801. Others have demonstrated that glutamate and NMDA also caused increases in whole cell currents in mature disaggregated rabbit osteoclasts, responses that were similarly antagonized in a voltage sensitive manner by Mg^{2+}, MK-801, and DEP (Espinosa *et al.*, 1999; Peet *et al.*, 1999; Itzstein *et al.*, 2000). Thus NMDA-type glutamate receptors expressed by osteoblasts and osteoclasts are electrophysiologically functional and have similar characteristics to their neuronal counterparts.

Significant differences exist between the electrophysiological responses to glutamate in clonal osteoblastic cells and primary rat osteoblasts. In primary cells, although application of NMDA results in a robust increase in whole cell membrane currents similar to that observed in clonal cells, there is little electrophysiological response to glutamate (Gu and Publicover, 2000).

Using a combined molecular, pharmacological, and electrophysiological approach, Gu and Publicover demonstrated that blockade of metabotropic glutamate receptors (mGluR) expressed by osteoblasts is required to induce equivalent responses from glutamate to those obtained with NMDA. Conversely, activation of these receptors decreased the response of NMDA to that seen with glutamate. It is apparent that cross talk occurs between metabotropic and NMDA-type glutamate receptors in osteoblasts, which is entirely consistent with the coexpression of NMDA receptors and mGluRs in these cells. Further experimental evidence supports a role for Group I metabotropic receptors in this process. Primary rat osteoblasts express mGluR1b, which couples to PLC, IP_3 production, and intracellular calcium mobilization, but not mGluR1a, -2, -3, -4a, or -5a or b. The involvement of PLC coupled mGluRs is supported by the effects of parathyroid hormone, which is similarly coupled to PLC signaling pathways and mimics the effect of mGluR receptor activation on the responses of NMDA receptors to glutamate in osteoblasts.

Expression of mGluRs has been the subject of further investigation in primary rat calvarial cells although results contrast with those obtained from primary rat femoral osteoblasts (Hinoi *et al.*, 2001). RT-PCR data indicate that calvarial osteoblasts expressed neither group I (mGluR1a or mGluR5) nor group II (mGluR2 or mGluR3) receptors, and while expression of some group III mGluR subunits was detected (mGluR4a/b and mGluR8), others appeared not to be expressed (mGluR6 and mGluR7). Group III receptors expressed by osteoblasts appear to be functional. Activation of receptors with AP-4 (L-(+)-2-amino-4-phosphonobutyric acid) resulted in the stimulation of inhibitory protein G, and attenuated forskolin induced cAMP production, an effect specifically blocked by the group III mGluR receptor antagonist CPPG ((*RS*)-α-cyclopropyl-4-phosphonophenylglycine).

In addition to functional roles identified for glutamate signaling in the control of osteoclastogenesis and bone resorption, evidence also exists to support an important role for glutamate in the control of osteoblastic bone formation and regulation of osteoprogenitor cell fate. Antagonism of osteoblastic NMDA receptors with MK-801 caused a dose-dependent inhibition of markers of bone formation in the rat bone marrow colony forming unit-fibroblastic (cfu-f) assay (Dobson and Skerry, 2000). Treatment of cfu-f cultures resulted in a significant reduction in the proportion of colonies that expressed the osteoblastic marker alkaline phosphatase and markedly decreased mineralization. In contrast to the effects of NMDA receptor antagonism on osteoclastogenesis, equivalent inhibitory effects were observed irrespective of when MK-801 was added, either early or late in culture. This suggested strongly that antagonism of NMDA receptors present in marrow stromal populations affected both preosteoblastic precursor cells and mature osteoblasts *in vitro*. Since the cfu-f assay is a measure of osteoblast progenitor numbers these findings implicated NMDA responsive cells, present in the bone marrow stroma, as potential pluripotent determinants of osteoblast differentiation. Dobson and Skerry also presented evidence to suggest that osteoblastic NMDA receptors were involved in the regulation of mesenchymal fate determination between adipocytic and osteoblastic lineages, which are derived from a common stromal precursor. MK-801 induced morphological changes in bone marrow stromal cells, which became more rounded, less fibroblastic, and developed perinuclear granules. These changes are similar to those observed during adipocyte differentiation suggesting that MK-801 may be acting to switch differentiation of precursors from osteoblasts to adipocytes. Pretreatment of cells with MK-801 before the addition of indomethacin and IBMX, which promote adipocyte differentiation, caused a significant increase in the number of lipid droplet-containing cells that stained with Oil Red-O, compared to cells without MK-801 pretreatment. Significantly, this effect was accompanied by reciprocal changes in gene expression associated with osteoblast and adipocyte differentiation.

Expression of markers of osteoblast differentiation, which include osteocalcin and type I collagen were reduced in MK-801 pretreated cells, while markers of adipocyte differentiation including peroxisome proliferating and activating receptor-γ (PPAR-γ) and lipoprotein lipase (LPL) were elevated (Dobson and Skerry, 2000).

5. Cellular Source of Glutamate in Bone

Studies on the expression and function of glutamate receptors in bone raised an important question regarding the source of agonist required to activate them. Recently, we have shown that osteoblasts spontaneously release glutamate in a regulated manner at levels equivalent to or greater than depolarized neurons (Genever and Skerry, 2001). FMI-143 labeling, a fluorescent-based microscopy technique employed in the study of intracellular trafficking, was used to demonstrate that osteoblasts contain vesicles that undergo repetitive recycling at peripheral sites, a process central to glutamate release at presynaptic sites of central glutamatergic neurons. Polar localization of vesicles have also been observed by electron microscopy at points of cell–cell contact in osteoblasts *in vitro* (P. G. Genever and P. S. Bhangu, personal communications). These studies suggested that the release of glutamate from osteoblasts was spontaneous, dependent on AMPA autoreceptors, and negatively regulated by depolarization induced voltage-dependent calcium entry. This markedly contrasts to glutamate release at central neurons where depolarization induced calcium influx provides the driving force for the initiation of glutamate exocytosis. Significantly, during osteoblast differentiation, levels of intracellular glutamate, exocytosed glutamate, and the susceptibility of glutamate release to depolarization-induced inhibition, increased as cells adopted a more osteoblastic phenotype suggesting that glutamatergic activity increased as cells differentiated.

In this study, the consequences of glutamate release inhibition were also investigated. Riluzole, a neuronal glutamate release inhibitor, inhibited the differentiation of osteoblasts and increased the relative proportion of cells undergoing apoptosis, as determined by viability assays, DNA fragmentation studies, and TUNEL positivity. Together, these data suggested that constitutive glutamate release was essential for osteoblast survival and differentiation. This is supported by evidence that the proapoptotic cytokines TNF-α and IFN-γ inhibited glutamate release while exogenously applied glutamate promoted the survival of osteoblasts grown under serum free conditions (Genever and Skerry, 2001).

Other studies have focused on determining whether osteoblasts contain the appropriate cellular machinery to support the vesicular release of glutamate. Using a combination of RT-PCR, immunohistochemistry, and Northern and Western blot analyses, we have demonstrated that osteoblasts also express the necessary exocytotic machinery required for regulated vesicular glutamate release from neurons (Bhangu *et al.*, 2001). Both primary and clonal osteoblasts express the v-SNARE VAMP and the t-SNAREs SNAP-25 and syntaxin, which together fulfill the minimum molecular requirement for the formation of the ternary exocytotic core complex essential for vesicular fusion and exocytosis (see Figure 12.1). Furthermore, the plasma-membrane t-SNARE SNAP-25, the vesicle specific protein synapsin I, and immunoreactive glutamate specifically colocalized, providing evidence for glutamate containing vesicles in osteoblasts. Expression of many other accessory and regulatory proteins involved in glutamate exocytosis have been identified in osteoblasts including Munc18, rSEC8, DOC2, syntaxin 6, Rab3A, and synaptophysin. Although analogous SNARE machinery is present in many non-neuronal cells (Bennet and Scheller, 1993), "synapse specific" proteins such as synapsin I and synaptophysin confer fast Ca^{2+}-dependent characteristics to standard trafficking systems, similar to that observed at central synapses (Bajjalieh *et al.*, 1999).

The vesicular release of glutamate from osteoblasts is not the only possible source of endogenous agonist for the activation of glutamate receptors in bone. A recent report based on immunohistochemical evidence and electron microscopy has described the presence of glutamate-containing nerve fibers in bone, which could supply glutamate for receptor activation (Serre *et al.*, 1999). It is difficult however, to envisage how static neurons could efficiently innervate highly motile and relatively short-lived bone cells, particularly osteoclasts that have a life span *in vivo* of approximately 3 weeks. However, it remains a possibility that both neuronal and non-neuronal cellular sources of glutamate exist in bone.

Our demonstrations of spontaneous glutamate release from osteoblasts may account for the lack of evidence for agonist effects in long-term bone cell cultures. In our experience, bone cells neither survive well nor differentiate in low-glutamate-containing medium making long-term consequences of glutamate receptor activation difficult to study (G.J. Spencer and P.G. Genever, unpublished observations).

A controversial report has recently been published that concluded that "glutamate does not play a major role in controlling bone growth" (Gray *et al.*, 2001). In this study, the authors tested the hypothesis that glutamate plays an important intercellular role in bone, controlling both the formative and resorptive activities of osteoblasts and osteoclasts. This hypothesis was based on previously published results describing the expression of glutamate receptors and transporters by bone cells, the effects of glutamate receptor inhibition on bone cell function *in vitro*, and the effects of mechanical loading on GLAST expression *in vivo*. Gray *et al.* examined the effects of glutamate receptor agonists and antagonists on bone formation and resorption *in vitro*, and analyzed the skeletal phenotype of glutamate transporter knockout mice ($GLAST^{-/-}$). The authors reported that while the noncompetitive NMDA receptor antagonist MK-801 reduced the number of resorption pits produced by disaggregated mature osteoclasts, the total area of resorption and volume of lacunae was unaffected. They also demonstrated that neither the competitive NMDA receptor antagonist APV nor the agonist NMDA affected the number of resorption pits, total area of resorption, or number of lacunae. Based on this evidence they concluded that the effect of MK-801 on osteoclast resorption was "unrelated to its action on glutamate receptors." Using a model of bone formation *in vitro*, Gray *et al.* also reported that neither NMDA nor APV significantly affected bone formation.

Further conclusions regarding the lack of the importance of glutamate in the skeleton were based on the examination of the phenotype of $GLAST^{-/-}$ mice. Macroscopically these mice displayed no differences in mandible or long bone size, morphology, trabeculation, number of resorption lacunae, nor areas of formation vs resorption, compared to wild-type siblings. However, the authors' conclusions regarding the lack of importance of glutamate signaling in the skeleton appear to be unnecessarily over interpreted, largely because they were based on relatively few experiments with limitations (Chenu *et al.*, 2001; Skerry *et al.*, 2001). We have found that the study of agonists and competitive antagonists (such as APV) in long-term cultures are notoriously problematic due to the significant contribution of exogenous glutamate concentrations present in the tissue-culture medium and serum, which are typically in the µM to mM range. The masking effect of exogenous glutamate in these experiments is also likely to be further compounded by endogenous glutamate spontaneously released from osteoblasts. Thus, intervention of glutamate signaling by either agonist or competitive antagonist would have to be perceived against a background of high glutamate concentrations. Therefore, it is not surprising that the competitive agents failed to affect either bone formation or bone resorption *in vitro*.

The lack of response of mature osteoclasts to APV and MK-801 in osteoclast resorption assays is consistent with other published data (Peet *et al.*, 1999), although MK-801 potently inhibited osteoclast differentiation and resorption in marrow/osteoblast cocultures when applied for very short durations (Peet *et al.*, 1999).

Surprisingly, on the basis of single knockout experiments Gray *et al.* implied that "any role that glutamate may play in the skeleton is minor" because of the absence of an obvious spontaneous skeletal phenotype. This is an unsafe assumption, since our demonstrations of multiple glutamate transporter expression in bone supports a potential for functional redundancy. In-depth functional analyses of GLAST$^{-/-}$ and other transgenic mice will help to resolve these important issues.

6. Glutamate Signaling in Bone Marrow

Bone marrow occupies the cylindrical cavities of the long bones. As well as being the site of blood production from hematopoietic stem cells, the marrow also interacts with bone cells, regulating the bone remodeling process. With such a divergent population of cells from both mesenchymal and hematopoietic lineages, sophisticated signaling systems within the bone marrow must operate to ensure lineage specificity of the differentiating stem cells. The observation that mononuclear bone marrow cells expressed the glutamate transporter GLT-1 (Mason *et al.*, 1997), suggested that glutamate signaling extended from bone surfaces and through the marrow cavity. Further analyses demonstrated that bone marrow megakaryocytes prominently expressed NMDA-type glutamate receptors, highlighting a multicellular glutamate signaling hierarchy in bone.

Megakaryocytes represent less than 0.5% of the cells in the bone marrow and they are sometimes described as "platelet precursors." Each megakaryocyte is capable of producing in the order of 3,000 platelets, leading to the release of 2×10^{11} platelets each day and therefore these cells play a pivotal role in hemostasis (Bruno and Hoffman, 1998; Italiano *et al.*, 1999). However, it is also clear that megakaryocytes are instrumental in regulating marrow cell populations by releasing numerous cytokines and growth factors that regulate hematopoietic and mesenchymal stem cell differentiation and transgenic manipulation of megakaryocyte differentiation *in vivo* frequently induces skeletal defects.

Earlier reports described the identification of a functional NMDA-type glutamate receptor on platelets. These studies demonstrated that NMDA antagonized the aggregating activity of archidonic acid and also inhibited ADP and platelet activating factor (PAF)-induced platelet aggregation (Franconi *et al.*, 1998). NMDA also increased intracellular calcium concentrations, completely inhibiting synthesis of the platelet aggregation promoter thromboxane B-12 (Franconi *et al.*, 1996). Changes in the sensitivity of these receptors in psychotic disorders has also lead to the possibility of platelets being used as peripheral markers of NMDA receptor dysfunction (Das *et al.*, 1995; Berk *et al.*, 2000). Using immunohistochemistry, RT-PCR, and northern blot analysis, we demonstrated that megakaryocytes expressed the NMDAR1 subunit of the NMDA receptor. Expression of the NMDAR2D subunit was also identified by RT-PCR of whole marrow and the megakaryoblastic cell line Meg-01 (Genever *et al.*, 1999). Interestingly, the level of glycosylation of the megakaryocytic NMDA receptors was significantly lower compared to NMDA receptors in the CNS. The exact purpose for glycosylation of the NMDA receptor remains unclear. However, it may aid the correct localization and orientation of NMDA receptors in the postsynaptic membrane, which may not be required for multidirectional glutamate signaling in bone marrow.

Radioligand binding with [3H]-MK-801 demonstrated that the megakaryocyte NMDA subunits coordinated as channel forming receptors, while inhibiting the receptors with MK-801 had profound effects on Meg-01 differentiation. Meg-01 cells treated with phorbol myristate acetate (PMA) undergo differentiation into a more megakaryocytic state, displaying increases in size, adhesion, cytoplasmic maturation, and the production of occasional platelet-like particles (Takeuchi *et al.*, 1991). Flow cytometric analyses demonstrated that PMA treatment increased CD41 (GPIIb-IIIa) expression, a cell surface marker of mature megakaryocytes. However, PMA-stimulated CD41 expression was inhibited by 30–55% when Meg-01 cells were treated with MK-801 (50–150 μM). PMA also activates many different types of integrins, including those implicated in megakaryocyte cell adhesion to both fibronectin and fibrinogen (Boudignon-Proudhon *et al.*, 1996; Tohyama *et al.*, 1998). Both MK-801 and APV inhibited Meg-01 adhesion in a dose-dependent manner. These experiments were repeated on another megakaryoblastic cell line CMK, with MK-801 also causing a significant decrease in PMA-mediated adhesion.

The identification of glutamate transporter protein GLT-1 in yet unidentified mononuclear marrow cells further supports the possibility of a glutamate-mediated signaling system within the bone marrow (Mason *et al.*, 1997). Furthermore, GLT-1 expression is prominent in selected cells located in the immediate vicinity of the megakaryocyte. This raises the possibility of these mononuclear cells being able to regulate the uptake and release of glutamate to the locality of the megakaryocyte. It is likely that characterization of the GLT-1 positive marrow cells will add another member to the ever-expanding multicellular glutamate-signaling family in bone tissues.

7. Summary

It is clear that bone and marrow cells, like neurons, have the capacity to receive and send glutamatergic signals. Osteoblasts express functional glutamate receptors, transporters, and release glutamate in a regulated manner through exocytotic machinery usually associated with neurons. Together these data indicate that glutamate signaling mechanisms exist within the bone microenvironment, which have fundamental analogies to synapses in the CNS. Despite widely differing physiological roles, glutamate signaling in bone cells and neurons may also regulate common cellular processes. In the CNS, glutamate signaling plays important roles in essential neuronal functions including learning and memory formation. Current evidence suggests that these activities are mediated through a complex sequence of cellular events that initiate long-lasting activity dependent changes in synaptic activity through a process of long-term potentiation (LTP). *In vitro* these processes can be mimicked by electrical stimulation using repetitive trains of high frequency electrical stimulation lasting only seconds that lead to a long-term increase in synaptic strength that can last for hours or days. The potency of LTP inducing stimuli in neurons has clear similarities to the osteogenic potency of mechanical loading, in which prolonged periods of new bone formation can be induced by a single brief period of loading. The precise cellular mechanism underlying the transduction of mechanical strain into an osteogenic response remains unknown, although is likely to depend on a form of cellular memory. Considering osteoblasts express the molecular architecture required to coordinate LTP and that mechanical loading regulates glutamate transporter expression, we hypothesize that the molecular events underlying memory formation in bone and brain may be similar or identical (Spencer and Genever, 2003). As such glutamate signaling may play a central role in the control of loading induced bone formation, the manipulation of which could be used to promote bone formation in diseases associated with low bone mass.

This is a rapidly advancing field of research. It is likely that future studies will be able to further clarify the functional role of glutamate signaling in bone and marrow tissues and ultimately determine if glutamate, a relatively simple amino acid with complex signaling capabilities, is indeed able to teach old bones new tricks.

8. Recent Developments

The identification of glutamine synthetase expression by osteoblastic cells provides further evidence of similarity between glutamate signalling in bone and the CNS (Olkku *et al.*, 2004). Glutamate transporters may act in concert with glutamine synthetase and mitochondrial glutaminase in bone cells to provide a glutamate/glutamine pathway that may act to recycle glutamate in the bone microenvironment. Other recent studies support the emerging regulatory role of glutamate signalling in the biology of osteoblasts, osteoclasts and megakaryocytes. Hinoi *et al.* (2002, 2003) have provided compelling data to complement our observations of regulated glutamate release from osteoblasts and further confirm the involvement of NMDA receptors in regulating osteoblast differentiation through the DNA binding activity of the key osteoblast transcription factor Cbfa-1. Merle *et al.* (2003) have recently described the expression of the NMDA receptor subunits NMDAR1, NMDAR2A, B and D in RAW 264.7 and mouse mononuclear marrow cells in a study aimed at determining the role of NMDA receptor activation in osteoclastogenesis. The NMDA receptor antagonists MK-801 and DEP caused a dose-dependent inhibition of osteoclast formation in the presence of the osteoclast differentiation factors RANKL and m-CSF, whilst activation induced translocation of NFkB, a key transcription factor required for osteoclast formation. These data, combined with those reported by Peet *et al.* (1999) support a functional role for NMDA receptors in the control of osteoclastogenesis through a direct effect on haematopoietic progenitor cells. We have recently further characterized the role of NMDA receptors in the regulation of human megakaryocytopoiesis, demonstrating that the NMDA receptor antagonist MK-801 inhibited proplatelet formation and induced marked ultrastructural changes in human megakaryocytes generated from $CD34^+$ progenitor cells, without affecting cell proliferation or apoptosis. These changes were accompanied by a decrease in the expression of the megakaryocyte markers CD61, CD41a and CD42a, consistent with an inhibitory effect on megakaryocyte differentiation (Hitchcock *et al.*, 2003).

Acknowledgments

We thank the Arthritis Research Campaign and British Heart Foundation for supporting our research.

References

Bajjalieh, S.M. (1999). Synaptic vesicle docking and fusion. *Curr. Opin. Neurobiol.* **9**, 321.

Bennett, M.K. and R.H. Scheller (1993). The molecular machinery for secretion is conserved from yeast to neurons. *Proc. Natl. Acad. Sci. USA* **90**, 2559.

Berk, M., H. Plein, and B. Belsham (2000). The specificity of platelet glutamate receptor supersensitivity in psychotic disorders. *Life Sci.* **66**, 2427.

Bhangu, P.S., P.G. Genever, G.J. Spencer, T.S. Grewal, and T.M. Skerry (2001). Evidence for targeted vesicular glutamate exocytosis in osteoblasts. *Bone* **29**, 16.

Boudignon-Proudhon, C., P.M. Patel, and L.V. Parise (1996). Phorbol ester enhances integrin alpha IIb beta 3-dependent adhesion of human erythroleukemic cells to activation-dependent monoclonal antibodies. *Blood* **87**, 968.

Bruno, E. and R. Hoffman (1998). Human megakaryocyte progenitor cells. *Semin. Hematol.* **35**, 183.

Chenu, C., C.M. Serre, C. Raynal, B. Burt-Pichat and P.D. Delmas (1998). Glutamate receptors are expressed by bone cells and are involved in bone resorption. *Bone* **22**, 295.

Chenu, C., C. Itzstein, and L. Espinosa (2001). Absence of evidence is not evidence of absence: Redundancy blocks determination of cause and effect. *J. Bone Miner. Res.* **16**, 1278.

Das, I., N.S. Khan, B.K. Puri, S.R. Sooranna, J. de Belleroche, and S.R. Hirsch (1995). Elevated platelet calcium mobilization and nitric oxide synthase activity may reflect abnormalities in schizophrenic brain. *Biochem. Biophys. Res. Commun.* **212**, 375.

Dobson, K.R. and T.M. Skerry (2000). The NMDA-type glutamate receptor antgonist MK-801 regulates differentiation of rat bone marrow osteoprogenitors and influences adipogenesis. *J. Bone Miner. Res.* **15**, S1, SA211.

Espinosa, L., C. Itzstein, H. Cheynel, P.D. Delmas, and C. Chenu (1999). Active NMDA glutamate receptors are expressed by mammalian osteoclasts. *J. Physiol.* **518**, 47.

Franconi, F., M. Miceli, M.G. De Montis, E.L. Crisafi, F. Bennardini, and A. Tagliamonte (1996). NMDA receptors play an anti-aggregating role in human platelets. *Thromb. Haemost.* **76**, 84.

Franconi, F., M. Miceli, L. Alberti, G. Seghieri, M.G. De Montis, and A. Tagliamonte (1998). Further insights into the anti-aggregating activity of NMDA in human platelets. *Br. J. Pharmacol.* **124**, 35.

Genever, P.G., D.J. Wilkinson, A.J. Patton, N.M. Peet, Y. Hong, and A. Mathur *et al.* (1999). Expression of a functional N-methyl-D-aspartate-type glutamate receptor by bone marrow megakaryocytes. *Blood* **93**, 2876.

Genever, P.G. and T.M. Skerry (2001). Regulation of spontaneous glutamate release activity in osteoblastic cells and its role in differentiation and survival: Evidence for intrinsic glutamatergic signaling in bone. *FASEB J.* **15**, 1586.

Gray, C., H. Marie, M. Arora, K. Tanaka, A. Boyde, and S. Jones (2001). Glutamate does not play a major role in controlling bone growth. *J. Bone Miner. Res.* **16**, 742.

Gu, Y. and S.J. Publicover (2000). Expression of functional metabotropic glutamate receptors in primary cultured rat osteoblasts, Cross-talk with N-methyl-D-aspartate receptors. *J. Biol. Chem.* **275**, 34252.

Hinoi, E., S. Fujimori, Y. Nakamura, and Y. Yoneda (2001). Group III metabotropic glutamate receptors in rat cultured calvarial osteoblasts. *Biochem. Biophys. Res. Commun.* **281**, 341.

Hinoi, E., S. Fujimori, T. Takarada, H. Taniura, and Y. Yoneda (2002). Facilitation of glutamate release by ionotropic glutamate receptors in osteoblasts. *Biochem. Biophys. Res. Commun.* **297**, 452.

Hinoi, E., S. Fujimori, and Y. Yoneda (2003). Modulation of cellular differentiation by N-methyl-D-aspartate receptors in osteoblasts. *FASEB J.* **17**, 1532.

Hitchcock, I.S., T.M. Skerry, M.R. Howard, and P.G. Genever (2003). NMDA receptor-mediated regulation of human megakaryocytopoiesis. *Blood* **102**, 1254.

Huggett, J., A. Vaughan-Thomas, and D. Mason (2000). The open reading frame of the Na(+)-dependent glutamate transporter GLAST-1 is expressed in bone and a splice variant of this molecule is expressed in bone and brain. *FEBS Lett.* **485**, 13.

Italiano, J.E., Jr., P. Lecine, R.A. Shivdasani, and J.H. Hartwig (1999). Blood platelets are assembled principally at the ends of proplatelet processes produced by differentiated megakaryocytes. *J. Cell Biol.* **147**, 1299.

Itzstein, C., L. Espinosa, P.D. Delmas, and C. Chenu (2000). Specific antagonists of NMDA receptors prevent osteoclast sealing zone formation required for bone resorption. *Biochem. Biophys. Res. Commun.* **268**, 201.

Itzstein, C., H. Cheynel, B. Burt-Pichat, B. Merle, L. Espinosa, P.D. Delmas *et al.* (2001). Molecular identification of NMDA glutamate receptors expressed in bone cells. *J. Cell Biochem.* **82**, 134.

Laketic-Ljubojevic, I., L.J. Suva, F.J. Maathuis, D. Sanders and T.M. Skerry (1999). Functional characterization of N-methyl-D-aspartic acid-gated channels in bone cells. *Bone* **25**, 631.

Manolagas, S.C. (2000). Birth and death of bone cells: Basic regulatory mechanisms and implications for the pathogenesis and treatment of osteoporosis. *Endocr. Rev.* **21**, 115.

Mason, D.J., L.J. Suva, P.G. Genever, A.J. Patton, S. Steuckle, R.A. Hillam *et al.* (1997). Mechanically regulated expression of a neural glutamate transporter in bone: A role for excitatory amino acids as osteotropic agents? *Bone* **20**, 199.

Merle, B., C. Itzstein, P.D. Delmas, and C. Chenu (2003). NMDA glutamate receptors are expressed by osteoclast precursors and involved in the regulation of osteoclastogenesis. *J. Cell Biochem.* **90**, 424.

Olkku, A., P.V. Bodine, A. Linnala-Kankkunen, and A. Mahonen (2004). Glucocorticoids induce glutamine synthetase expression in human osteoblastic cells: a novel observation in bone. *Bone* **34**, 320.

Patton, A.J., P.G. Genever, M.A. Birch, L.J. Suva, and T.M. Skerry (1998). Expression of an N-methyl-D-aspartate-type receptor by human and rat osteoblasts and osteoclasts suggests a novel glutamate signaling pathway in bone. *Bone* **22**, 645.

Peet, N.M., P.S. Grabowski, I. Laketic-Ljubojevic, and T.M. Skerry (1999). The glutamate receptor antagonist MK801 modulates bone resorption in vitro by a mechanism predominantly involving osteoclast differentiation. *FASEB J.* **13**, 2179.

Rodan, G.A. and T.J. Martin (2000). Therapeutic approaches to bone diseases. *Science* **289**, 1508–1514.

Serre, C.M., D. Farlay, P.D. Delmas, and C. Chenu (1999). Evidence for a dense and intimate innervation of the bone tissue, including glutamate-containing fibers. *Bone* **25**, 623.

Skerry, T.M., P.G. Genever, A.F. Taylor, K. Dobson, D. Mason, and L.J. Suva (2001). Absence of evidence is not evidence of absence: The shortcomings of the GLAST knockout mouse. *J. Bone Miner. Res.* **16**, 1729.

Spencer, G.J. and P.G. Genever (2003). Long-term potentiation in bone—a role for glutamate in strain-induced cellular memory? *BMC Cell Biology* **4**, 8.

Storck, T., S. Schulte, K. Hofmann, and W. Stoffel (1992). Structure, expression, and functional analysis of a Na(+)-dependent glutamate/aspartate transporter from rat brain. *Proc. Natl. Acad. Sci. USA* **89**, 10955.

Takeuchi, K., M. Ogura, H. Saito, M. Satoh, and M. Takeuchi (1991). Production of platelet-like particles by a human megakaryoblastic leukemia cell line (MEG-01). *Exp. Cell Res.* **193**, 223.

Tohyama, Y., K. Tohyama, M. Tsubokawa, M. Asahi, Y. Yoshida, and H. Yamamura (1998). Outside-In signaling of soluble and solid-phase fibrinogen through integrin alphaIIbbeta3 is different and cooperative with each other in a megakaryoblastic leukemia cell line, CMK. *Blood* **92**, 1277.

13

Expression and Function of Metabotropic Glutamate Receptors in Liver

Marianna Storto, Maria Pia Vairetti, Francesc X. Sureda,
Barbara Riozzi, Valeria Bruno, and Ferdinando Nicoletti

1. Introduction

Glutamate is the major excitatory neurotransmitter in the Central Nervous System (CNS) where its actions are mediated by different types of membrane receptors classified into "ionotropic" (iGluR) and "metabotropic" (mGluR) glutamate receptors (reviewed by Nakanishi, 1992). iGlu receptors (NMDA, N-methyl-D-aspartate; AMPA, α-amino-3-hydroxy-5-methyl-4-isoxazole propionic acid, and kainate receptors) form membrane ion channels permeable to mono- and divalent cations, and their activation is responsible for fast excitatory synaptic transmission; however, this class of receptors, particularly NMDA receptors, has been widely implicated in the pathophysiology of neuronal damage in most of acute and chronic neurodegenerative disorders (Choi, 1992; Doble, 1999). A sustained activation of NMDA receptors induces excessive and prolonged increases in intracellular free Ca^{2+} levels. Ca^{2+} activates a series of intracellular enzymes, including phospholipase A_2, xanthine oxidase, and constitutive nitric oxide synthase, which in turn trigger a cascade of events leading to a form of cell death, termed "excitotoxicity" (Choi, 1992). mGlu receptors are coupled to G proteins and regulate the activity of a variety of membrane enzymes and ion channels (reviewed by Pin and Duvoisin, 1995). mGlu receptors form a family of eight subtypes, which are subdivided into three groups on the basis of sequence similarities and transduction pathways. Group-I includes mGlu1 and mGlu5 receptors, which are coupled to polyphosphoinositide (PPI) hydrolysis and can also regulate the activity of different types of K^+ channels. Group-II (mGlu2 and mGlu3) and group-III (mGlu4, -6, -7, -8) receptors are negatively coupled to adenylate cyclase activity in heterologous expression systems and, with the exception of mGlu6 receptors, share the ability to modulate different types of voltage-sensitive

Marianna Storto, Barbara Riozzi, Valeria Bruno, and Ferdinando Nicoletti • I.N.M. Neuromed, Pozzilli, Italy. **Maria Pia Vairetti** • Department of Internal Medicine and Medical Therapy, University of Pavia, Pavia. **Francesc X. Sureda** • Pharmacology Unit, School of Medicine, University of Rovira i Virgili, Reus, Spain. **Valeria Bruno and Ferdinando Nicoletti** • Department of Human Physiology and Pharmacology, University of Rome "La Sapienza," Rome, Italy.

Glutamate Receptors in Peripheral Tissue, edited by Santokh Gill and Olga Pulido.
Kluwer Academic / Plenum Publishers, New York, 2005.

Ca^{2+} channels when expressed in their native environment. Both iGlu and mGlu receptors have an established role in the CNS, and, in particular, mGlu receptors are involved in various aspects of CNS physiology and pathology, including modulation of excitatory synaptic transmission, developmental plasticity, learning and memory processes, and neurodegeneration (Nakanishi, 1994; Conn and Pin, 1997; Bruno *et al.* 2001).

There is a growing interest on the expression and function, in cells outside the CNS, of receptor subtypes normally present in the CNS, as several neurotransmitter receptors have been found in peripheral organs. Many components of glutamatergic synapses, such as postsynaptic and presynaptic receptors (reviewed by Gill and Pulido, 2001; Skerry and Genever, 2001) and/or glutamate transporters (Howell *et al.*, 2001), are expressed in tissues that are remote both functionally and embryologically from the CNS.

We have recently focused our interest on the expression and function of peripheral mGlu receptors and have found evidence for the existence of mGlu receptors in thymus, testis, melanocytes, and liver (Sureda *et al.*, 1997; Frati *et al.*, 2000; Storto *et al.*, 2000a, b, 2001, 2003). The presence of peripheral mGlu receptors is also provided by several studies which have identified this class of receptors in osteoblasts, keratinocytes, bone marrow, heart, taste buds, gastrointestinal tract, ovary, kidney, and pancreas (reviewed in Gill and Pulido, 2001; Skerry and Genever, 2001).

2. Expression and Function of mGlu Receptors in Liver

Several metabotropic subtypes of neurotransmitter receptors have been recently identified in isolated or cultured hepatocytes, hepatocytoma cell lines, liver membranes, and the whole liver, both in embryos and in adult animals, such as $\alpha 1b$ and $\beta 2$ adrenergic receptors (Auman *et al.*, 2002; Castrejon-Sosa *et al.*, 2002), m3 cholinergic receptors (Vatamaniuk *et al.*, 2003), P2Y purinergic receptors (Suarez-Huerta *et al.*, 2001; Feranchak and Fitz, 2002; Junankar *et al.*, 2002), A2A and A2B adenosine receptors (Harada *et al.*, 2001), and GABAB receptors (Castelli *et al.*, 1999; Biju *et al.*, 2002), supporting an involvement of these receptors in liver physiology and pathology.

The first evidence for a glutamatergic receptor in liver comes from studies of the distribution of radiolabelled NMDA channel blockers (Nasstrom *et al.*, 1993; Lin *et al.*, 1996; Samnick *et al.*, 1998); since then, the iGlu receptor subunits NMDAR1, GluR2/3, and KA2 have been localized in liver (reviewed by Gill *et al.*, 2000), but evidence for the presence of mGlu receptors was missing.

We have first identified a functional mGlu receptor belonging to group-I in cultured rat hepatocytes as indicated by stimulation of PPI hydrolysis induced by the group-I mGluR agonists, 1*S*,3*R*-ACPD (ACPD) and quisqualate, which is inhibited by the group-I mGlu receptor antagonist, MCPG (Sureda *et al.*, 1997) (Table 13.1). We therefore investigated whether the presence of these receptors may have, in liver, physiological and/or pathological implications, as, in the CNS, activation of group-I mGlu receptors facilitates the development of excitotoxic and hypoxic neuronal damage and blockade of this class of receptors is neuroprotective (Nicoletti *et al.*, 1996; Bruno *et al.*, 2001). A similar function in hepatocytes could therefore offer new strategies for the treatment of liver damage by ischemia/reperfusion, which may develop during severe episodes of hypotension and is an unavoidable complication of liver transplantation. In isolated hepatocytes deriving from both neonate and adult rat liver, we have found the presence of the group-I mGlu receptor subtype, mGlu5, as assessed by

Table 13.1. Functional Activity of Group-I mGlu Receptors in
Cultured Hepatocytes

Days *in vitro*	[³H] InsP formation (cpm/mg protein)
2 days *in vitro*	
Basal	6075 ± 202
ACPD (100 μM)	11325 ± 150^a
Quisqualate (100 μM)	10800 ± 675^a
ACPD + MCPG (500 μM)	7025 ± 650^b
5 days *in vitro*	
Basal	4162 ± 225
ACPD (100 μM)	5887 ± 350^a
Quisqualate (100 μM)	5650 ± 120^a
ACPD + MCPG (500 μM)	3900 ± 150^b

Stimulation of PPI hydrolysis by mGlu receptor agonists in cultured hepatocytes at 2 or 5 days
in vitro. Values are means $\pm$ SEM of 4–8 determinations.
[a]$p < 0.05$ vs basal values.
[b]$p < 0.05$ vs ACPD (one-way ANOVA + Fisher's PLSD).

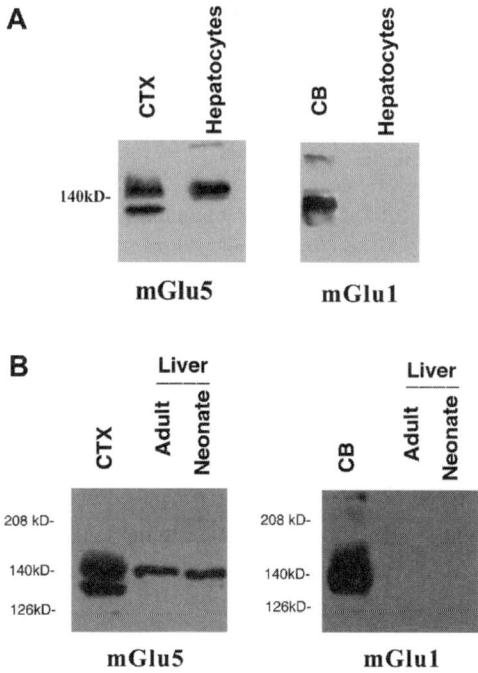

Figure 13.1. Expression of mGlu5 receptor protein in (A) isolated hepatocytes and (B) adult or neonate (8-day-old) rat liver. Western blot analysis shows the presence of a band corresponding to the monomeric form of mGlu5 receptors. Note the lack of expression of mGlu1 receptor protein. In membranes from isolated hepatocytes high molecular size aggregates are present in addition to the receptor monomer. The adult rat cortex (CTX) is used as a reference tissue for mGlu5 receptors, whereas adult rat cerebellum is used as reference tissue for mGlu1 receptors. Twenty micrograms of proteins have been loaded per lane.

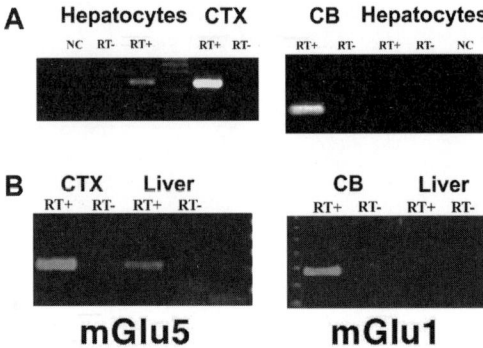

Figure 13.2. Reverse-transcription polymerase chain reaction (PCR) analysis of mGlu5 receptor mRNA in (A) isolated hepatocyte extracts and (B) adult rat liver. Note the lack of mGlu1 receptor mRNA. The adult rat cortex (CTX) is used as a reference tissue for mGlu5 receptors, whereas adult rat cerebellum is used as reference tissue for mGlu1 receptors. RNA samples have been incubated in the presence (+) or absence (−) of reverse transcriptase (RT). Negative control (NC) has been loaded with dH_2O.

Western blot (Figure 13.1) and polymerase chain reaction (PCR) (Figure 13.2). This pattern of expression is different than in brain tissue, in which mGlu5 expression is high early after birth and then declines in the adult life. In contrast, we could not detect mGlu1 receptors. As the same expression profile was observed in the whole liver (Figures 13.1 and 13.2), hepatocytes may represent the major source of mGlu5 receptor expression in the rat liver, although we cannot exclude that other cell types, such as Kupffer cells, endothelial cells, and lypocytes also express mGlu5 receptors. mGlu5 receptors were also observed in HepG2 hepatoma cells, which are commonly used for the study of hepatocyte metabolism. Activation of mGlu5 receptors in heterologous expression systems leads to the formation of inositoltrisphosphate and DAG, which in turn produce oscillatory increases in cytosolic free Ca^{2+} and activation of PKC (Kawabata *et al.*, 1996). These intracellular processes mediate the physiological actions of mGlu5 receptors, but may become detrimental in cells subjected to environmental stresses, such as hypoxia, ischemia, or other pathological conditions (Choi *et al.*, 1992). When isolated hepatocytes are subjected to an anoxic insult, a decline in hepatocyte viability occurs in a time-dependent fashion, which is accelerated by the addition of the group-I mGlu receptor agonists, ACPD, and quisqualate, as well as by the addition of glutamate (used at micromolar concentrations that are physiologically found in the portal blood). So it can be speculated that, *in vivo*, glutamate reaching the liver via the portal circulation may play a permissive role in the development of hypoxic damage, and high levels of glutamate are present in the portal blood after ingestion of food rich in proteins or monosodium glutamate. The effect of ACPD and quisqualate is reversed in the presence of the noncompetitive mGlu5 receptor antagonist, MPEP; moreover, MPEP is highly protective by itself, rescuing a substantial proportion of hepatocytes from anoxic death, and facilitates cell recovery after re-oxygenation (Table 13.2); this suggests that mGlu5 receptors are endogenously activated during the induction of anoxia, as this experimental condition induces an increase in extracellular glutamate to levels which are sufficient to activate mGlu5 receptors (Storto *et al.*, 2000b). Thus, endogenous activation of mGlu5 receptors can be translated into a death signal in cells that are made vulnerable by hypoxia, and MPEP could exert its protective effect by preventing the permissive function of mGlu5 receptors activated by glutamate released from anoxic hepatocytes. Therefore, in isolated hepatocytes, mGlu5 receptors are endogenously activated and may be involved in several

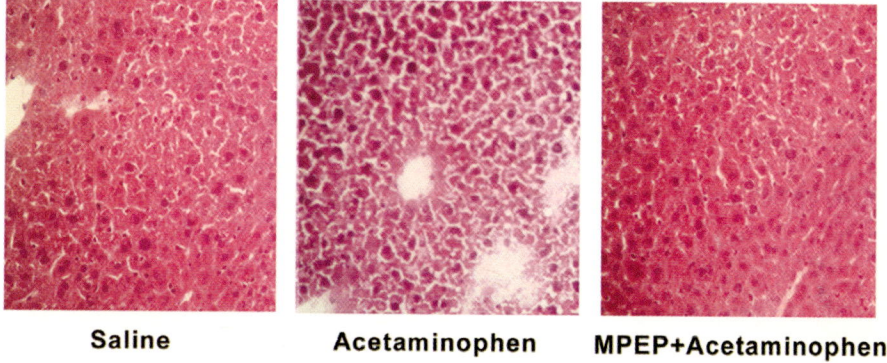

Saline   **Acetaminophen**   **MPEP+Acetaminophen**

Figure 13.3.   Hematoxylin-eosin staining in liver sections of mice treated with a single injection of acetaminophen (300 mg/kg, i.p.) and sacrificed 24 hr later. Note the areas of centrilobular necrosis in the acetaminophen condition. Objective = 40×.

Table 13.2. The mGlu5 Receptor Antagonist, MPEP, is Protective Against Hypoxic and Oxidative Damage in Isolated Hepatocytes

Treatment	Cell viability (%)
O_2	78 ± 0.9
N_2	42 ± 2.5
+ ACPD (100 μM)	25 ± 2.1[a]
+ Quisqualate (50 μM)	18 ± 1.7[a]
+ Glutamate (100 μM)	24 ± 0.8[a]
+ ACPD + MPEP	60 ± 3.4[b]
+ MPEP (30 μM)	48 ± 5.6[a]
Control	86 ± 2.2
t-BuOOH (0.5M)	64 ± 2.6[a]
t-BuOOH + MPEP (30 μM)	72 ± 1.7[b]

Cell death was assessed by measuring lactate dehydrogenase, released by damaged hepatocytes, 60 min after induction of hypoxia or addition of t-BuOOH in the perfusion fluid. Values are mean ± SEM of 8 determinations.
[a] $p < 0.05$ vs hypoxia (N_2) or control.
[b] $p < 0.05$ vs ACPD or t-BuOOH (one-way ANOVA + Fisher's PLSD).

aspects of liver physiology and pathology. Accordingly, mGlu5 receptor antagonists are also able to protect isolated hepatocytes against oxidative damage produced by *tert*-butylhydroperoxide (*t*-BuOOH) (Table 13.2), by reducing the formation of radical oxygen species (Storto *et al.*, 2003). The treatment with *t*-BuOOH induces in hepatocytes a perturbation in intracellular Ca^{2+} homeostasis and free-radical formation, which are direct consequence of *t*-BuOOH metabolism by the glutathione peroxidase reductase system (Bellomo *et al.*, 1982, 1984). Hence, activation of mGlu5 receptors by endogenous glutamate, by increasing intracellular Ca^{2+} levels, could act synergistically with oxidizing agents in the induction of liver damage. The same hepatoprotective effect is observed *in vivo*, in animals treated with a single i.p. injection of high doses of acetaminophen, which induce centrilobular liver necrosis (Figure 13.3, see color insert), due to the bioactivation of acetaminophen by cytochrome P-450 leading to the formation of a highly reactive metabolite, which forms chemical adducts with intracellular proteins (Hinson *et al.*, 1981). In this case, the protective effect induced by MPEP is due not only to the reduction in the formation of reactive oxygen species and in the increase of G6PDH activity, but also to the reduction in the enhanced iNOS expression and activity, induced by acetaminophen. We have also assessed the hepatoprotective effect of mGlu5 receptor antagonists in a more drastic model of liver toxicity, in which acetaminophen is administered for 6 days; even under this experimental condition, MPEP reduces the extent of centrilobular necrosis and prevents the increase in reactive oxygen species and G6PDH activity induced by acetaminophen treatment (Storto *et al.*, 2003).

It could be speculated that mGlu5 receptors can become therapeutic targets for the development of drugs useful for prophylaxis or treatment of liver toxicity induced by hepatotoxic agents, although information on the safety profile of mGlu5 receptor antagonists are still lacking. In our studies, mice treated with mGlu5 receptor antagonists showed neither signs of liver toxicity nor remarkable changes in spontaneous behavior, therefore, the possibility that mGlu5 receptors can be targeted by novel drugs endowed with protective activity against liver toxicity is certainly exciting and worth to be investigated.

3. Conclusions

There is compelling evidence for the expression and function of glutamate receptors in peripheral organs, suggesting that glutamate can act as a signaling molecule also outside the CNS. In addition to endogenous glutamate, other naturally occurring substances endowed with glutamate-like excitatory properties and whose neurotoxic effects have been described, such as monosodium glutamate, β-*N*-oxalylamino-L-alanine (BOAA), β-*N*-methylamino-L-alanine (BMAA), and domoic acid (Spencer *et al.*, 1986; Zautcke *et al.*, 1986; Weiss *et al.*, 1989; Peng *et al.*, 1994), present in food as additives or contaminants, may enter the body and reach non-neuronal tissue sites, especially the liver through the portal blood, and affect peripheral organs expressing glutamate receptors. Therefore, peripheral glutamate receptor targeting may become valuable therapeutically, and awaits the development of subtype-selective and tissue-targeted drugs.

References

Auman, J.T., F.J. Seidler, C.A. Tate, and T.A. Slotkin (2002). Are developing beta-adrenoceptors able to desensitise? Acute and chronic effects of beta-agonists in neonatal heart and liver. *Am. J. Physiol. Regul. Integr. Comp. Physiol.* **283**, R205–R217.

Bellomo, G., S.A. Jewell, H. Thor, and S. Orrenius (1982). Regulation of intracellular calcium compartmentation: Studies with isolated hepatocytes and t-butyl hydroperoxide. *Proc. Natl. Acad. Sci. USA* **79**, 6842–6846.

Bellomo, G., A. Martino, P. Richelmi, G.A. Moore, S.A. Jewell, and S. Orrenius (1984). Pyridine-nucleotide oxidation. Ca^{2+} cycling and membrane damage during tert-butyl hydroperoxidase metabolism by rat-liver mitochondria. *Eur. J. Biochem.* **140**, 1–6.

Biju, M.P., S. Pyroja, N.V. Rajeshkumar, and C.S. Paulose (2002). Enhanced GABAB receptor in neoplastic rat liver: Induction of DNA synthesis by baclofen in hepatocyte cultures. *J. Biochem. Mol. Biol. Biophys.* **6**, 209–214.

Bruno, V., G. Battaglia, A. Copani, M. D'Onofrio, P. Di Iorio, A. De Blasi *et al.* (2001). Metabotropic glutamate receptor subtypes as targets for neuroprotective drugs. *J. Cereb. Blood Flow Metab.* **21**, 1013–1033.

Castelli, M.P., A. Inganni, E. Stefanini, and G.L. Gessa (1999). Distribution of GABAB receptor mRNAs in the rat brain and peripheral organs. *Life Sci.* **64**, 1321–1328.

Castrejon-Sosa, M., R. Villalobos-Molina, R. Guinzberg, and E. Pina (2002). Adrenaline (via α1B-adrenoceptors) and ethanol stimulate OH* radical production in isolated rat hepatocytes. *Life Sci.* **71**, 2469–2474.

Choi, D.W. (1992). Excitotoxic cell death. *J. Neurobiol.* **23**, 1261–1276.

Conn, P.J. and J.P. Pin (1997). Pharmacology and function of metabotropic glutamate receptors. *Ann. Rev. Pharmacol. Toxicol.* **37**, 205–237.

Doble, A. (1999). The role of excitotoxicity in neurodegenerative diseases: Implications for therapy. *Pharmacol. Ther.* **81**, 163–221.

Feranchak, A.P. and J.G. Fitz (2002). Adenosine triphosphate release and purinergic regulation of cholangiocyte transport. *Semin. Liver Dis.* **22**, 251–262.

Frati, C., C. Marchese, G. Fisichella, A. Copani, M.R. Nasca, M. Storto *et al.* (2000). Expression of functional mGlu5 metabotropic glutamate receptors in human melanocytes. *J. Cell. Physiol.* **183**, 364–372.

Gill, S.S. and O.M. Pulido (2001). Glutamate receptors in peripheral tissues: Current knowledge, future research, and implications for toxicology. *Toxicol. Pathol.* **29**, 208–223.

Gill, S.S., O.M. Pulido, R.W. Mueller, and P.F. McGuire (2000). Potential target sites in peripheral tissues for excitatory neurotransmission and excitotoxicity. *Toxicol. Pathol.* **28**, 277–284.

Harada, H., O. Asano, T. Kawata, T. Inoue, T. Horizoe, N. Yasuda *et al.* (2001). 2-Alkynyl-8-aryladenines possessing an amide moiety: Their synthesis and structure-activity relationships of effects on hepatic glucose production induced via agonism of the A2B adenosine receptor. *Bioorg. Med. Chem.* **9**, 2709–2726.

Hinson, J.A., L.R. Pohl, T.J. Monks, and J.R. Gillette (1981). Acetaminophen-induced hepatotoxicity. *Life Sci.* **29**, 107.

Howell, J.A., A.D. Matthews, K.C. Swanson, D.L. Harmon, and J.C. Matthews (2001). Molecular identification of high-affinity glutamate transporters in sheep and cattle forestomach, intestine, liver, kidney, and pancreas. *J. Anim. Sci.* **79**, 1329–1336.

Junankar, P.R., A. Karjalainen, and K. Kirk (2002). The role of P2Y1 purinergic receptors and cytosolic Ca^{2+} in hypotonically activated osmolyte efflux from a rat hepatoma cell line. *J. Biol. Chem.* **277**, 40324–40334.

Kawabata, S., R. Tsutsumi, A. Kohara, T. Yamaguchi, S. Nakanishi, and M. Okada (1996). Control of calcium oscillations by phosphorylation of metabotropic glutamate receptors. *Nature* **383**, 89–92.

Lin, Y.J., S. Bovetto, J.M. Carver, and T. Giordano (1996). Cloning of the cDNA for the human NMDA receptor NR2C subunit and its expression in the central nervous system and periphery. *Brain Res. Mol. Brain Res.* **43**, 57–64.

Nakanishi, S. (1992). Molecular diversity of glutamate receptors and implications for brain function. *Science* **258**, 597–602.

Nakanishi, S. (1994). Metabotropic glutamate receptors: Synaptic transmission, modulation, and plasticity. *Neuron* **13**, 1031–1037.

Nasstrom, J., E. Boo, M. Stahlberg, and O.G. Berge (1993). Tissue distribution of 2 NMDA receptor antagonists, [^{3}H]CGS 19755 and [^{3}H]MK801, after intrathecal injection in mice. *Pharmacol. Biochem. Behav.* **44**, 9–15.

Nicoletti, F., V. Bruno, A. Copani, G. Casabona, and T. Knöpfel (1996). Metabotropic glutamate receptors: A new target for the therapy of neurodegenerative disorders? *Trends Neurosci.* **19**, 267–271.

Peng, Y.G., T.B. Taylot, R.E. Finch, R.C. Switzer, and J.S. Ramsdell (1994). Neuroexcitatory and neurotoxic actions of the amnesic shellfish poison, domoic acid. *NeuroReport* **5**, 981–985.

Pin, J.P., and R. Duvoisin (1995). The metabotropic glutamate receptors: Structure and functions. *Neuropharmacology* **34**, 1–26.

Samnick, S., S. Ametamey, K.L. Leenders, P. Vontobel, G. Quack, C.G. Parsons *et al.* (1998). Electrophysiological study, biodistribution in mice, and preliminary PET evaluation in a rhesus monkey of 1-amino-3-[^{18}F]-fluoromethyl-5-methyl-adamantane (^{18}F-MEM): A potential radioligand for mapping the NMDA-receptor complex. *Nucl. Med. Biol.* **25**, 323–330.

Skerry, T.M. and P.G. Genever (2001). Glutamate signalling in non-neuronal tissues. *Trends Pharmacol. Sci.* **22**, 174–181.

Spencer, P.S., D.N. Roy, A. Ludolph, J. Hugon, M.P. Dwivedi, and H.H. Schaumburg (1986). Lathyrism: Evidence for role of the neuroexcitatory amino acid BOOA. *Lancet* **ii**, 1066–1067.

Storto, M., U. De Grazia, G. Battaglia, M.P. Felli, M. Maroder, A. Gulino *et al.* (2000a). Expression of metabotropic glutamate receptors in murine thymocytes and thymic stromal cells. *J. Neuroimmunol.* **109**, 112–120.

Storto, M., U. De Grazia, T. Knöpfel, P.L. Canonico, A. Copani, P. Richelmi *et al.* (2000b). Selective blockade of mGlu5 metabotropic glutamate receptors protects rat hepatocytes against hypoxic damage. *Hepatology* **31**, 649–655.

Storto, M., M. Sallese, L. Salvatore, R. Poulet, D.F. Condorelli, P. Dell'Albani *et al.* (2001). Expression of metabotropic glutamate receptors in the rat and human testis. *J. Endocrinol.* **170**, 71–78.

Storto, M., R.T. Ngomba, G. Battaglia, I. Freitas, P. Griffino, P. Richelmi *et al.* (2003). Selective blockade of mGlu5 metabotropic glutamate receptors is protective against acetaminophen hepatotoxicity in mice. *J. Hepatol.* **38**, 179–187.

Suarez-Huerta, N., V. Pouillon, J. Boeynaems, and B. Robaye (2001). Molecular cloning and characterization of the mouse P2Y4 nucleotide receptor. *Eur. J. Pharmacol.* **416**, 197–202.

Sureda, F., A. Copani, V. Bruno, T. Knöpfel, G. Meltzger, and F. Nicoletti (1997). Metabotropic glutamate receptor agonists stimulate polyphosphoinositide hydrolysis in primary cultures of rat hepatocytes. *Eur. J. Pharmacol.* **338**, R1–R2.

Vatamaniuk, M.Z., O.V. Horyn, O.K. Vatamaniuk, and N.M. Doliba (2003). Acetylcholine affects rat liver metabolism via type 3 muscarinic receptors in hepatocytes. *Life Sci.* **72**, 1871–1882.

Weiss, J.H., J.-Y. Koh, and D.W. Choi (1989). Neurotoxicity of β-N-methylamino-L-alanine (BMAA) and β-oxalyl-amino-L-alanine (BOAA) on cultured cortical neurons. *Brain Res.* **497**, 64–71.

Zautcke, J.L., J.A. Schwartz, and E.J. Mueller (1986). Chinese restaurant syndrome: A review. *Ann. Emerg. Med.* **15**, 1210–1213.

14

Neuroexcitatory Signaling in Immune Tissues

Helga S. Haas and Konrad Schauenstein

1. Introduction

The concept of mutual immuno–neuro–endocrine interactions came up in the early 1970s, after Besedovsky and colleagues (Besedovsky *et al.*, 1975) had described the neuroendocrine feedback regulation of immune responses via glucocorticoid hormones. A transient increase in serum glucocorticoids at the time of maximal antibody production after an antigenic challenge is mediated by central activation of the hypothalamic–pituitary–adrenal axis (HPA axis). It contributes to the specificity of immune reactions (Besedovsky *et al.*, 1979), and defects of this feedback predispose to unwanted immune reactions, as suggested by data in animal models with spontaneous (Schauenstein *et al.*, 1987), as well as experimentally induced autoimmune diseases (Sternberg *et al.*, 1989a, b; Mason *et al.*, 1990; Hu *et al.*, 1993) In addition, it is well established that the sympathetic nervous system via noradrenaline conveys information from the brain to the peripheral immune system (Friedman and Irwin, 1997; Schorr and Arnason, 1999; Kohm and Sanders, 2000), and several other neurotransmitters, such as dopamine, serotonin, acetylcholine, opioids, as well as the major excitatory and inhibitory transmitters, glutamate and gamma-aminobutyric acid (GABA), have been shown to be critically involved in central nervous system (CNS)—immune communications (Antonica *et al.*, 1996; Roy and Loh, 1996; Song *et al.*, 1996; Bergeret *et al.*, 1998; Mossner and Lesch, 1998; Rinner *et al.*, 1998a; Espey and Basile, 1999; Hegg and Thayer, 1999; Basu and Dasgupta, 2000). In the past years, the new discipline "Psychoneuroimmunology" has expanded rapidly (Watkins, 1997; Rabin, 1999; Sternberg, 2000), and it became obvious that the interactions with the immune system involve not only neuroendocrine signaling routes and synaptic activity within the whole brain, but also an intimate paracrine cross talk with peripheral nerve terminals as well as autocrine communications between immunocompetent cells themselves (Bergquist *et al.*, 1994; Ader *et al.*, 1995; Kaplan, 1996; Haas and Schauenstein, 1997, 2001). Back in the 1980s, noradrenergic nerve terminals were first

Helga S. Haas and Konrad Schauenstein • Department of Pathophysiology, Medical University Graz, Heinrichstrasse 31A, A-8010 Graz, Austria.

Glutamate Receptors in Peripheral Tissue, edited by Santokh Gill and Olga Pulido.
Kluwer Academic / Plenum Publishers, New York, 2005.

described to form "synaptic-like" contacts with immune cells in the spleen (Felten and Olschowka, 1987). The detailed molecular mechanisms underlying these communications remained, however, largely speculative in those times. Furthermore, noradrenergic, peptidergic, and/or cholinergic nerve fibers have been shown to distribute in other lymphatic tissues, such as bone marrow, thymus, lymph nodes, as well as gut-associated lymphoid tissue (GALT), although in the case of the thymus, the cholinergic innervation is still somewhat unclear (Bulloch and Moore, 1981; Felten et al., 1985, 1987; Fatani et al., 1986; Nance et al., 1987; Micic et al., 1994; Romano et al., 1994; Niijima 1995). During the last decades, the core of research in psychoneuroimmunology has focused at describing the functions and regulation of this complex communication system. It has been demonstrated that peripheral immune cells express functional receptor molecules for neurotransmitters/-peptides as well as several channel proteins, such as binding sites for norepinephrine (Heijnen and Kavelaars, 1999; Schauenstein et al., 2000), acetylcholine (Kawashima and Fujii, 2000; Schauenstein et al., 2000), serotonin (Stefulj et al., 2000, 2001), dopamine (Amenta et al., 1999; Levite et al., 2001), as well as different K^+, Ca^{2+}, Cl^-, and Na^+ channels similar to those observed in the CNS (Lai et al., 2000; Lepple-Wienhues et al., 2000; Blunck et al., 2001; Lewis, 2001). Furthermore, lymphocytes and macrophages themselves have been found to be equipped to produce neurotransmitters and a number of peptide hormones, which have all been described along with their effects on both immune as well as nerval functions (Weigent and Blalock, 1997; Bergquist et al., 1998; Rinner et al., 1998b; Kawashima and Fujii, 2000). So far, relatively little attention has been paid to the peripheral distribution pattern and function of glutamate receptors/transporters in immune tissues. This is somewhat surprising as components of the central glutamatergic signaling system have been identified in several other peripheral tissues. Furthermore, in view of the pleiotropic effects of the dicarboxylic amino acid glutamate (metabolic agent, taste stimulus, neurotransmitter, neurotoxin), it is reasonable to assume that the precise control of the glutamatergic system is not only crucial for normal brain functions but also for cellular responses in peripheral tissues. Here we want to review data suggesting that local autocrine/paracrine glutamatergic signaling mechanisms participate in peripheral neuroimmunomodulation, providing novel, fast-acting signal transduction pathways between nerval and immune tissues.

2. Distribution of Glutamate Receptors/Transporters in Immune Tissues

Initial evidence for regulatory effects of neuroexcitatory amino acids on human peripheral mononuclear cells came from a study by Malone and colleagues (Malone et al., 1986). In their experiments, they describe a chemotactic response of human monocytes to 4-carboxyglutamic acid (an analogue of glutamate) as well as kainate. The maximal response that occurred at nanomolar concentrations, was stereospecific and cell-type specific, that is, negative results were obtained with neutrophils and fetal bovine fibroblasts. Unexpectedly, however, glutamic acid itself acted as a competitive antagonist in this experimental system without any agonist activity. That the human blood leukocytes may contain glutamate or glutamate-like receptors was indirectly suggested by data of Whitlock et al. (1996), who primarily intended to characterize a novel drug-binding site for sigma and opioid ligands on human and rat leukocytes, and incidentally found high-affinity binding of compounds capable to act at the N-methyl-D-aspartate (NMDA) subtype of glutamate receptors. Furthermore, a functional polyamine site associated with the

NMDA receptor complex was suggested on rat peritoneal mast cells (Purcell *et al.*, 1996). This was indicated by the findings that the natural polyamine spermine triggers histamine secretion from mast cells in a concentration-dependent manner with an EC_{50} value of 200 μM. Additionally, this spermine-induced histamine release was totally dependent on the presence of calcium in the external milieu, and could be inhibited by the noncompetitive polyamine antagonist Ifenprodil as well as the NMDA receptor channel blocker dizoclipine = (+)-5-methyl-10,11-dihydro-5H-dibenzo[a,d]cyclohepten-5,10-imine (MK-801).

The presence of specific glutamate-binding sites on human blood lymphocytes was first described in a series of Russian studies by Kostanyan and coworkers (Kostanyan *et al.*, 1997). ^{3}H-labeled glutamate was found to bind specifically and with considerable affinity ($K_d = 2.36 \times 10^{-7}$M) to T lymphocytes separated from normal human blood. The specificity of this glutamate receptor was verified by competition experiments with several dipeptides (Glu–Ala, Ala–Glu, and Glu–Glu) and the broad-spectrum glutamate receptor agonist quisqualate. Furthermore, by using bulky conjugates of radiolabeled and unlabeled glutamate with dextran it was demonstrated that conjugated as well as free glutamate interacts with this binding site expressed on the cell membrane of human T cells. In a following study (Kostanyan *et al.*, 1998a), a similar glutamate-binding site was described on cells of the human leukemia line HL-60 during their differentiation. However, although these experiments suggested that human lymphocytes express quisqualate-sensitive glutamate receptors, these studies did not define receptor subtypes, since quisqualate is known to activate both ionotropic (iGluR) and metabotropic (mGluR) glutamate receptors. Furthermore, it is debatable if such binding experiments alone are sufficient to define a given surface receptor molecule. More recent results (Storto *et al.*, 2000) provided direct molecular evidence that group I (mGluR1 and mGluR5) as well as group II (mGluR2 and mGluR3) mGluR are present in whole thymus, isolated murine thymocytes, and the thymic stromal cell line TC-1S. Using RT-PCR, mGluR1, 2, 3, and 5 transcripts were all found in whole thymus of adult mice, but appeared to be differentially expressed in thymocytes vs thymic stromal cells. mGluR 2 mRNA, for example, was exclusively found in TC-1S cells, whereas mGluR1, 3, and 5 transcripts were expressed by both TC-1S cells as well as thymocytes. The authors additionally investigated mGluR cell surface protein expression by immunoprecipitation and Western blot analysis. The dimeric form of mGluR5 as well as a 100-kDa band corresponding to monomeric mGlu2/3 receptors were stained in immunoprecipitates from whole thymus, thymocytes, and TC-1S cells. mGluR2/3 receptor dimers, however, were detected in TC-1S cells only, whereas mGluR1 protein was not detectable. Furthermore, the expression pattern of the different mGluR subtypes was found to change during thymocyte differentiation. mGluR2/3 expression was observed in immature double negative (DN) CD4$^-$CD8$^-$ T cell precursors, intermediate CD4$^+$CD8$^+$ double positive (DP) cells, as well as mature single positive (SP) CD4$^+$ T cells. Consistent with the immunoblot analysis, mGluR1 expression was lowest in mature cells as compared to immature DN cells. mGlu5 receptors were virtually absent in DN thymocytes, but were found in a high percentage of intermediate DP as well as mature SP cells, suggesting that this subtype is associated with later stages of thymocyte differentiation. Functionality of these glutamate-binding sites in the thymic microenvironment was confirmed by measuring agonist-stimulated polyphosphoinositide hydrolysis and inhibition of forskolin-induced cAMP formation. Interestingly, TC-1S thymic stromal cells obviously share further similarities with neural tissue, since they were also found to express synapsin I, a synaptic vesicle-associated neuronal phosphoprotein (Screpanti *et al.*, 1992). Taken together, Storto's study indicated that peripheral mGluRs are developmentally regulated, analogous to what has been observed within the CNS (Ghosh *et al.*, 1997; Heck *et al.*, 1997; Liu *et al.*, 1998). The data also suggest that novel autocrine/paracrine glutamatergic signaling mechanisms may participate in intrathymic lymphocyte–stroma interactions, and affect processes such as thymocyte apoptosis, similar to what our group has previously described for intrathymic non-neuronal acetylcholine (Rinner *et al.*, 1999). Besides mGluR, peripheral lymphocytes also express members of the iGluR family. This is demonstrated in a recent study (Lombardi *et al.*, 2001), showing that L-glutamate, receptor agonists (NMDA, AMPA = α-amino-3-hydroxy-5-methylisoxazole-4-propionic acid, kainate) as well as antagonists (D(−)-2-amino-5-phosphonopentanoic acid = D-AP5, a competitive NMDA receptor antagonist; MK-801, the noncompetitive NMDA receptor channel blocker; 2,3-dihydroxy-6-nitro-7-sulfamoyl-benzo(F)-quinoxaline (NBQX), a competitive AMPA/kainate receptor antagonist; and kynurenic acid, a broad spectrum iGluR antagonist), are functionally operating as modulators of human lymphocyte activation (see also below).

As to the glutamate transport systems, it is generally known that macrophages express the cystine/glutamate antiporter (system X_c^-) (Watanabe and Bannai, 1987). This Na^+-independent glutamate transport system, originally described in the plasma membrane of human fibroblasts (Bannai, 1986), represents an "anion-exchanging agency" and transports extracellular cystine in exchange for intracellular glutamate. The induction of this carrier system increases the intracellular cysteine pool, which consequently elevates the synthesis of glutathione (Bannai and Tateishi, 1986). Glutathione, in turn, is the main intracellular antioxidant, and importantly modulates several immune functions (Dröge *et al.*, 1994). Macrophages also possess members of the high-affinity Na^+/K^+-dependent glutamate uptake system (system X_{AG}^- = the excitatory amino acid transporter "EAATs"), as was recently demonstrated by Rimaniol's group (Rimaniol *et al.*, 2000). This transport system consists of five subtypes (EAAT1-5) and is driven by gradients of both sodium and potassium (for review see Danbolt, 2001). It represents not only the major glutamate removal mechanism in the CNS, but has also been shown to be critically involved in the fine-tuning of synaptic transmission, and was recently identified in several peripheral non-neural tissues including heart, intestine, kidney, pancreas, and bone. Rimaniol and colleagues (2000) showed that human monocyte-derived macrophages, besides expressing the cystine/glutamate antiporter, do also express subtypes of the high-affinity glutamate transporter family (EAATs). RT-PCR analyses revealed that EAAT1 (GLAST) and EAAT2 (GLT-1) gene expression began to increase after 1 hr in culture, reached a maximum by day 2, and then slowly decreased until day 12. Freshly isolated tissue macrophages apparently do not possess functional EAATs, since efficient glutamate transport was observed only in cultured human splenic macrophages, macrophages activated by adhesion to plastic, and in cytokine (tumor necrosis factor-α = TNF-α) stimulated cells. Similarly, as described in glial cells (Hu *et al.*, 2000), this latter observation argues for a role of inflammatory mediators as modulators of high-affinity glutamate transport in macrophages, and opens up exciting aspects for future research in neuroimmunomodulation as well as toward novel strategies to control neuroinflammatory conditions. To further confirm that monocyte-derived macrophages express EAATs and that extracellular glutamate is predominantly taken up by this Na^+/K^+-dependent high-affinity carrier system, competitive inhibitors specific for EAATs were used and compared with EAA analogues capable to block the cystine/glutamate antiporter (system X_c^-). These results revealed the higher affinity of glutamate for EAATs (EAAT1 over EAAT2) and suggest that peripheral macrophages may be equally efficient as neuronal cells in clearing extracellular glutamate (Rimaniol *et al.*, 2000).

Finally, from a more general view, it should be mentioned that immune cells have been shown to be equipped with molecules involved in vesicular traffic, that is, fusion and exocytosis, similar to those required for the fusion of synaptic vesicles at the presynaptic membrane. For example, synaptosome-associated protein-25 (SNAP-25), known to be ubiquitously distributed in nerval tissues, was found in human neutrophils (Nabokina *et al.*, 1997) as well as eosinophils (Hoffman *et al.*, 2001), and implicated in the regulation of exocytosis. Furthermore, the process of phagocytosis requires a series of specific fusion events including coordinated interactions between the membrane of the phagocyte and surface molecules of the pathogen, as well as the docking and fusion of intracellular vesicles with the phagosomal membrane. In this context, members of the SNARE family (soluble *N*-ethylmaleimide-sensitive fusion [NSF] factor attachment protein receptor), known to be involved in vesicular traffic in the central as well as peripheral nervous system, were detected in murine peritoneal macrophages, the murine macrophage cell line J774, as well as in the human macrophage cell lines U937 and THP-1 (Hackam *et al.*, 1996). However, even though recent studies on intracellular trafficking (Bajjalieh, 2001; Bruns and Jahn, 2002) have led to striking advances in the understanding of processes, such as vesicle docking and fusion, the regulatory mechanisms mediating these processes are still far from clear. Presently, it is not known whether glutamate receptors expressed on immune cells may be functionally involved in, for example, granule secretion or glutamate release itself. Nevertheless, the involvement of peripheral glutamate receptors in the modulation of glutamate secretion via presynaptic-like exocytosis has recently been demonstrated in pinealocytes and osteoblasts (Yatsushiro *et al.*, 2000; Bhangu *et al.*, 2001). These results altogether indicate that immune cells share by far more features with nerval cells than previously assumed. Clarifying the ways by which this neuroimmune communication system is organized and regulated will significantly strengthen the acceptance of "Psychoneuroimmunology" as a major concept in human health and disease.

3. Neuroimmune Regulation via Peripheral Glutamatergic Signaling Pathways

It has long been established that immune cells metabolize and secrete glutamate both in the absence of any stimuli as well as—strongly induced—following stimulation with inflammatory mediators including phorbol myristoyl acetate (PMA), lipopolysaccharide (LPS), and cytokines (Brand *et al.*, 1987; Ardawi, 1988; Curi *et al.*, 1997; Klegeris *et al.*, 1997). Furthermore, macrophages synthesize and release quinolinic acid, a metabolite formed along the kynurenine pathway of tryptophan metabolism, but also known as a potent NMDA receptor agonist and immunomodulator (Espey *et al.*, 1995; Koennecke *et al.*, 1999; Chiarugi *et al.*, 2001). Disturbances in the metabolism of glutamate combined with abnormally high venous plasma glutamate levels have been reported in several diseases, such as cancer, HIV infection, autism, or a number of neurodegenerative disorders (Westall *et al.*, 1980; Dröge *et al.*, 1987, 1988; Plaitakis and Caroscio, 1987; Moreno-Fuenmayor *et al.*, 1996), and were associated, at least in part, with a decreased glutamate uptake capacity by peripheral muscle tissue (Hack *et al.*, 1996). Elevated glutamate levels, however, can also be found more locally at sites of inflammation. Toward this, the group around Westlund (Lawand *et al.*, 2000) conducted extensive research to investigate the local release of excitatory amino acids into the knee joint of rats during the course of inflammation. Within 10 min after induction of joint inflammation the concentrations of glutamate, as well as the nitric oxide (NO) metabolites arginine and citrulline doubled and remained elevated for at least 2 hr. The local application of an anesthetic, lidocaine, was effective in blocking this glutamate release. Furthermore, following induction of arthritis in monkeys a doubling in the percentage of myelinated glutamate-containing axons has been reported in the medial articular nerve innervating the affected joint (Westlund *et al.*, 1992). Increased glutamate and aspartate levels have also been found in the synovial fluid of patients with arthritis, whereby according to McNearney's study (McNearney *et al.*, 2000), excitatory amino acid levels were highest in Reiter's infectious arthropathies and systemic lupus erythematosus. Since all these disease states are commonly associated with dysregulated immune functions and/or pain processes, it is assumed that elevated glutamate, besides being a neurotoxin, is either also a proinflammatory or an "immunotoxic" agent. One line of studies addressing this issue apparently confirms this view. Elevated plasma glutamate concentrations were quantitatively correlated with a reduced mitogenic activity of lymphocytes as well as with higher death rates in tumor patients (Dröge *et al.*, 1987, 1988; Eck *et al.*, 1989a; Sommer *et al.*, 1994). Furthermore, increased glutamate levels in HIV-1-infected persons, as well as the glutamate receptor agonists quisqualate, NMDA, and kainate were shown to impair immune functions in macrophages as well as lymphocytes (Eck *et al.* 1989b). The fact that glutamate also competitively inhibits the import of cystine by the cystine/glutamate antiporter (system X_c^-), which consequently results in decreased intracellular cysteine levels (Bannai, 1986; Watanabe and Bannai, 1987), generally served as one of the putative explanations of these results. However, in view of the high variety of glutamate receptors as well as transport proteins, and the diversity of signal transduction cascades involved, glutamate may have other effects besides to suppress immune functions. Furthermore, elevated glutamate is not necessarily associated with inflammatory states. For example, an increase in plasma glutamate can also be found in healthy human subjects during senescence (Hack *et al.*, 1996), and, by using a specific microdialysis technique, high local concentrations of glutamate can be found *in vivo* in tendons of patients suffering from chronic tendinopathies, but without any biochemical signs of inflammation (normal PGE_2 levels) (Alfredson *et al.*, 2000, 2001; Alfredson and Lorentzon, 2002). As to lymphocyte proliferation, Pawlikowski and Kunert-Radek (1995) reported that neither NMDA nor quisqualate affected the tritiated thymidine incorporation into DNA of mouse spleen cells. Lombardi and colleagues (2001), on the other hand, confirmed the commonly reported glutamate-induced inhibition of mitogen-stimulated lymphocyte proliferation. No change in the proliferation rate of resting lymphocytes was noted even following administration of increasing doses of glutamate. Additionally, this group demonstrated the enhancing effect of glutamate, receptor agonists, and antagonists on intracellular calcium concentrations $[Ca^{2+}]_i$ in human lymphocytes (Fura-2 method), suggesting a receptor-mediated mechanism for glutamate regulation of lymphocyte functions. Again, neither glutamate nor receptor agonists modified basal $[Ca^{2+}]_i$ in resting lymphocytes. However, glutamate significantly potentiated the effects of anti-CD3 monoclonal antibody (mAb) or phytohemagglutinin (PHA)-induced $[Ca^{2+}]_i$ rises. Similar to what has been observed with glutamate, also NMDA, AMPA, and kainate produced a potentiation of $[Ca^{2+}]_i$ responses in activated lymphocytes, whereby AMPA and kainate receptor agonists were most effective if compared to NMDA or glutamate itself. In accordance, the non-NMDA receptor antagonists, NBQX and kynurenic acid completely abolished the effects of glutamate on $[Ca^{2+}]_i$, whereas competitive and noncompetitive NMDA receptor antagonists (D-AP5 and MK-801) inhibited but did not abolish the glutamate response. Together, Lombardi's results (2001) clearly indicate that human peripheral lymphocytes express functional iGluR which, depending on the subtype involved, may selectively affect immune-activated lymphocyte functions. Glutamate was also found to affect differentiation and functions of nonspecific immune cells. At a concentration of 0.1 μM glutamate was shown to induce the differentiation of promyelocytic HL-60 cells into granulocytes or neutrophils (Kostanyan et al., 1998b). It also potentiated the TNF-α-induced differentiation of this cell line (Astapova et al., 1999). In addition, glutamate was found to selectively modulate cytokine binding and receptors on HL-60 cells. Whereas it completely inhibited the high-affinity binding of ^{125}I-labeled interleukin (IL)-1β without changing the receptor number, it increased the number of receptors for TNF-β (Kostanyan et al., 1998b). Last not least, in neutrophil respiratory burst activation, known to be dependent on Ca^{2+} cascade mechanisms, also a potentiating effect of glutamate and NMDA has been observed (Kim-Park et al., 1997).

As seen from the hitherto discussed data, glutamate receptor signaling may either suppress or enhance immune functions, and/or exert even cytoprotective effects analogous to what has been described in the CNS, where activation of specific mGluR protects neuronal cells against anoxia, NO toxicity, and post-traumatic cell death (Maiese et al., 1996; Faden et al., 1997). In this regard, Belokrylov's group (Belokrylov et al., 1999) observed that glutamate, dipeptides such as GluTrp, as well as the mixture of these amino acids markedly stimulated phagocytosis of murine granulocytes and protected splenocytes from benzene toxicity. From these data, it was concluded that certain amino acids may be promising therapeutic agents in preventing diseases associated with chronic benzene intoxication. Referring to the concept of immunoenhancement via glutamine supplementation, and with the knowledge of the continuous metabolic conversion of glutamate vs glutamine (Alexander, 1993; Klimberg and McClellan, 1996; Castell, and Newsholme, 1997), a recent study (Lin et al., 1999) aimed to investigate the effect of dietary glutamate on immune functions in rats recovering from chemotherapy by feeding a glutamine/glutamate-free amino acid diet. Two in vivo immune tests were performed: (a) the delayed-type hypersensitivity (DTH) assay, which evaluates the degree of skin-contact sensitivity after challenge with 2,4-dinitrofluorobenzene (DNFB), and (b) the popliteal lymphoproliferation (PLP) assay, whereby following subcutaneous injection of sheep red blood cells (SRBC) the proliferative response of lymphocytes from the draining (popliteal) lymph node is measured. Furthermore, several dietary treatment groups with different feeding durations (23 days/44 days) were tested. The results demonstrate that dietary glutamate dose-dependently increases both types of immune response, and that the effect is clearly more pronounced after longer duration of dietary intake. In addition, plasma glutamine, glutamate, and glutathione levels were measured, but were found only transiently increased by dietary glutamate at the immediate postprandial state. Although further studies are required to elucidate the mechanisms underlying these immunoenhancing effects of glutamate, such investigations yield first exciting insights into the potential of peripheral excitatory amino acid-induced effector responses.

As follows from the above, it seems likely that besides glutamate receptors also the peripheral high-affinity transporter molecules (EAATs) participate in neuroimmunomodulation, and may be involved in normal immune functions as well as in immune dysregulation. Considering Rimaniol's data (2000), which provided clear evidence that human macrophages derived from monocytes have both the Na^+-independent glutamate carrier system (X_c^-) as well as the high-affinity Na^+/K^+-coupled transporters (EAATs = system X_{AG}^-), and in view of the accumulation of extracellular glutamate found in several pathological conditions (see above), a high-speed peripheral glutamate removal system in macrophages should be of outstanding importance. Thus, Rimaniol and coworkers performed additional experiments to investigate (a) whether EAAT activity in monocyte-derived macrophages is capable of protecting neuronal cultures from glutamate toxicity, and (b) if peripheral EAATs may be involved in the regulation of glutathione synthesis. The results revealed that macrophages time-dependently protected primary mouse neuronal cultures from toxicity of different gluta-mate concentrations (Rimaniol *et al.*, 2000). V_{max} for this Na^+-dependent glutamate uptake by macrophages was 3,000 pmol/mg protein/min. As to glutathione synthesis, this work as well as a following report by the same group (Rimaniol *et al.*, 2001) reveals that EAATs on macrophages, in concert with the cystine/glutamate antiporter, are indeed involved in the reg-ulation of intracellular glutathione concentration: extracellular cystine is transported into macrophages via the cystine/glutamate antiporter (system X_c^-) leading to increased extracel-lular glutamate, which in turn activates EAATs to fuel the intracellular glutamate pool. This leads to enhanced cystine uptake via system X_c^- (called "trans-stimulation"), and consequently increases glutathione synthesis. A second mechanism by which EAATs on macrophages regulate intracellular glutathione is that EAATs also provide intracellular glutamate for direct insertion into glutathione.

In view of the well known control of phagocyte functions by neural activity (Baciu, 1988; Zeev-Brann *et al.*, 1998), and the facts that defects in proliferation, differentiation, cell signaling, secretory functions, as well as phagocytosis should critically affect nonspecific immunity (Douglas, 1999), the aim of our own preliminary studies is to precisely examine the role of glutamate receptor signaling in effector functions of phagocytic cells. By using flow cytometric analysis and a commercially available phagocytosis Phagotest® (Orpegen Pharma, Heidelberg, Germany), we measured changes in the phagocytic activity of human granulocytes and monocytes after incubation of whole blood from healthy volunteers with various concentrations of (6-cyano-7-nitroquinoxaline-2,3-dione) (CNQX), an antagonist at AMPA/kainate glutamate receptors, as well as with the receptor agonists AMPA and kainate. The results so far indicate that kainate exerted up to 90% stimulation, whereas CNQX caused significant ($p < 0.05$) suppression of percent phagocytotic monocytes as well as granulo-cytes. No significant effects were observed on the mean fluorescence intensity per cell. Since these preliminary experiments with whole blood preparations revealed a high interindividual variation in phagocytic activity, we presently try to reproduce the above mentioned results with the human monocytic cell line U937. Further, we initiated studies on the role of gluta-mate signaling in releasing the chemokine monocyte chemotactic protein-1 (MCP-1) from mouse macrophages. Chemokines have recently been implicated in central glutamate neuro-transmission, as chemokine signaling has been described to trigger glutamate release from astrocytes via a complex signaling cascade also involving cytokine (TNF-α) signal transduc-tion pathways (Allan and Attwell, 2001; Bezzi *et al.*, 2001). Vice versa, intrahippocampal injections of NMDA induced MCP-1 expression in the CNS, and systemic administration of kainate stimulated the expression of the CCR5 chemokine receptor in forebrain structures (Szaflarski *et al.*, 1998; Galasso *et al.*, 2000; Mennicken *et al.*, 2002). In our experiments, the levels of MCP-1 released from cells of the mouse macrophage line RAW 264 were measured with a commercial enzyme-linked immunosorbent assay (ELISA) (Pharmingen, CA, USA). Cells were incubated for 4 hr at 37°C with or without LPS and different concentrations of CNQX and kainate. Compared to controls, CNQX ($p < 0.05$) significantly inhibited basal MCP-1 release, without affecting LPS-stimulated cells. In contrast, kainate significantly enhanced MCP-1 release from LPS-stimulated cells, without any effect on basal secretion. These first data indicate that functional activities of human and murine phagocytes are affected by agonists and antagonist of AMPA/kainate receptors, and hence suggest the existence of peripheral glutamate signaling pathways in these cells, similar to what has been described for other peripheral tissues.

4. Conclusions

It is accepted that glutamate is the major excitatory neurotransmitter in the mammalian CNS. As is apparent from the published literature cited in this review, glutamate is also a peripheral neuroimmunomodulator acting in a receptor-/transporter-mediated manner

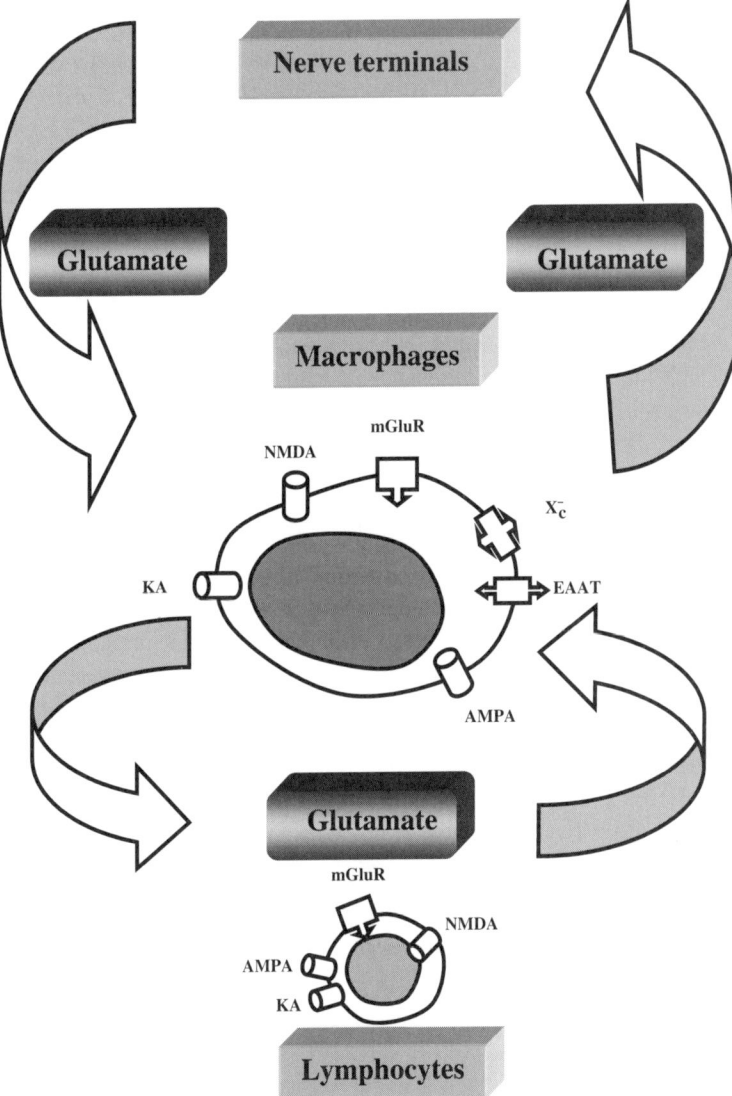

Figure 14.1. Autocrine and paracrine neuroimmunomodulation via glutamatergic signaling mechanisms.

similarly as observed within the CNS. As such, glutamate is essential for many of the cell's activity, and may be a central element in the regulation and homeostasis of immune responses, not the least due to its extremely fast signaling activity via ion channel receptors (Figure 14.1). However, in spite of exciting new results in this field of research, many pieces are still missing from the puzzle. At present, only a few subtype-specific glutamate receptor compounds are available, which prevent to, precisely, define receptor distribution and function, or to treat diseases with anti-excitotoxic agents. These problems are mainly due to the complexity of the glutamate-system itself with its 25-receptor subtypes including the less mentioned δ receptors and the 5 high affinity transporters as well as all glutamate carrier mechanisms other than EAATs. As a further obstacle, glutamate-binding sites within mGluRs of the same group share high similarity, which will make the design of subtype selective drugs extremely difficult. In addition, it has to be recognized that a receptor molecule localized in peripheral tissues is not necessarily regulated in the same ways as in the CNS. Even though the complete solution of the whole puzzle may remain an illusion, every step toward a better understanding of the major excitatory neurotransmitter glutamate in context with immunity will lead to a better understanding of psycho-neuro-immune communications.

Acknowledgments

Glutamate research in the authors' laboratory is supported by a grant from the Österreichischen Nationalbank, Project Nr. 9365. Further, we wish to thank Mrs. Elfgard Heintz for her excellent technical contribution.

References

Ader, R., N. Cohen, and D. Felten (1995). Psychoneuroimmunology: Interactions between the nervous system and the immune system. *Lancet* **345**, 99–103.

Alexander, J.W. (1993). Immunoenhancement via enteral nutrition. *Arch. Surg.* **128**, 1242–1245.

Alfredson, H. and R. Lorentzon (2002). Chronic tendon pain: No signs of chemical inflammation but high concentrations of the neurotransmitter glutamate. Implications for treatment? *Curr. Drug Targets* **3**, 43–54.

Alfredson, H., B.O. Ljung, K. Thorsen, and R. Lorentzon (2000). In vivo investigation of ECRB tendons with microdialysis technique—no signs of inflammation but high amounts of glutamate in tennis elbow. *Acta Orthop. Scand.* **71**, 475–479.

Alfredson, H., S. Forsgren, K. Thorsen, and R. Lorentzon (2001). In vivo microdialysis and immunohistochemical analyses of tendon tissue demonstrated high amounts of free glutamate and glutamate NMDAR1 receptors, but no signs of inflammation, in Jumper's knee. *J. Orthop. Res.* **19**, 881–886.

Allan, N.J. and D. Attwell (2001). A chemokine-glutamate connection. *Nat. Neurosci.* **4**, 676–678.

Amenta, F., E. Bronzetti, L. Felici, A. Ricci, and S.K. Tayebati (1999). Dopamine D2-like receptors on human peripheral blood lymphocytes: A radioligand binding assay and immunocytochemical study. *J. Auton. Pharmacol.* **19**, 151–159.

Antonica, A., E. Ayroldi, F. Magni, and N. Paolocci (1996). Lymphocyte traffic changes induced by monolateral vagal denervation in mouse thymus and peripheral lymphoid organs. *J. Neuroimmunol.* **64**, 115–122.

Ardawi, M.S.M. (1988). Glutamine and glucose metabolism in human peripheral lymphocytes. *Metabolism* **37**, 99–103.

Astapova, M.V., V.M. Lipkin, M.V. Askhipova, S.G. Andreeva, S.M. Dranitsyna, M.I. Merkulova *et al.* (1999). L-glutamic acid—a modulator of the physiological status of myeloid series blood cells. *Bioorganicheskaya Khimiya* **25**, 816–820.

Baciu, I. (1988). Nervous control of the phagocytic system. *Int. J. Neurosci.* **41**, 127–141.

Bajjalieh, S. (2001). SNAREs take the stage: A prime time to trigger neurotransmitter secretion. *Trends Neurosci.* **24**, 678–680.

Bannai, S. (1986). Exchange of cystine and glutamate across plasma membrane of human fibroblasts. *J. Biol. Chem.* **261**, 2256–2263.

Bannai, S. and N. Tateishi (1986). Role of membrane transport in metabolism and function of glutathione in mammals. *J. Membr. Biol.* **89**, 1–8.

Basu, S. and P.S. Dasgupta (2000). Dopamine, a neurotransmitter, influences the immune system. *J. Neuroimmunol.* **102**, 113–124.

Belokrylov, G.A., O. Ya Popova, and E.I. Sorochinskaya (1999). Immuno-, phagocytosis-modulating and antitoxic properties of dipeptides are defined by the activity of their constituent amino acids. *Int. J. Immunopharm.* **21**, 879–883.

Bergeret, M., M. Khrestchatisky, E. Tremblay, A. Bernard, A. Gregoire, and C. Chany (1998). GABA modulates cytotoxicity of immunocompetent cells expressing GABA A receptor subunits. *Biomed. Pharmacother.* **52**, 214–219.

Bergquist, J., A. Tarkowski, R. Ekman, and A. Ewing (1994). Discovery of endogenous catecholamines in lymphocytes and evidence for catecholamine regulation of lymphocyte function via an autocrine loop. *Proc. Natl. Acad. Sci. USA* **91**, 12912–12916.

Bergquist, J., A. Tarkowski, A. Ewing, and R. Ekman (1998). Catecholaminergic suppression of immunocompetent cells. *Immunol. Today* **19**, 562–567.

Besedovsky, H.O., E. Sorkin, M. Keller, and J. Müller (1975). Changes in blood hormone levels during the immune response. *Proc. Soc. Exp. Biol. Med.* **150**, 466–470.

Besedovsky, H.O., A. Del Rey, and E. Sorkin (1979). Antigenic competition between horse and sheep red blood cells is a hormone-dependent phenomenon. *Clin. Exp. Immunol.* **37**, 106–113.

Bezzi, P., M. Domercq, L. Brambilla, R. Galli, D. Schols, E. De Clercq *et al.* (2001). CXCR4-activated astrocyte glutamate release via TNF-a: Amplification by microglia triggers neurotoxicity. *Nat. Neurosci.* **4**, 702–710.

Bhangu, P.S., P.G. Genever, G.J. Spencer, T.S. Grewal, and T.M. Skerry (2001). Evidence for targeted vesicular glutamate exocytosis in osteoblasts. *Bone* **29**, 16–23.

Blunck, R., O. Scheel, M. Muller, K. Brandenburg, U. Seitzer, and U. Seydel (2001). New insights into endotoxin-induced activation of macrophages: Involvement of a K+ channel in transmembrane signaling. *J. Immunol.* **166**, 1009–1015.

Brand, K., J. von Hintzenstern, K. Langer, and W. Fekl (1987). Pathways of glutamine and glutamate metabolism in resting and proliferating rat thymocytes: Comparison between free and peptide-bound glutamine. *J. Cell. Physiol.* **132**, 559–564.

Bruns, D. and Jahn, R. (2002). Molecular determinants of exocytosis. *Pflügers Arch.- Eur. J. Physiol.* **443**, 333–338.

Bulloch, K. and R.Y. Moore (1981). Innervation of the thymus gland by brain stem and spinal cord in mouse and rat. *Am. J. Anat.* **162**, 157–166.

Castell, L.M. and E.A. Newsholme (1997). The effects of oral glutamine supplementation on athletes after prolonged, exhaustive exercise. *Nutrition* **13**, 738–742.

Chiarugi, A., M. Calvani, E. Meli, E. Traggiai, and F. Moroni (2001). Synthesis and release of neurotoxic kynurenine metabolites by human monocyte-derived macrophages. *J. Neuroimmunol.* **120**, 190–198.

Curi, T.C., M.P. De Melo, R.B. De Azevedo, T.M. Zorn, and R. Curi (1997). Glutamine-utilization by rat neutrophils: Presence of phosphate-dependent glutaminase. *Am. J. Physiol.* **273**, C1124–C1129.

Danbolt, N.C. (2001). Glutamate uptake. *Prog. Neurobiol.* **65**, 1–105.

Douglas, S.D. (1999). Monocytes/macrophages in diagnosis and immunopathogenesis. *Clin. Diagn. Lab. Immunol.* **6**, 283–285.

Dröge, W., H.P. Eck, M. Betzler, and H. Naher (1987). Elevated plasma glutamate levels in colorectal carcinoma patients and in patients with acquired immunodeficiency syndrome (AIDS). *Immunobiology* **174**, 473–479.

Dröge, W., H.-P. Eck, M. Betzler, P. Schlag, P. Drings, and W. Ebert (1988). Plasma glutamate concentration and lymphocyte activity. *J. Cancer Res. Clin. Oncol.* **114**, 124–128.

Dröge, W., K. Schulze-Osthoff, S. Mihm, D. Galter, H. Schenk, H.P. Eck *et al.* (1994). Functions of glutathione and glutathione disulfide in immunology and immunopathology. *FASEB J.* **8**, 1131–1138.

Eck, H.P., P. Drings, and W. Dröge (1989a). Plasma glutamate levels, lymphocyte reactivity and death rate in patients with bronchial carcinoma. *J. Cancer Res. Clin. Oncol.* **115**, 571–574.

Eck, H.P., H. Frey, and W. Dröge (1989b). Elevated plasma glutamate concentrations in HIV-1-infected patients may contribute to loss of macrophage and lymphocyte functions. *Int. Immunol.* **1**, 367–372.

Espey, M.G. and A.S. Basile (1999). Glutamate augments retrovirus-induced immunodeficiency through chronic stimulation of the hypothalamic-pituitary-adrenal axis. *J. Immunol.* **162**, 4998–5002.

Espey, M.G., J.R. Moffett, and M.A. Namboodiri (1995). Temporal and spatial changes of quinolinic acid immunoreactivity in the immune system of lipopolysaccharide-stimulated mice. *J. Leukoc. Biol.* **57**, 199–206.

Faden, A.I., S.A. Ivanova, A.G. Yakovlev, and A.G. Mukhin (1997). Neuroprotective effects of group III mGluR in traumatic neuronal injury. *J. Neurotrauma* **14**, 885–895.

Fatani, J.A., M.A. Quayyum, L. Mehta, and U. Singh (1986). Parasympathetic innervation of the thymus: A histochemical and immunocytochemical study. *J. Anat.* **147**, 115–119.

Felten, D.L., S.Y. Felten, S.L. Carlson, J.A. Olschowka, and S. Livnat (1985). Noradrenergic and peptidergic innervation of lymphoid tissue. *J. Immunol.* **135**, Suppl.2, 755–765.

Felten, S.Y. and J. Olschowka (1987). Noradrenergic sympathetic innervation of the spleen: II. Tyrosine hydroxylase (TH)-positive nerve terminals form synapticlike contacts on lymphocytes in the splenic white pulp. In J.R. Perez-Polo, K. Bulloch, R.H. Angeletti, G.A. Hashim, and J. de Vellis (eds), *Neuroimmunomodulation*, Alan R. Liss, Inc., New York, pp. 70–74.

Felten, D.L., S.Y. Felten, D.L. Bellinger, S.L. Carlson, K.D. Ackerman, K.S. Madden *et al.* (1987). Noradrenergic sympathetic neural interactions with the immune system: Structure and function. *Immunol. Rev.* **100**, 225–260.

Friedman, E.M. and M.R. Irwin (1997). Modulation of immune cell function by the autonomic nervous system. *Pharmacol. Ther.* **74**, 27–38.

Galasso, J.M., M.J. Miller, R.M. Cowell, J.K. Harrison, J.S. Warren, and F.S. Silverstein (2000). Acute excitotoxic injury induces expression of monocyte chemoattractant protein-1 and its receptor, CCR2, in neonatal rat brain. *Exp. Neurol.* **165**, 295–305.

Ghosh, P.K., N. Baskaran, and A.N. van den Pol (1997). Developmentally regulated gene expression of all eight metabotropic glutamate receptors in hypothalamic suprachiasmatic and arcuate nuclei—a PCR analysis. *Dev. Brain Res.* **102**, 1–12.

Haas, H.S. and K. Schauenstein (1997). Neuroimmunomodulation via limbic structures—The neuroanatomy of psychoimmunology. *Prog. Neurobiol.* **51**, 195–222.

Haas, H.S. and K. Schauenstein (2001). Immunity, hormones, and the brain. *Allergy* **56**, 470–477.

Hack, V., O. Stutz, R. Kinscherf, M. Schykowski, M. Kellerer, E. Holm *et al.* (1996). Elevated venous glutamate levels in (pre)catabolic conditions result at least partly from a decreased glutamate transport activity. *J. Mol. Med.* **74**, 337–343.

Hackam, D.J., O.D. Rotstein, M.K. Bennett, A. Klip, S. Grinstein, and M.F. Manolson (1996). Characterization and subcellular localization of target membrane soluble NSF attachment protein receptors (t-SNAREs) in macrophages. Syntaxins 2, 3, and 4 are present on phagosomal membranes. *J. Immunol.* **156**, 4377–4383.

Heck, S., R. Enz, C. Richter-Landsberg, and D.H. Blohm (1997). Expression of eight metabotropic glutamate receptor subtypes during neuronal differentiation of P19 embryocarcinoma cells: A study by RT-PCR and in situ hybridization. *Dev. Brain Res.* **101**, 85–91.

Hegg C.C. and Thayer S.A. (1999). Monocytic cells secrete factors that evoke excitatory synaptic activity in rat hippocampal cultures. *Eur. J. Pharmacol.* **385**, 231–237.

Heijnen, C.J. and A. Kavelaars (1999). The importance of being receptive. *J. Neuroimmunol.* **100**, 197–202.

Hoffmann, H.J., T. Bjerke, M. Karawajczyk, R. Dahl, M.A. Knepper, and S. Nielsen (2001). SNARE proteins are critical for regulated exocytosis of ECP from human eosinophils. *Biochem. Biophys. Res. Commun.* **282**, 194–199.

Hu, S., W.S. Sheng, L.C. Ehrlich, P.K. Peterson, and C.C. Chao (2000). Cytokine effects on glutamate uptake by human astrocytes. *Neuroimmunomodulation* **7**, 153–159.

Hu, Y., H. Dietrich, M. Herold, P.C. Heinrich, and G. Wick (1993). Disturbed immuno-endocrine communication via the hypothalamo–pituitary–adrenal axis in autoimmune disease. *Int. Arch. Allergy Immunol.* **102**, 232–241.

Kaplan, D. (1996). Autocrine secretion and the physiological concentration of cytokines. *Immunol. Today* **17**, 303–304.

Kawashima, K. and T. Fujii (2000). Extraneuronal cholinergic system in lymphocytes. *Pharmacol. Ther.* **86**, 29–48.

Kim-Park, W.K., M.A. Moore, Z.W. Hakki, and M.J. Kowolik (1997). Activation of the neutrophil respiratory burst requires both intracellular and extracellular calcium. *Ann. N. Y. Acad. Sci.* **832**, 394–404.

Klegeris, A., D.G. Walker, and P.L. McGeer, (1997). Regulation of glutamate in cultures of human monocytic THP-1 and astrocytoma U-373 MG cells. *J. Neuroimmunol.* **78**, 152–161.

Klimberg, V.S. and J.L. McClellan, (1996). Claude H. Organ, Jr. Honorary Lectureship. Glutamine, cancer, and its therapy. *Am. J. Surg.* **172**, 418–424.

Koennecke, L.A., M.A. Zito, M.G. Proescholdt, N. van Rooijen, and M.P. Heyes (1999). Depletion of systemic macrophages by liposome-encapsulated clodronate attenuates increases in brain quinolinic acid during CNS-localized and systemic immune activation. *J. Neurochem.* **73**, 770–779.

Kohm, A.P. and V.M. Sanders (2000). Norepinephrine: A messenger from the brain to the immune system. *Immunol. Today* **21**, 539–542.

Kostanyan, I.A., M.I. Merkulova, E.V. Navolotskaya, and R.I. Nurieva (1997). Study of interaction between L-glutamate and human blood lymphocytes. *Immunol. Lett.* **58**, 177–180.

Kostanyan, I.A., R.I. Nurieva, T.N. Lepikhova, M.V. Astapova, E.V. Navolotskaya, V.P. Zavyalov *et al.* (1998a). Appearance of glutamate receptors on the surface of HL-60 cells upon differentiation. *Bioorganicheskaya Khimiya* **24**, 468–470.

Kostanyan, I.A., R.I. Nurieva, E.V. Navolotskaya, M.V. Astapova, S.M. Dranitsyna, V.P. Zavyalov *et al.* (1998b). Effect of L-glutamic acid on the reception of cytokines by HL-60 cells. *Bioorganicheskaya Khimiya* **24**, 3–9.

Lai, Z.-F., Y.-Z. Chen, Y. Nishimura, and K. Nishi (2000). An amiloride-sensitive and voltage-dependent Na+ channel in an HLA-DR-restricted human T cell clone. *J. Immunol.* **165**, 83–90.

Lawand, N.B., T. McNearney, and K.N. Westlund (2000). Amino acid release into the knee joint: Key role in nociception and inflammation. *Pain* **86**, 69–74.

Lepple-Wienhues, A., I. Szabò, U. Wieland, L. Heil, E. Gulbins, and F. Lang (2000). Tyrosine kinases open lymphocyte chloride channels. *Cell. Physiol. Biochem.* **10**, 307–312.

Levite, M., Y. Chowers, Y. Ganor, M. Besser, R. Hershkovits, and L. Cahalon (2001). Dopamine interacts directly with its D3 and D2 receptors on normal human T cells, and activates beta1 integrin function. *Eur. J. Immunol.* **31**, 3504–3512.

Lewis, R.S. (2001). Calcium signaling mechanisms in T lymphocytes. *Ann. Rev. Immunol.* **19**, 497–521.

Lin, C.-M., S.F. Abcouwer, and W.W. Souba (1999). Effect of dietary glutamate on chemotherapy-induced immunosuppression. *Nutrition* **15**, 687–696.

Liu, X.B., A. Munoz, and E.G. Jones (1998). Changes in subcellular localization of metabotropic glutamate receptor subtypes during postnatal development of mouse thalamus. *J. Comp. Neurol.* **395**, 450–465.

Lombardi, G., C. Dianzani, G. Miglio, P.L. Canonico, and R. Fantozzi (2001). Characterization of ionotropic glutamate receptors in human lymphocytes. *Br. J. Pharmacol.* **133**, 936–944.

Maiese, K., M. Swiriduk, and M. TenBroeke (1996). Cellular mechanisms of protection by metabotropic glutamate receptors during anoxia and nitric oxide toxicity. *J. Neurochem.* **66**, 2419–2428.

Malone, J.D., M. Richards, and A.J. Kahn (1986). Human peripheral monocytes express putative receptors for neuroexcitatory amino acids. *Proc. Natl. Acad. Sci. USA* **83**, 3307–3310.

Mason, D., I. MacPhee, and F. Antoni (1990). The role of the neuroendocrine system in determining genetic susceptibility to experimental allergic encephalomyelitis in the rat. *Immunology* **70**, 1–5.

McNearney, T., D. Speegle, N. Lawand, J. Lisse, and K.N. Westlund (2000). Excitatory amino acid profiles of synovial fluid from patients with arthritis. *J. Rheumatol.* **27**, 739–745.

Mennicken, F., J.G. Chabot, and R. Quirion (2002). Systemic administration of kainic acid in adult rat stimulates expression of the chemokine receptor CCR5 in the forebrain. *Glia* **37**, 124–138.

Micic, M., G. Leposavic, and N. Ugresic (1994). Relationship between monoaminergic and cholinergic innervation of the rat thymus during aging. *J.Neuroimmunol.* **49**, 205–212.

Moreno-Fuenmayor, H., L. Borjas, A. Arrieta, V. Valera, and L. Socorro-Candanoza (1996). Plasma excitatory amino acids in autism. *Invest. Clin.* **37**, 113–128.

Mossner, R. and K.P. Lesch (1998). Role of serotonin in the immune system and in neuroimmune interactions. *Brain Behav. Immun.* **12**, 249–271.

Nabokina, S., G. Egea, J. Blasi, and F. Mollinedo (1997). Intracellular location of SNAP-25 in human neutrophils. *Biochem. Biophys. Res. Commun.* **239**, 592–597.

Nance, D.M., D.A. Hopkins, and D. Bieger (1987). Re-investigation of the innervation of the thymus gland in mice and rats. *Brain Behav. Immun.* **1**, 134–147.

Niijima, A. (1995). An electrophysiological study on the vagal innervation of the thymus in the rat. *Brain Res. Bull.* **38**, 319–323.

Pawlikowski, M. and J. Kunert-Radek (1995). Failure of excitatory amino acids receptor agonists NMDA and quiscalate to affect the cell proliferation. *Pol. J. Pharmacol.* **47**, 185–187.

Plaitakis, A. and J.T. Caroscio (1987). Abnormal glutamate metabolism in amyotrophic lateral sclerosis. *Ann. Neurol.* **22**, 575–579.

Purcell, W.M., K.M. Doyle, C. Westgate, and C.K. Atterwill (1996). Characterisation of a functional polyamine site on rat mast cells: Association with a NMDA receptor macrocomplex. *J. Neuroimmunol.* **65**, 49–53.

Rabin, B.S. (1999). *Stress, Immune Function, and Health: The Connection.* Wiley-Liss, Inc., New York.

Rimaniol, A.-C., S. Haïk, M. Martin, R. Le Grand, F.D. Boussin, N. Dereuddre-Bosquet *et al.* (2000). Na$^+$-dependent high-affinity glutamate transport in macrophages. *J. Immunol.* **164**, 5430–5438.

Rimaniol, A.-C., P. Mialocq, P. Clayette, D. Dormont, and G. Gras (2001). Role of glutamate tarnsporters in the regulation of glutathione levels in human macrophages. *Am. J. Physiol.* **281**, C1964–C1970.

Rinner, I., P. Felsner, P. Liebmann, D. Hofer, A. Woelfler, A, Globerson et al. (1998a). Adrenergic/cholinergic immunomodulation in the rat model—in vivo veritas? Dev. Immunol. 6, 245–252.

Rinner, I., K. Kawashima, and K. Schauenstein (1998b). Rat lymphocytes produce and secrete acetylcholine in dependence of differentiation and activation. J. Neuroimmunol. 81, 31–37.

Rinner, I., A. Globerson, L. Kawashima, W. Korsatko, and K. Schauenstein (1999). A possible role for acetylcholine in the dialogue between thymocytes and thymic stroma. Neuroimmunomodulation 6, 51–55.

Romano, T.A., S.Y. Felten, J.A. Olschowka, and D.L. Felten (1994). Noradrenergic and peptidergic innervation of lymphoid organs in the beluga, Delphinapterus leucas: An anatomical link between the nervous and immune systems. J. Morphol. 221, 243–259.

Roy, S. and H.H. Loh (1996). Effects of opioids on the immune system. Neurochem. Res. 21, 1375–1386.

Schauenstein, K., R. Faessler, H. Dietrich, S. Schwarz, G. Kroemer and G. Wick (1987). Disturbed immune-endocrine communication in autoimmune diseases. Lack of corticosterone response to immune signals in Obese Strain chickens with spontaneous autoimmune thyroiditis. J. Immunol. 139, 1830–1833.

Schauenstein, K., P. Felsner, I. Rinner, P.M. Liebmann, J.R. Stevenson, J. Westermann et al. (2000). In vivo immunomodulation by peripheral adrenergic and cholinergic agonists/antagonists in rat and mouse models. Ann. N. Y. Acad. Sci. 917, 618–627.

Schorr, E.C. and B.G. Arnason (1999). Interactions between the sympathetic nervous system and the immune system. Brain Behav. Immun. 13, 271–278.

Screpanti, I., D. Meco, S. Scarpa, S. Morrone, L. Frati, A. Gulino et al. (1992). Neuromodulatory loop mediated by nerve growth factor and interleukin-6 in thymic stromal cell cultures. Proc. Natl. Acad. Sci. USA 89, 3209–3212.

Sommer, M.H., M.H. Xavier, M.B. Fialho, C.M. Wannmacher, and M. Wajner (1994). The influence of amino acids on mitogen-activated proliferation of human lymphocytes in vitro. Int. J. Immunopharmacol. 16, 865–872.

Song, D.K., H.W. Suh, J.S. Jung, M.B. Wie, J.H. Song, and Y.H. Kim (1996). Involvement of NMDA receptor in the regulation of plasma interleukin-6 levels in mice. Eur. J. Pharmacol. 316, 165–169.

Stefulj, J., B. Jernej, L. Cicin-Sain, I. Rinner and K. Schauenstein (2000). mRNA expression of serotonin receptors in cells of the immune tissues of the rat. Brain Behav. Immun. 14, 219–224.

Stefulj, J., L. Cicin-Sain, K. Schauenstein, and B. Jernej (2001). Serotonin and immune response: Effect of the amine on in vitro proliferation of rat lymphocytes. Neuroimmunomodulation 9, 103–108.

Sternberg, E.M., J.M. Hill, G.P. Chrousos, T. Kamilaris, S.J. Listwak, P.W. Gold et al. (1989a). Inflammatory mediator-induced hypothalamic-pituitary-adrenal axis activation is defective in streptococcal cell wall arthritis-susceptible Lewis rats. Proc. Natl. Acad. Sci. USA 86, 2374–2378.

Sternberg, E.M., W.S. Young, R. Bernadini, A.E. Calogero, G.P. Chrousos, P.W. Gold et al. (1989b). A central nervous system defect in biosynthesis of corticotropin releasing hormone is associated with susceptibility to streptococcal cell wall-induced arthritis in Lewis rats. Proc. Natl. Acad. Sci. USA 86, 4771–4775.

Sternberg, E.M. (2000). The Balance Within: The Science Connecting Health and Emotions. W.H. Freeman and Co., New York.

Storto, M., U. de Grazia, G. Battaglia, M.P. Felli, M. Maroder, A. Gulino, et al. (2000). Expression of metabotropic glutamate receptors in murine thymocytes and thymic stromal cells. J. Neuroimmunol. 109, 112–120.

Szaflarski, J., J. Ivacko, X.H. Liu, J.S. Warren, and F.S. Silverstein (1998). Excitotoxic injury induces monocyte chemoattractant protein-1 expression in neonatal rat brain. Mol. Brain Res. 55, 306–314.

Watanabe, H. and S. Bannai (1987). Induction of cystine transport activity in mouse peritoneal macrophages. J. Exp. Med. 165, 628–640.

Watkins, A. (1997). Mind–Body Medicine. A Clinician's Guide to Psychoneuroimmunology. Churchill Livingstone, New York.

Weigent, D.A. and J.E. Blalock (1997). production of peptide hormones and neurotransmitters by the immune system. Chem. Immunol. 69, 1–30.

Westall, F.C., A. Hawkins, G.W. Ellison and L.W. Myers (1980). Abnormal glutamic acid metabolism in multiple sclerosis. J. Neurol. Sci. 47, 353–364.

Westlund, K.N., Y.C. Sun, K.A. Sluka, P.M. Dougherty, L.S. Sorkin and W.D. Willis (1992). Neural changes in acute arthritis in monkeys. II. Increased glutamate immunoreactivity in the medial articular nerve. Brain Res. Rev. 17, 15–27.

Whitlock, B.B., Y. Liu, S. Chang, P. Saini, B.K. Ha, T.W. Barrett et al. (1996). Initial characterization and autoradiographic localization of a novel sigma/opioid binding site in immune tissues. J. Neuroimmunol. 67, 83–96.

Yatsushiro, S., H. Yamada, M. Hayashi, A. Yamamoto and Y. Moriyama (2000). Ionotropic glutamate receptors trigger microvesicle-mediated exocytosis of L-glutamate in rat pinealocytes. J. Neurochem. 75, 288–297.

Zeev-Brann, A.B., O. Lazarov-Spiegler, T. Brenner, and M. Schwartz (1998). Differential effects of central and peripheral nerves on macrophages and microglia. Glia 23, 181–190.

15

Platelet Glutamate Receptors as a Window into Psychiatric Disorders

Michael Berk

1. Introduction

1.1. Overview of Platelet Biochemistry and Psychiatric Illness

Access to pathophysiological processes is replete with difficulty in psychiatry. It is obviously not possible to directly access brain tissue in the living subject. In response to this, ranges of options for the study of pathophysiological processes have been developed. Brain imaging has been useful, although the study of psychiatric conditions has been limited by the lack of overt structural changes in these conditions and technical issues pertaining to resolution. Newer techniques that allow imaging of neurochemical and pharmacological parameters such as positron emission tomography (PET) scanning are in ascendancy. Postmortem studies are the only way that brain tissue can be accessed, but there are substantial practical issues, including postmortem changes and issues of retrospective diagnosis. Pharmacological challenge paradigms, where an agonist or antagonist is given that causes a measurable central nervous system (CNS)-mediated change, usually to a neurohormone such as a growth hormone, is a well-utilized research strategy.

The development of valid peripheral markers of psychiatric illness may offer a potentially valuable tool for understanding the pathophysiology of these disorders. Peripheral markers offer access to a window that has many advantages, being simple, relatively non-invasive, affordable, and practical. A major challenge however is confronting the face validity issue that peripheral marker studies such as the use of platelets have in psychiatric disorders. In order to attempt to justify such limitations, it is necessary to discuss a number of issues. First, physiological similarities between platelets and neuronal tissue needs to be established. Second, changes in pathophysiological processes in parallel between platelets and central findings need to be demonstrated. Last, state related changes in symptomatology that are reflected by peripheral changes would augment the validity of the model.

Michael Berk • Professor of Psychiatry, Barwon Health and Geelong Clinic, Department of Clinical and Biomedical Sciences, University of Melbourne and Community and Mental Health, Barwon Health Swanston Centre, PO Box 281, Geelong, Victoria 3220, Australia.

Glutamate Receptors in Peripheral Tissue, edited by Santokh Gill and Olga Pulido.
Kluwer Academic / Plenum Publishers, New York, 2005.

1.2. Similarities between Platelets and Neuronal Tissue

Platelets have a number of common biochemical and morphological features with CNS neurons, allowing for comparisons of both structure and function. This makes them attractive candidate models for studying CNS receptors (Stahl, 1977; Da Prada *et al.*, 1988). The similarities between the platelet and neuron are greatest with the serotonergic neuron, and it is serotonergic function that has the longest tradition of research using platelet models, although noradrenaline, dopamine, adenosine, and glutamate have been studied using platelet models (Sneddon, 1973; Stahl, 1977; Da Prada *et al.*, 1988; Berk *et al.*, 1994; Berk *et al.*, 2001). Both platelets and neurons have subcellular storage systems, a cytoplasmic membrane with an active transport system for a number of neurotransmitter systems. Both have analogous binding sites for drugs and neurotransmitters. The kinetic characteristics of these systems are similar, and both physiological and pharmacological effects can be determined and potentially extrapolated as reflecting processes occurring in the brain (Healy and Leonard, 1987). Platelets in addition express a range of receptors. These include adrenoceptors (alpha-2 and beta-2), benzodiazepine, *N*-methyl-D-aspartate (NMDA), 5HT2, and adenosine receptors, which appear analogous to their CNS relatives.

Platelets are derived from megakaryocytes, which in turn are derived from special stem cells, colony-forming-granulocyte-erythroid-macrophage-megakaryocyte cells, which create promegakaryocytes. Once megakaryocytes leave bone marrow, they lodge in the pulmonary vasculature and fragment to form platelets that thereby enter the circulation. There are a number of important organelles in the platelet cytoplasm. Alpha granules are the most numerous and these contain a range of proteins and other factors secreted by the platelet during activation, including PF-4, beta-TG, adhesive glycoproteins such as von Willebrand factor and fibronectin, coagulation factors including factors V, XI, and XIII, protein S, and mitogenic factors including alpha-2, plasmin inhibitor-1, plasminogen activator inhibitor-1, and P-selectin. Dense bodies contain calcium, adenosine diphosphate (ADP), adenosine triphosphate (ATP), and serotonin. These bodies, like vesicles in the terminal boutons, act as storage sites for serotonin that is released by a calcium dependent excitation coupling mechanism. A significant difference is that unlike the nerve terminal, the platelet is dependent on systemic uptake of serotonin as is cannot synthesize serotonin directly (Pearse, 1986; Camacho and Dimsdale, 2000).

There are however major differences between the platelet and the neuron. Perhaps the most significant difference is that nerve terminals are part of an interconnected network with a vast number of neurons, and interact with a range of neurotransmitters and neuromodulators, whereas platelets are physically unconnected with other cell types. Platelets also do not express the full range of receptors present in the brain, expressing, for example, only a subset of serotonin receptors. Signaling is primarily chemical in platelets, while in neuronal tissue, chemical and electrical signaling operates (Leonard, 1992).

1.3. Psychiatric Nosology

The classification of psychiatric disorders is essentially on the basis of phenomenology. Other factors such as genetics, longitudinal course, and treatment effects also inform classifications. This is a substantial limitation, as in most areas of medicine, once the pathophysiology of a disorder is established, it becomes clear that individual pathologies can have pleomorphic phenomenology, and conversely, many pathologies can have similar clinical presentations. Perhaps the greatest problem in psychiatry is the absence of clear pathology for almost all disorders, as well as an understanding of normal physiology for many psychological processes. The development of markers that are capable of providing insight into the pathology of these disorders is thus of primary importance. There is an ongoing debate as to the diagnostic boundaries of many of the major psychiatric disorders.

Schizophrenia is defined as a chronic psychotic disorder of unknown etiology. Structural and functional changes are evident in imaging studies, and a significant genetic component is clear. Psychosis is defined as a loss of reality testing, and symptoms of psychosis include abnormal thought content including delusions, abnormal form of thought including illogical thought processes, and alterations in perception such as hallucinations. In schizophrenia, symptoms of psychosis occur together with other features such as a decline in functioning, and blunted or inappropriate mood. Depression is a syndrome consisting of a range of symptoms. These include depressed mood, loss of interest, changes in sleep, appetite, energy, concentration, and thoughts of suicide, guilt, and worthlessness. Patients can have recurrent episodes of depression in a depressive disorder. In bipolar disorder, the hallmark is the presence of episodes of both mania and depression. Mania is the converse of depression, and symptoms include inflated self-esteem or grandiosity, decreased need for sleep, increased energy, drive, thought speed, and goal-directed behavior, and is typically associated with impairment. Psychotic symptoms can occur in both depression and mania. In the presence of dominant mood symptoms, a mood disorder diagnosis is made even if psychotic symptoms are present (American Psychiatric Association, 2000).

1.4. Platelet Studies

Platelets have been used as models of neuronal tissue in a number of psychiatric disorders, particularly schizophrenia and depression. Parallels between processes in the platelet and brain have been shown in some neurological conditions such as Huntington's Disease, where increased aspartate in both brain and platelet have been shown (Reilmann *et al.*, 1994), validating this approach.

A number of differing methods have been used to study platelet parameters. Platelet specific factors such as PF4 and beta-TG have been examined to quantify platelet function. Radioligand binding can be used to study platelet receptors such as the alpha-2 adrenoceptor. Platelet monoamine oxidase (MAO) activity can be studied using radioenzymatic assay. Whole blood aggregation can be measured. Analysis of specific membrane receptors can be done via flow cytometry. Spectrofluorometry can be used to examine basal and receptor-regulated shifts in intracellular calcium (Camacho and Dimsdale, 2000).

A widely used approach to the understanding of receptor pathophysiology centered on binding of specific ligands to receptors (Arora and Meltzer, 1989; Biegon *et al.*, 1990; Butler *et al.*, 1992; Arora and Meltzer, 1993; Bakish *et al.*, 1997). Targets of study have included platelet aggregation, platelet MAO activity, and binding to the serotonin-2 receptor, the serotonin uptake site and the alpha-2 receptor. Receptor-regulated second messenger responses may represent a more physiological methodology, measuring transduction of receptor stimuli into parallel second messenger signals. Intracellular calcium, as a second messenger response may serve as an index of the functional capability of the receptor-regulated neurotransmitter response. This may facilitate understanding of an important neurobiological process underpinning these disorders. Calcium as a second messenger is involved in the regulation of many processes implicated in affective disorders (Dubovsky *et al.*, 1989).

NMDA receptors are present on platelet membranes. Agonist stimulation of NMDA receptors has been shown to increase intracellular free calcium in non-neuronal cell lines has been linked to the NMDA receptor complex. Similarities between the kinetic properties of glutamate uptake in platelets and brain slices have been reported (Mangano and Schwarcz, 1981; Almazov et al., 1988; Grant et al., 1997). NMDA has an inhibitory role in platelet aggregation. NMDA appears to antagonize the aggregation induced by ADP and arachidonic acid. NMDA also increases cyclic AMP as well as being an inhibitor of thromboxane B2 synthesis from arachidonic acid (Franconi et al., 1998). The receptor however is distinct from neuronal NMDA in that the platelet receptor is not potentiated by glycine, as is the case with neurons (Franconi et al., 1996).

1.5. Glutamate and Schizophrenia

The amino acid glutamate functions as the dominant excitatory neurotransmitter in the human cortex. It has a number of receptors, which form two families, ionotropic and metabotropic. The ionotropic receptors are represented by NMDA, α-amino-3-hydroxy-5-methyl-4-isoxazolepropionic acid (AMPA), and kainate receptors. Glutamate and, in particular, the NMDA receptor have received increased attention regarding its potential role in a range of psychiatric disorders, particularly schizophrenia.

Kim et al. (1980) put forward the glutamate theory of schizophrenia, on the basis of reduced levels of the excitatory amino acid, glutamate, in the cerebrospinal fluid of patients with schizophrenia. Subsequent studies have added to the literature suggesting that schizophrenia may be caused by a hypoglutamatergic state (Tsai and Coyle, 2002). It was observed that phencyclidine, a glutamate receptor antagonist, causes a psychosis resembling both the positive and negative symptoms of schizophrenia (Javitt and Zukin, 1991). This data is augmented by an increasing number of postmortem studies in schizophrenia that have described changes in glutamate receptor binding in a number of brain areas including the putamen, temporal lobe, and frontal areas (Ulas and Cotman, 1993; Olney and Farber, 1995; Aparicio-Legarza et al., 1998; Gluck et al., 2002; Heckers and Konradi, 2002). In addition, a number of antipsychotics, such as haloperidol and clozapine have been shown to have effects on the glutamatergic system (Goff et al., 2002). Interactions between both dopamine and serotonin and glutamate may play a role in schizophrenia (Breese et al., 2002). Imaging studies have further demonstrated abnormal glutamate metabolism in the brains of schizophrenic patients (Tsai et al., 1995).

2. Peripheral Markers in Psychiatric Illness

2.1. Platelets and Schizophrenia

There is a paucity of research available on effective peripheral markers in schizophrenia. While dopamine receptors are expressed on the platelet, the platelet intracellular calcium response to dopamine has not been studied in schizophrenia. A single study in bipolar disorder failed to show any differentiation from controls (Berk et al., 1994). Serotonin is increasingly implicated in schizophrenia. The platelet intracellular calcium response to serotonin in schizophrenia has unfortunately not proved to be a useful marker in this disorder (Mikuni et al.,

1992; Konopka et al., 1996). Therefore, other targets need to be investigated with regard to the use of the platelet as a peripheral marker for schizophrenia. The first study to look at peripheral excitatory amino acids in psychiatric populations was that of Altamura et al. (1993), who investigated platelet levels of excitatory amino acids in platelets and plasma from patients with schizophrenia, anxiety, and mood disorders. Higher plasma levels of glutamate were found in mood disorders, a trend not seen in platelet levels. No significant trend in schizophrenia was found however with regard to glutamate.

Das and colleagues (1995) demonstrated the effect of neuroleptics on the glutamatergic system by utilizing platelet receptors to show a significant alteration in intracellular calcium and nitric oxide synthase in response to thrombin stimulation in patients with schizophrenia. This research demonstrated an increase in intracellular calcium that was replicated by Ripova and colleagues (1997). Neuroleptic treatment reduced intracellular calcium, but not to control levels. Both studies utilized drug naive patients. Increased turnover of phosphatidic acid has been reported suggesting alterations in the phosphoinositide signaling process in schizophrenia (Ripova et al. 1999). Das and colleagues (1995) reported increased nitric oxide synthase activity in patients who were drug naive, compared to both healthy controls and schizophrenic patients currently taking neuroleptic medication. Treatment with neuroleptic medication returned these elevations to control levels. Nitric oxide is generated from nitric oxide synthase in response to NMDA receptor stimulation by glutamate therefore supporting the involvement of glutamate in schizophrenia.

2.2. Methodology of the Platelet Intracellular Calcium Response to Glutamate

The methodology for the following three studies was specifically developed. In brief summary, platelet-rich-plasma was obtained, and the platelets were incubated with a calcium sensitive dye, fura-2-AM. Fura-2-AM fluoresces at different wavelengths when bound and unbound, allowing levels of intracellular calcium to be determined. Thereafter, glutamate concentrations of 0–100 mM were added sequentially, and a dose response curve of intracellular calcium against glutamate was obtained. In order to confirm the role of the NMDA receptor in these results, platelets were incubated with 100 μM dizoclipine (MK-801). Dizoclipine is a noncompetitive NMDA receptor antagonist that cannot be displaced by glutamate. This resulted in the complete blockade of the agonist effect of glutamate in this model, confirming the role of the NMDA receptor in the model.

2.3. Platelet Intracellular Calcium Response to Glutamate in Psychosis

The first study utilizing this method (Berk et al., 1999) compared 15 drug-free patients diagnosed on structured interview as suffering from schizophrenia, with 15 age and sex-matched controls. A highly significant difference between the dose response curves of patients and controls was found, with a far greater increase in intracellular calcium responses in the patients with schizophrenia than control subjects. It was thought that this finding might reflect secondary upregulation of NMDA receptor sensitivity in the face of decreased glutamatergic function. This was of particular interest, given the paucity of peripheral markers of clinical or theoretical utility in schizophrenia.

The next study attempted to answer the question of whether the supersensitivity seen in the first study was specific to schizophrenia, or if it was a reflection of psychosis. The answer was unexpected, and confirmed neither of the above. In the second study, three groups of patients with psychotic features were compared; schizophrenia, mania with psychotic features, and depression with psychotic features. Both the schizophrenia and the depression with psychotic features group demonstrated statistically significant platelet NMDA receptor supersensitivity compared to the control group. Surprisingly, the mania with psychotic features group did not differ significantly from controls (Berk et al., 2000). This suggested that the response seen was not a nonspecific finding in psychosis. Unipolar major depressive disorder became the next focus of study. The next question was if the marker was specific to psychosis or if it was to be found in nonpsychotic depression.

2.4. Platelet Intracellular Response to Glutamate in Depression

There is increasing interest in the excitatory amino acids in depression. Higher plasma and platelet levels of glutamate in a cohort of depressed patients compared to controls have been described (Mauri et al., 1998). There are negative studies, in which no difference in serum levels of glutamate between patients with depression and age and sex-matched controls were found (Maes et al., 1998). However, changes in plasma glutamate have been described in other conditions such as migraine (Ferrari et al., 1990). Traditional antidepressants have significant effects on glutamate. Both acute changes in glutamate neurons as well as chronic changes in the NMDA receptor complex are described with antidepressant administration (Pangalos et al., 1992; Nowak, 1996; Bouron and Chatton, 1999). Peripheral glutamate function is also altered by antidepressants. Antidepressant therapy appears to reduce plasma levels of glutamate (Maes, 1998).

Therapeutic response to NMDA antagonists is a powerful argument for the involvement of glutamate in depression. Ketamine, an NMDA antagonist, has been demonstrated to show antidepressant properties in a double-blind trial (Berman, 2000). Similarly, lamotrigine is an anticonvulsant with glutamate antagonist properties, in that it inhibits the excessive release of glutamate. Lamotrigine has been clearly shown to be efficacious in the depressive phase of bipolar disorder in a large placebo controlled design (Calabrese et al., 1999). Newer unpublished data by the same author suggests that lamotrigine may have prophylactic effects in bipolar disorder that are greater in depression than mania. This is intriguing given that the platelet marker data suggests that glutamate sensitivity may occur in depression rather than mania. It would be fascinating to study if glutamate sensitivity was a predictor of lamotrigine response.

In order to characterize the status of the platelet NMDA receptor in depression, a further study was conducted. In this study of 15 drug-free nonpsychotic patients suffering from unipolar major depression and 17 controls, clear differences were seen between patients and controls. The depression group showed a significantly greater platelet intracellular calcium response to glutamate stimulation than the control group. This suggested that platelet NMDA receptors may be supersensitive in depression. It also confirmed that the platelet might be a possible peripheral marker of glutamate function in depression.

3. Conclusion

Much research remains to be done. The state trait nature of the marker needs to be elucidated. It would be illuminating to know if the marker settled with improvement in clinical condition. Treatment specificity with regard to the marker would also be interesting. It would be illuminating to know if the marker had any predictive value in terms of treatment. More research in a wider range of psychiatric illnesses would be interesting. It is also necessary to look at studies to correlate the peripheral marker with central markers to validate the use of peripheral markers.

It is hoped that platelet intracellular second messenger responses to glutamate will reflect the pathogenesis of the disease processes. Further development of these accessible and practical markers may allow for a better understanding of these disorders, and may guide in the rational development of treatments of these conditions.

References

Almazov, V.A., Y.G. Popov, A.I. Gorodinsky, I.A. Mikhailova, S.A. Dambinova, and V.S. Gurevich (1988). The sites of high affinity binding of L-[^{3}H] glutamic acid in human platelets: A new type of platelet receptor? *Biokhimiia* **53**(5), 848–852.

Altamura, C.A., M.C. Mauri, A. Ferrara, A.R. Moro, G. D'Andrea, and F. Zamberlan (1993). Plasma and platelet excitatory amino acids in psychiatric disorders. *Am. J. Psychiatry* **150**, 1731–1733.

American Psychiatric Association. Diagnostic and Statistical Manual of Mental Disorders, text revision, 4th ed. American Psychiatric Association 2000, Washington DC.

Aparicio-Leagarza, M.I., B. Davis, P.H. Hutson, and G.P. Reynolds (1998). Increased density of glutamate/N-methyl-D-aspartate receptors in putamen from schizophrenic patients. *Neurosci. Lett.* **241**, 143–146.

Arora, R.C. and H.Y. Meltzer (1989). Increased serotonin 2 receptor binding as measured by ^{3}H-LSD binding in blood platelets of depressed patients. *Life Sci.* **44**, 725–734.

Arora, R.C. and H.Y. Meltzer (1993). Serotonin 2 receptor binding in blood platelets of schizophrenic patients. *Psychiatry Res.* **47**, 111–119.

Bakish, D., P. Cavazzoni, J. Chudzik *et al.* (1997). Effects of serotonin reuptake inhibitors on platelet serotonin parameters in major depressive disorder. *Biol. Psychiatry* **41**, 184–190.

Berk, M., W. Bodemer, T. Van Oudenhove *et al.* (1994). Dopamine increases platelet intracellular calcium in bipolar disorder and controls. *International Clin. Pharmacol.* **9**, 291–293.

Berk, M., H. Plein, and B. Belsham (2000). The specificity of the platelet glutamate receptor supersensitivity in psychotic disorders. *Life Sci.* **66**, 2427–2432.

Berk, M., H. Plein, and T. Czismadia (1999). Supersensitive platelet glutamate receptors as a possible peripheral marker in schizophrenia. *International Clin. Psychopharmacol.* **14**, 119–122.

Berk, M., H. Plein, and D. Ferreira (2000). Platelet glutamate supersensitivity in depression. *Int. J. Neuropsychopharmacol.* **3**(Suppl 1), S112.

Berk, M., H. Plein, D. Ferreira, and B. Jersky (2001). Blunted adenosine A2a receptor function in platelets in patients with major depression. *Eur. Neuropsychopharmacol.* **11**(2), 183–186.

Berman, R.M., A. Cappiello, A. Anand, D.A. Oren, G.R. Heninger, D.S. Charney *et al.* (2000). Antidepressant effects of ketamine in depressed patients. *Biol. Psychiatry* **47**, 351–354.

Biegon, A., A. Grinspoon, B. Blumenfeld *et al.* (1990). Increased serotonin 5HT2 receptor binding on blood platelets of suicidal men. *Psychopharmacology (Berl)* **100**(2), 165–167.

Bouron, A and J.Y. Chatton (1999). Acute application of the tricyclic antidepressant desipramine presynaptically stimulates the exocytosis of glutamate in the hippocampus. *Neuroscience* **90**, 729–736.

Breese, G., D. Knapp, and S. Moy (2002). Integrative role for serotonergic and glutamatergic receptor mechanisms in the action of NMDA antagonists: Potential relationships to antipsychotic drug actions on NMDA antagonist responsiveness. *Neurosci. Biobehav. Rev.* **26**(4), 441.

Butler, J., A. O'Halloran, and B.E. Leonard (1992). The Galway study of panic disorder II changes in some peripheral markers of noradrenergic and serotonergic function in DSM III-R panic disorder. *J. Affect Disord.* **26**(2), 89–99.

Calabrese, J.R., C.I. Bowden, G.S. Sacks, J.A. Ascher, E. Monaghn, and G.D. Rudd (1999). A double blind placebo controlled study of lamotrigine monotherapy in outpatients with bipolar 1 depression. *J. Clin. Psychiatry* **60**, 79–88.

Camacho, A. and J.E. Dimsdale (2000). Platelets and psychiatry: Lessons learned from old and new studies. *Psychosom. Med.* **62**, 326–336.

Da Prada, M., A.M. Cesura, J.M. Launay *et al.* (1988). Platelets as a model for neurones? *Experentia* **44**, 115–126.

Das, I., N.S. Khan, B.K. Puri, S.R. Sooranna, J. de Belleroche, and S.R. Hirsch (1995). Elevated platelet calcium mobilization and nitric oxide synthase activity may reflect abnormalities in schizophrenic brain. *Biochem. Biophys. Res. Commun.* **212**(2), 375–380.

Dubovsky, S.L., J. Christiano, L.C. Daniell *et al.* (1989). Increased platelet intracellular calcium concentration in patients with bipolar affective disorders. *Arch. Gen. Psychiatry* **46**, 632–638.

Ferrari, M.D., J. Odink, K.D. Bos, M.J. Malessy, and G.W. Bruyn (1990). Neuroexitatory amino acids are elevated in migraine. *Neurology* **40**, 1582–1586.

Franconi, F., M. Miceli, L. Alberti, G. Seghieri, M.G. De Montis, and A. Tagliamonte (1998). Further insights into the anti aggregating activity of NMDA in human platelets. *Br. J. Pharmacol.* **124**, 35–40

Franconi, F., M. Miceli, M.G. De Montis, E.L. Crisafi, F. Bennardini, and A. Tagliamonte (1996). NMDA receptors play an anti aggregating role in human platelets. *Thromb. Haemost.* **76**, 84–87.

Gluck, M.R., R.G. Thomas, K.L. Davis, and V. Haroutunian (2002). Implications for altered glutamate and GABA metabolism in the dorsolateral prefrontal cortex of aged schizophrenic patients. *Am. J. Psychiatry* **159**(7), 1165–1173.

Goff, D.C., J. Hennen, I.K. Lyoo, G. Tsai, L.L. Wald, A.E. Evins *et al.* (2002). Modulation of brain and serum glutamatergic concentrations following a switch from conventional neuroleptics to olanzapine. *Biol. Psychiatry* **51**(6), 493–497.

Grant, E.R., B.J. Bacskai, D.E. Pleasure *et al.* (1997). N-methyl-D-aspartate receptors expressed in a non-neuronal cell line mediate subunit-specific increases in free intracellular calcium. *J. Biol. Chem.* **272**, 647–656.

Healy, D. and B.E. Leonard (1987). Monoamine transport in depression: Kinetics and dynamics. *J. Affect. Disord.* **12**(2), 91–103.

Heckers, S. and C. Konradi (2002). Hippocampal neurons in schizophrenia. *J. Neural. Transm.* **109**(5–6), 891–905.

Javitt, D.C. and S.R. Zukin (1991). The role of excitatory amino acids in neuropsychiatric illness. *Am. J. Psychiatry* **148**, 1301–1308.

Kim, J.S., H.H. Kornhuber, W. Schmid- Burgk *et al.* (1980). Low cerebrospinal fluid glutamate in schizophrenic patients and a new hypothesis on schizophrenia. *Neurosci. Lett.* **20**, 379–382.

Konopka, L.M., R. Cooper, and J.W. Crayton (1996). Serotonin-induced increases in platelet cytosolic calcium in depressives, schizophrenic, and substance abuse patients. *Biol. Psychiatry* **39**, 708–713.

Leonard, B.E. (1992). *Fundamentals of Psychopharmacology.* John Wiley Publishers, Chichester.

Maes, M., R. Verkerk, E. Vandoolaeghe, A. Lin, and S. Scharpe (1998). Serum levels of excitatory amino acids, serine glyceine, histidine, threonine, taurine, alanine and arginine in treatment resistant depression: Modulation by treatment with antidepressants and prediction of clinical responsivity. *Acta. Psychiatr. Scand.* **97**, 302–308.

Mangano, R.M. and R. Schwarcz (1981). The human platelet as a model for the glutamatergic neuron: Platelet uptake of L-glutamate. *J. Neurochem.* **36**, 1067–1076.

Mauri, M.C., A. Ferrara, L. Boscati, S. Bravin, F. Zamberlan, M. Alecci *et al.* (1998). Plasma and platelet amino acid concentrations in patients affected by major depression and under fluvoxamine treatment. *Neuropsychobiology* **37**, 124–129.

Mikuni, M., A. Kagaya, K. Takahashi *et al.* (1992). Serotonin but not norepinephrine induced calcium mobilization of platelets is enhanced in affective disorders. *Psychopharmacology* **106**, 311–314.

Nowak, G., Y. Li, and I.A. Paul (1996). Adaptation of cortical but not hippocampal NMDA receptors after chronic citalopram treatment. *Eur. J. Pharmacol.* **295**, 75–85.

Olney, J.E. and N.B. Farber (1995). Glutamate receptor dysfunction and schizophrenia. *Arch Gen Psychiatry* **52**, 998–1007.

Pangalos, M.N., A.I. Malizia, P.T. Francis, S.I. Lowe, P.H. Bertolucci, A.W. Procter *et al.* (1992). Effects of psychotropic drugs on excitatory amino acids in patients undergoing psychosurgery for depression. *Br. J. Psychiatry* **160**, 638–642.

Pearse, A.G.E. (1986). The diffuse neuroendocrine system: Peptides, amines, placodes and the APUD theory. *Prog Brain Res.* **68**, 25–31.

Reilmann, R., L.H. Rolf, and H.W. Lange, (1994). Huntington's disease: The neuroexcitotoxin aspartate is increased in platelets and decreased in plasma. *J. Neurol. Sci.* **127**, 48–53.

Ripova, D., A. Strunecka, V. Nemcova, and I. Farska (1997). Phospholipids and calcium alteration in platelets of schizophrenic patients. *Physiol. Res.* **46**, 59–68.

Ripova, D., A. Strunecka, V. Platilova, and C. Hoschl, (1999). Phosphoinositide signalling system in platelets of schizophrenic patients and the effect of neuroleptic therapy. *Prostaglandins, Leukot. Essent. Fatty Acids* **61**, 125–129.

Sneddon, J.M. (1973). Blood platelets as a model for monoamine containing neurons. *Prog. Neurobiol.* **1**, 151–198.

Stahl, S.M. (1977). The human platelet: A diagnostic and research tool for the study of biogenic amines in psychiatric and neurologic disorders. *Arch. Gen. Psychiatry* **34**, 509–516.

Tsai, G. and Coyle, J.T. (2002). Glutamatergic mechanisms in schizophrenia. *Annu. Rev. Pharmacol. Toxicol.* **42**, 165–179.

Tsai, G., L.A. Passani, B.S. Slusher *et al.* (1995). Abnormal excitatory neurotransmitter metabolism in schizophrenic brains. *Arch. Gen. Psychiatry* **52**, 829–836.

Ulas, J. and C.W. Cotman (1993). Excitatory amino acid receptors in schizophrenia. *Schizophr. Bulle.* **19**(1), 105–113.

Part III

Non-Mammalian Organisms

Analysis of Glutamate Receptor Genes in Plants: Progress and Prospects

Joanna C. Chiu, Eric D. Brenner, Rob DeSalle, Nora M. Barboza, and Gloria M. Coruzzi

1. Introduction

Glutamate receptor-like genes (*GLR*) have been described in a number of plant species including *Arabidopsis thaliana* (Lam *et al.*, 1998; Kim *et al.*, 2001; Lacombe *et al.*, 2001) and *Brassica napus* (Genbank accession number AF109392). These *GLR* genes have similarity in primary sequence and domain organization to ionotropic glutamate receptors (iGluRs) in animals. At the moment, plant genes that have sequence similarity to the other major animal glutamate receptor gene family, metabotropic glutamate receptor, have not been found. However, a study by Turano *et al.* (2001) suggests an evolutionary link between *Arabidopsis* glutamate receptor genes (*AtGLRs*) and seven transmembrane G-protein-linked receptors, which includes GABA receptors as well as metabotropic glutamate receptors, based on primary sequence similarity in the N-terminus of the proteins. Like animals, plants appear to have multiple *GLR* genes, and 20 genes have been uncovered in the complete *Arabidopsis* genome. In this chapter, we will focus on works that examine the glutamate receptor gene family from *Arabidopsis (AtGLR)*, including forward and reverse genetic approaches and other studies that aim to determine the *in vivo* function of *GLRs* in plants.

2. Sequence and Phylogenetic Analysis of *Arabidopsis* GLR Genes

With the completion of the *A. thaliana* genome-sequencing project (AGI, 2000), all the ionotropic glutamate receptor-like genes (*AtGLRs*) in this model plant have been uncovered. A nomenclature system (Table 16.1) has been established for the 20 members of the *AtGLR*

Joanna C. Chiu, Eric D. Brenner, Nora M. Barboza, and Gloria M. Coruzzi • New York University, Department of Biology, New York, NY. **Rob DeSalle** • Americas Museum of Natural History, Division of Invertebrate Zoology, New York, NY.

Glutamate Receptors in Peripheral Tissue, edited by Santokh Gill and Olga Pulido.
Kluwer Academic / Plenum Publishers, New York, 2005.

Table 16.1 Genbank Accession Numbers and Nomenclature for the *Arabidopsis GLR* Gene Family (Reprinted from Chiu *et al.*, 2002, Copyrighted by the Society of Molecular Biology and Evolution)

			cDNA		Genomic	
		Previous name	Full-length	Splice-variants	BAC	Protein ID
I	AtGLR1.1	AtGLR1[a]	AF079998		AC016829	AAF26802.1
	AtGLR1.2		AY072064[b]	AY072065[b]	AB020745.1	BAA96960.1
	AtGLR1.3				AB020745.2	BAA96961.2
	AtGLR1.4		AY072066[b]	AY072067[b]	AC009853	AAF02156.1
II	AtGLR2.1	AtGLR3[a]			AF007271	AAB61068.1
	AtGLR2.2		AY072068[b]		AC007266.1	AAD26895.1
	AtGLR2.3				AC007266.2	AAD26894.1
	AtGLR2.4				AL031004	CAA19752.1
	AtGLR2.5				AL360314.1	CAB96656.1
	AtGLR2.6				AL360314.2	CAB96653.1
	AtGLR2.7		AY072069[b]		AC005315.1	AAC33239.1
	AtGLR2.8	Glur9	AJ311495		AC005315.2	AAC33237.1
	AtGLR2.9				AC005315.3	AAC33236.1
III	AtGLR3.1	AtGLR2[a], ACL1[c]	AF079999	AF038557	AC002329	AAF63223.1
	AtGLR3.2	AtGluR2[d]	AF159498		AL022604	CAA18740.1
	AtGLR3.3				AC025815	AAG51316.1
	AtGLR3.4	AtGLR4[a], GLUR3[e]	AF167355	AY072070[b]	AC000098	AAB71458.1
	AtGLR3.5	GLR6	AF170494		AC005700.1	AAC69939.1
	AtGLR3.6				AL133452	CAB63012.1
	AtGLR3.7	GLR5	AF210701		AC005700.2	AAC69938.1

[a]Lam *et al.* (1998)
[b]Chiu *et al.* (2002)
[c]*ACL1* corresponds to the splice variant AF038557.
[d]Kim *et al.* (2001)
[e]*GLur3* is deposited into Genbank as AF167355.

gene family based on a preliminary parsimony analysis that separates the gene family into three clades (Lacombe *et al.*, 2001). This nomenclature system helps to avoid future confusion in the naming of the *AtGLR* genes as more *AtGLR* complementary DNAs (cDNAs) are being isolated. With the complete set of *AtGLR* gene sequences, phylogenetic analyses have been conducted to examine: (a) sequence similarity between animal iGluRs and *AtGLR* genes, and (b) phylogenetic relationships between animal iGluRs and *AtGLRs*.

2.1. Sequence Similarity between Animal iGluR and *Arabidopsis GLR* Genes

Based on sequence similarity, *Arabidopsis GLR* genes encode all the signature domains of animal iGluR, including the two glutamate-binding domains (GlnH1 and GlnH2), the three transmembrane domains (M1, M2, and M3), and the putative pore region (P) between M1 and M2. The highest level of sequence similarity between *AtGLRs* and animal iGluRs is observed in M2, in which the percentage identity is above 60% (Chiu *et al.*, 1999). The sequence similarity between animal iGluR and *AtGLR* genes span all the important functional domains defined in animal iGluRs (Figure 16.1). Many of the amino acid residues that are conserved between animal iGluRs, *AtGLRs*, as well as *Synechocystis GluR0*, have been identified as

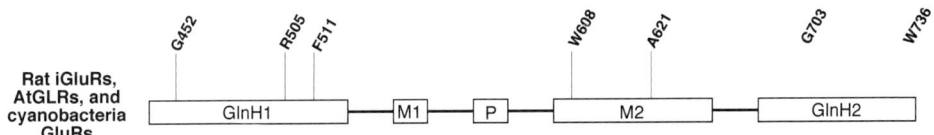

Figure 16.1. Amino acid residues invariant to animal, plant, and cyanobacterial iGluRs. Schematic diagram illustrating the functional domains of the ionotropic glutamate receptor (iGluR) protein and the locations of invariant amino acid residues between rat iGluRs, *AtGLRs*, and cyanobacterial iGluRs. All amino acid numbers correspond to the sequence of *AtGLR1.1* (accession number AF079998) (adapted from Chiu *et al.*, 2002, material is copyrighted by the Society of molecular biology and evolution).

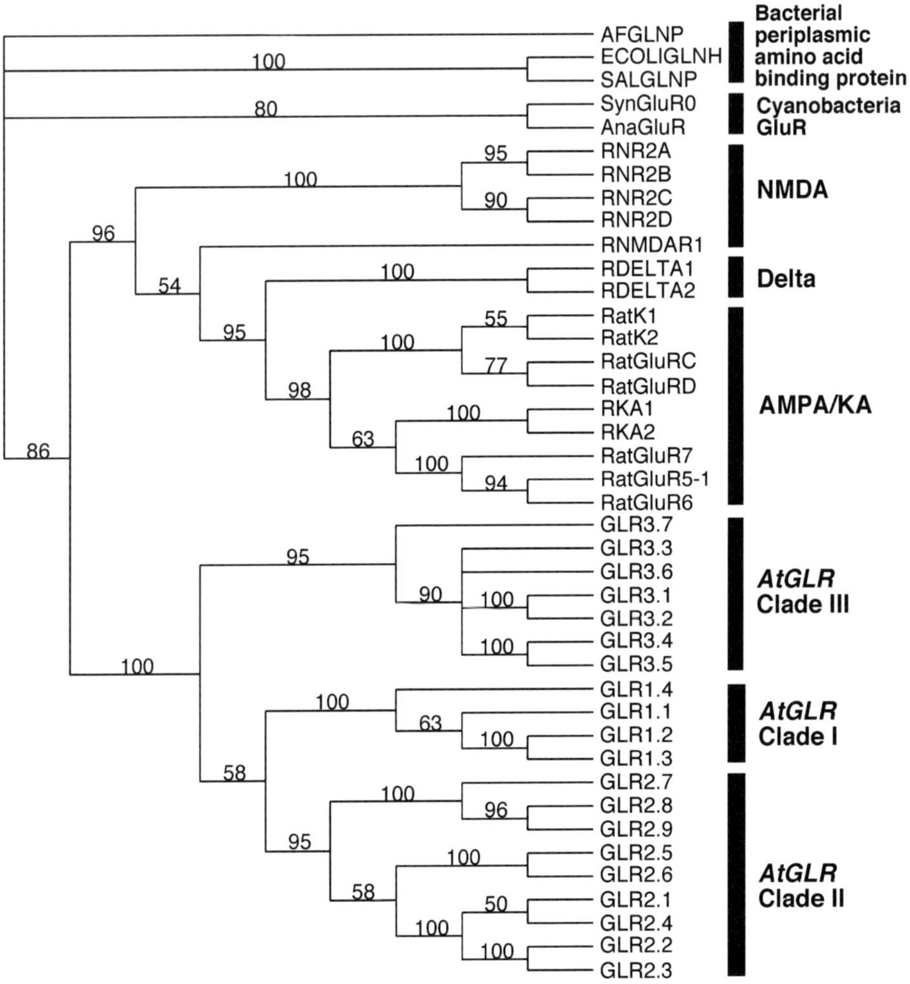

Figure 16.2. *Arabidopsis GLR* genes group into three distinct clades. Bootstrap consensus tree resulting from parsimony analysis illustrating the phylogenetic relationships between amino acid sequences of rat ionotropic glutamate receptors, *AtGLRs*, and two prokaryotic cyanobacterial iGluRs using three bacterial periplasmic amino acid binding protein sequences as outgroups (reprinted from Chiu *et al.*, 2002, material is copyrighted by the Society of molecular biology and evolution).

residues required for the proper functioning in animal iGluRs (Chiu *et al.*, 2002). In addition to primary sequence similarity, hydropathy plot analysis predicted that the transmembrane topologies of *AtGLRs* are similar to animal iGluRs (Lam *et al.*, 1998).

2.2. Phylogenetic Analysis of Animal iGluR and *Arabidopsis* *GLR* Genes

Chiu *et al.* (2002) performed a parsimony analysis of the iGluR gene family, which includes all 20 *Arabidopsis GLR* genes, all rat iGluRs (except NR3 genes), and two prokaryotic iGluRs (*Synechocystis GluR0* and *Anabaena* iGluR), using bacterial periplasmic amino acid binding proteins as outgroups (Figure 16.2). Results of this analysis support a previous finding (Chiu *et al.*, 1999) suggesting that the divergence of animal iGluRs and *AtGLRs* precedes that divergence of the animal iGluR classes (AMPA, KA, NMDA, and Delta). Based on this phylogenetic inference, it is unclear whether *AtGLR* genes may encode functional ion channel subunits, as in the case of animal iGluRs. However, the fact that *Synechocystis GluR0*, a gene that diverged from *Arabidopsis GLRs* and animal iGluRs early on, has been shown to encode a functional ion channel subunit (Chen *et al.*, 1999) suggests that *AtGLR* genes are likely to encode ion channel subunits. Future electrophysiological experiments on the isolated *AtGLR* genes are needed to confirm this prediction. However, *in planta* electrophysiological measurements suggest the existence of glutamate-gated ion channels in *Arabidopsis* (Dennison and Spalding, 2000; see section on *in planta* electrophysiology).

Since the rat iGluR gene family has been separated into functional classes based on biochemistry and electrophysiology, it represents an excellent test case for using phylogenetic analysis to separate a large gene family into potential functionally distinct protein classes. According to the results of parsimony analysis (Chiu *et al.*, 2002), animal iGluR clades defined phylogenetically coincide with animal iGluR classes (AMPA, KA, NMDA, and Delta) that were previously established based on biochemical properties such as ligand selectivity. Therefore, the separation of the 20 *AtGLR* genes into three clades by parsimony analysis suggests that the *AtGLR* gene family may also contain genes that encode proteins with distinct biochemical properties and physiological functions. Future experiments are necessary to establish true functional *AtGLR* protein classes.

3. Expression Analysis of *Arabidopsis GLR* Genes

Based on the phylogenetic analysis of rat iGluRs and *AtGLRs* (Chiu *et al.*, 2002), rat iGluR genes form different clades that coincide with functional iGluR classes defined by biochemistry and electrophysiology. Due to the lack of physiological and biochemical data at the current time, it is impossible to determine whether the three *AtGLR* clades defined by the same phylogenetic analysis also represent functionally distinct protein classes. As a first attempt to examine the functional significance of the *AtGLR* clade division, Chiu *et al.* (2002) performed a comprehensive RT-PCR analysis on all 20 members of the *AtGLR* gene family. Since animal iGluR channels are homomultimers or heteromultimers assembled from proteins encoded by iGluR genes from the same functional classes, this may also be the case for *AtGLR* proteins. In order for *AtGLR* proteins to coassemble, it is likely that their mRNA will colocalize in the same organs and cell types. Assuming that the three *AtGLR* clades represent functionally distinct protein classes, it is therefore hypothesized that *AtGLR* genes from the same clade may be expressed in the same organs, and that the three *AtGLR* clades might have distinct expression profiles. To increase the resolution of the RT-PCR analysis, Chiu *et al.* (2002) examined the cell-type expression of one representative gene from each *AtGLR* clade using a promoter–reporter gene system.

3.1. Examining Organ-Level Expression of *AtGLRs* by RT-PCR

By using RT-PCR, Chiu *et al.* (2002) presented the first comprehensive organ-level expression profile of all 20 members of the *Arabidopsis GLR* gene family. Figure 16.3 shows the representative expression patterns for each *AtGLR* clade (see Chiu *et al.*, 2002 for expression patterns for all 20 genes). In 8-week old *Arabidopsis* plants, genes from clade I are expressed in all organs tested, namely leaf, root, flower, and silique. This expression pattern also holds true for all genes from clade III. On the other hand, expression of *AtGLR* genes from clade II are less ubiquitous. Five of the nine clade II genes (*AtGLR2.1, AtGLR2.2, AtGLR2.3, AtGLR2.6,* and *AtGLR2.9*) are root-specific in 8-week old *Arabidopsis* plants. *AtGLR2.4* is detected in siliques in addition to roots, while *AtGLR2.7* and *AtGLR2.8* are detected in all organs tested except flowers. Out of all clade II genes, *AtGLR2.5* is the only one that is detected in all organs tested.

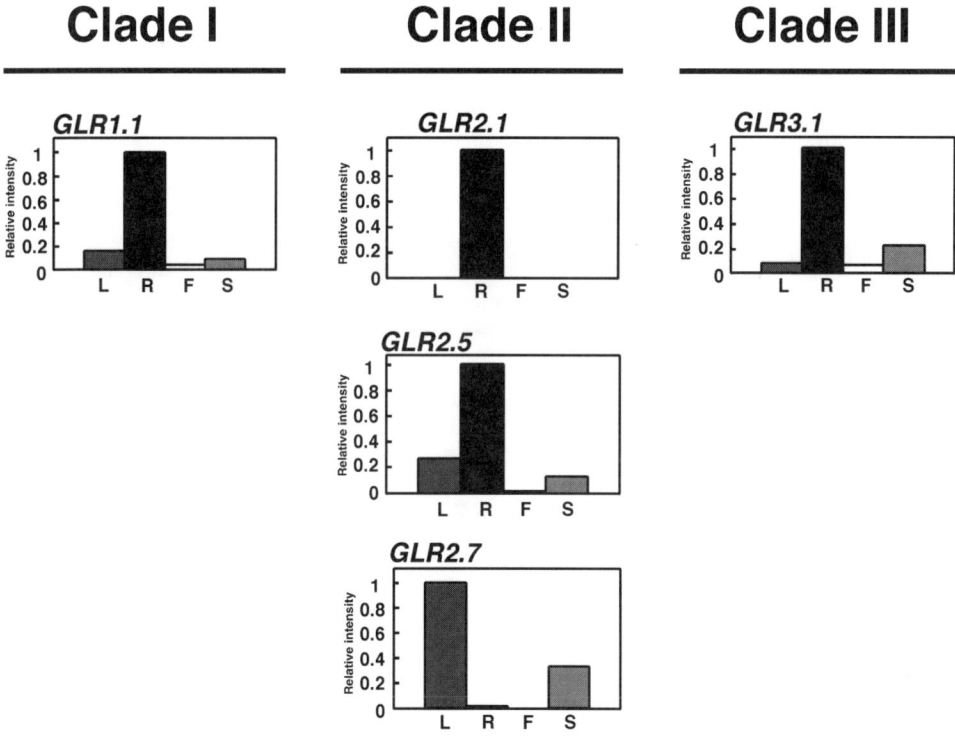

Figure 16.3. Representative RT-PCR mRNA expression patterns for each of the three *AtGLR* clades. Graphic representation of the data from RT-PCR analysis of the *AtGLR* gene family using mRNA generated from four different *Arabidopsis* organs: leaf (L), root (R), flower (F), and silique (S). Each graph shows the relative intensity of the four *AtGLR* bands normalized against a control gene *TUB5* (accession number M84702; adapted from Chiu *et al.*, 2002, material is copyrighted by the Society of molecular biology and evolution).

The RT-PCR results show that although there is similarity in organ-level expression patterns between *AtGLR* genes in the same clade, there is also overlap in expression profiles between genes from different clades. For example, *AtGLR* genes from clades I and III are all expressed ubiquitously with strongest expression in roots. These results show that the three *AtGLR* clades are not distinct based on organ-level expression, therefore suggesting that the three *AtGLR* clades may contain genes with overlapping function *in vivo*.

In addition to exploring the functional significance of *AtGLR* clade division, RT-PCR analysis also showed that almost all *AtGLR* genes are expressed at high levels in roots (except *GLR2.7*). Their strong expression in roots suggest that *AtGLRs* may be involved in regulating ion-uptake from soil, especially if they indeed function as ion channels (Chiu *et al.*, 2002).

3.2. Examining Cell-Type Level Expression of AtGLRs Using Promoter–Reporter Gene Expression System

Although *AtGLR* genes from different clades are expressed in the same organs, for example, genes from clades I and III, it is possible that they may be expressed in different cell types within the same organ. Chiu *et al.* (2002) examined the cell-type expression of a representative *GLR* gene from each *AtGLR* clade (*AtGLR1.1, AtGLR2.1,* and *AtGLR3.1*) by examining transgenic plants that have been transformed with constructs in which the putative promoter of each of the three *AtGLR* genes was fused to the reporter gene β-glucuronidase respectively (GUS; Jefferson, 1989).

These studies showed that although *AtGLR1.1* and *AtGLR3.1* are both expressed in all organs tested as shown by RT-PCR (Chiu *et al.*, 2002), they are observed in different cell types according to GUS expression in transgenic plants. Whereas *GLR3.1* expression is concentrated in the vasculature, *GLR1.1* expression is not detected in the vasculature (Figure 16.4, see color insert; see Chiu *et al.*, 2002 for more complete GUS expression patterns). The vasculature-specific expression of clade III genes is also supported by the fact that *AtGLR3.2*, a second clade III gene, is also strongly expressed in the vasculature (Kim *et al.*, 2001). Future experiments are necessary to determine if all *AtGLR* clade III genes are vasculature-specific, while all *AtGLR* clade I genes are expressed in cell types similar to *AtGLR1.1*. If it turns out that *AtGLR* genes from clades I and III are expressed in distinct cell types, it is likely that clades I and III represent distinct functional protein classes.

4. Functional Analysis of *Arabidopsis GLR* Genes *In Vivo*

The ultimate goal for researchers studying plant *GLR* genes is to determine the function of these genes *in vivo*. The question concerning the function of *GLR* genes in plants is particularly intriguing since plants lack a nervous system in which animal iGluRs are known to be important. By using *A. thaliana* as a model plant, researchers have used different approaches to attempt to unlock the *in vivo* function of *GLR* genes in plants.

4.1. *In Planta* Electrophysiology

Despite the sequence similarity between animal iGluRs and *AtGLRs*, it is currently unknown whether *AtGLRs*, like animal iGluRs, encode functional ligand-gated ion channel subunits. By performing *in planta* electrophysiology, Dennison and Spalding (2000)

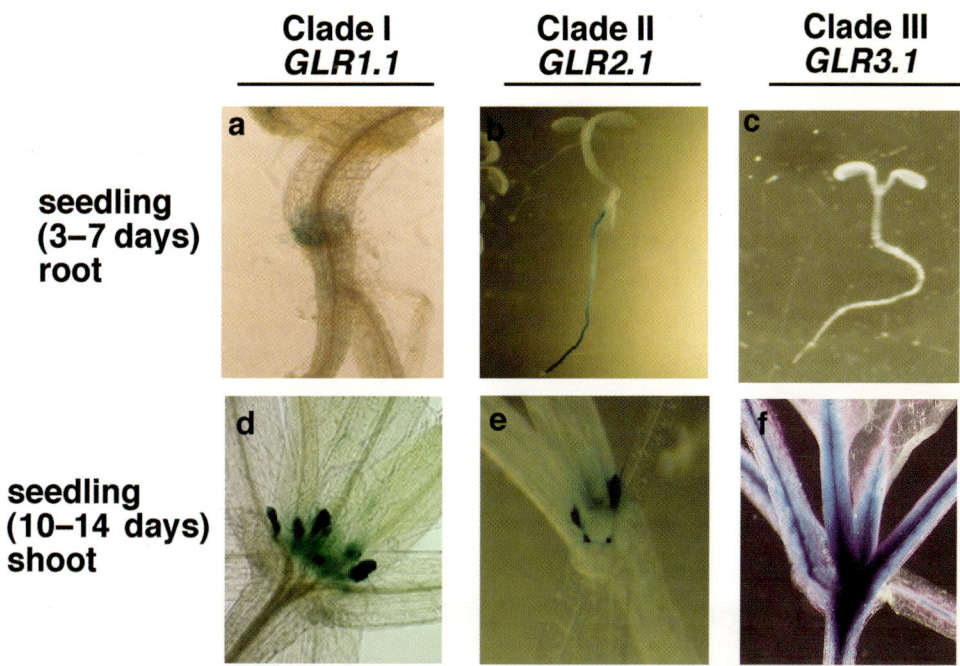

Clade I *GLR1.1*	Clade II *GLR2.1*	Clade III *GLR3.1*

**seedling
(3–7 days)
root**

a   b   c

**seedling
(10–14 days)
shoot**

d   e   f

Figure 16.4. Representative genes from each of the three *AtGLR* clades exhibit cell-specific expression patterns in young seedlings. Expression of *GLR1.1*, *GLR2.1*, and *GLR3.1* as seen in transgenic *Arabidopsis* seedlings harboring promoter::GUS fusions. Weak expression of *GLR1.1* can be seen in the shoot–root junction (a). *GLR2.1* shows strong expression in all cell types of the root except the apex between 3 and 5 days after germination (b). *GLR3.1* expression cannot be detected at this early stage (c). In more mature seedlings, *GLR1.1* and *GLR2.1* are expressed in stipules (d and e), whereas *GLR3.1* expression is found in the vasculature-related tissues (f) (adapted from Chiu *et al.*, 2002, material is copyrighted by the Society of molecular biology and evolution).

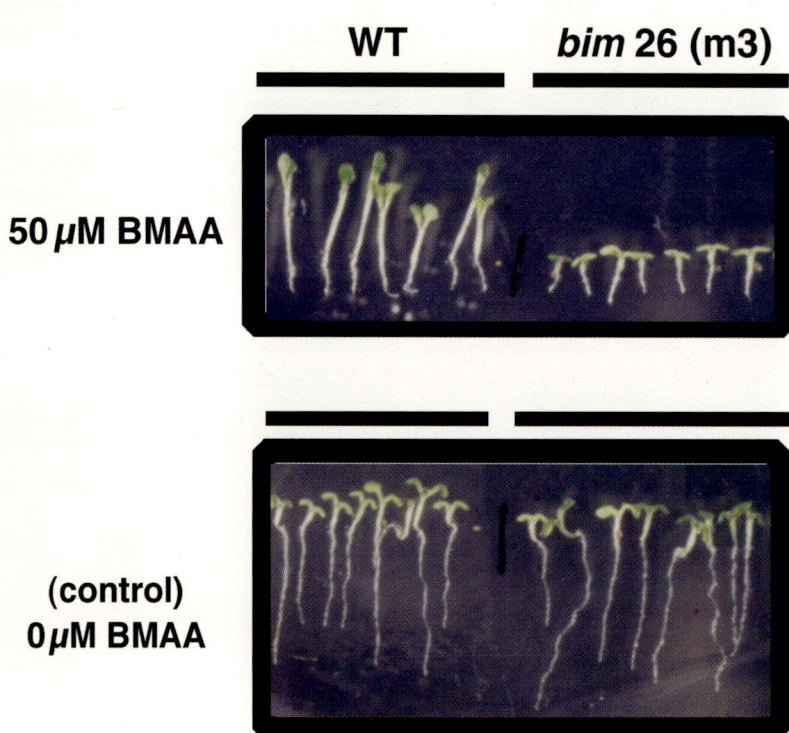

Figure 16.6. Effects of BMAA on *Arabidopsis* seedlings and the isolation of BMAA-insensitive (*bim*) mutants. Wild-type seedlings have increased hypocotyl length and decreased cotyledon-opening when treated with BMAA. On the other hand, *bim* mutants are insensitive to BMAA and resemble wild-type seedlings without BMAA treatment.

presented the first evidence suggesting the presence of glutamate-gated ion channels in *Arabidopsis* plants. With the use of transgenic seedlings expressing acquorin, a calcium sensitive luminescent protein (Knight *et al.*, 1991), Dennison and Spalding (2000) monitored [Ca^{2+}] in the cytoplasm before and after glutamate application. They observed a large, transient increase in [Ca^{2+}]$_{cyt}$ immediately after glutamate application (Figure 16.5).

In separate experiments, Dennison and Spalding (2000) monitored the membrane potential (V_m) of *Arabidopsis* root cells by inserting intracellular microelectrodes into root apices of *Arabidopsis* seedlings. When the bath chamber medium is switched from 1 mM KCl to 1 mM K-Glu (addition of glutamate), a large and rapid membrane depolarization is observed (average peak change = 55 mV; Figure 16.5). This suggests that the large, transient increase in [Ca^{2+}]$_{cyt}$ observed in transgenic seedlings expressing aequorin upon glutamate application is due to the opening of glutamate-gated calcium-permeable channels at the plasma membrane. Dennison and Spalding (2000) pointed out that other activities in addition to glutamate-gated inward calcium current across the plasma membrane might also contribute to the positive shift in membrane potential. These include (a) secondary effects of increased [Ca^{2+}]$_{cyt}$ on other ion transporters, (b) activities of glutamate-gated chloride ion channels, and (c) electrogenic glutamate-uptake mechanism.

To prove that the increase in [Ca^{2+}]$_{cyt}$ induced by glutamate application is at least partially due to the calcium influx through the plasma membrane, Dennison and Spalding (2000) made use of La^{3+}, a calcium channel blocker, and EGTA, a calcium ion chelator. Application of either of the two modulators inhibits the [Ca^{2+}]$_{cyt}$ spike normally induced by glutamate application. Combining all the evidence, their results suggest that glutamate application led to an influx of

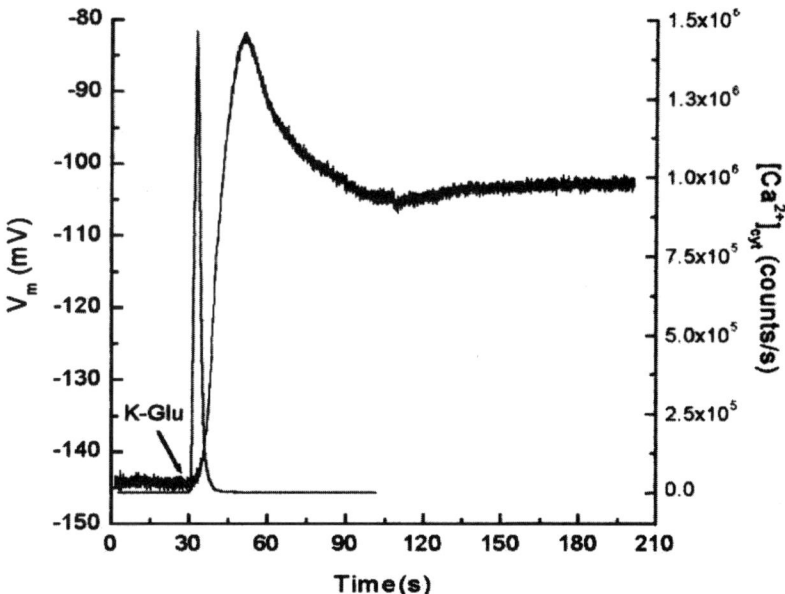

Figure 16.5. Glutamate-gated calcium fluxes and membrane depolarization in *Arabidopsis* seedlings. The left axis measures plasma membrane potential (V_m) while the right axis measures calcium concentration in the cytoplasm as represented by calcium dependent luminescence from aequorin-expressing seedlings. Glutamate application (indicated by arrow) triggers a calcium spike as well as a positive shift in membrane potential in separate experiments (reprinted from Dennison and Spalding (2000) with permission; material is copyrighted by the American society of plant biologists).

calcium ions across the plasma membrane, thereby increasing $[Ca^{2+}]_{cyt}$. An increase in $[Ca^{2+}]_{cyt}$ may then activate downstream signaling pathways involved in plant development and metabolism.

In addition to glutamate, the native ligand of animal iGluRs, the ability of other non-native animal iGluR agonists, such as AMPA and NMDA, to induce the increase in $[Ca^{2+}]_{cyt}$ were also tested (Dennison and Spalding, 2000). These non-native agonists were not effective in triggering a positive response that is comparable to the glutamate-induced response. This is not surprising since the divergence of animal iGluR subclasses (AMPA, KA, NMDA, and Delta) occurs after the divergence of plant and animal iGluRs (Chiu *et al.*, 1999). As a matter of fact, these non-native agonists also failed to activate *Synechocystis GluR0* (Chen *et al.*, 1999), which diverged from plant and animal iGluRs early on (Chiu *et al.*, 2002).

Although the observation of glutamate-gated calcium influxes and membrane depolarization is exciting, and suggests the presence of glutamate-gated calcium channels in *Arabidopsis*, future experiments are necessary to establish a link between these phenomena and *Arabidopsis GLR* genes.

4.2. Pharmacological Approaches to Examine the Function of *GLR* Genes in Plants

In an attempt to study the *in vivo* role of *GLR* genes in plants, *Arabidopsis* seedlings were grown in the presence of DNQX, an antagonist which blocks the function of animal non-NMDA type iGluRs (Lam *et al.*, 1998). This was an attempt to block the function of native *GLRs* in *Arabidopsis*, assuming that DNQX affects *AtGLRs* in the same manner as in animals. Seedlings grown in the presence of DNQX phenocopy *Arabidopsis* long hypocotyl (hy) mutants (Koornneef *et al.*, 1980), which are impaired in light signal transduction. Normally, wild type seedlings that are grown in the light have short hypocotyls and open green cotyledons, whereas those that are grown in the dark have much longer hypocotyls and closed yellow cotyledons. Seedlings grown in the presence of DNQX behave as if they cannot sense light, therefore resulting in their long hypocotyl phenotype. DNQX-induced hypocotyl-elongation is shown to be light dependent as well as dose dependent. Results of this experiment suggest that *GLRs* may be involved in light signal transduction in plants.

A similar pharmacological experiment supporting the importance of *AtGLRs* in light signal transduction and photomorphogenesis was performed using BMAA, a non-native animal glutamate receptor agonist derived from cycad plants (Brenner *et al.*, 2000). Wild-type *Arabidopsis* seedlings grown in the presence of BMAA show a light-dependent long-hypocotyl phenotype (Figure 16.6, see color insert). In addition, BMAA also inhibits cotyledon opening, another light-dependent photomorphogenic response. Interestingly, the aforementioned BMAA-dependent phenotype can be reversed by the addition of glutamate. This finding that BMAA, an agonist of animal GluRs, has the same effects on *Arabidopsis* seedlings as DNQX, an animal iGluR antagonist, is puzzling. However, it has been shown that agonists of animal glutamate receptors can sometimes act as neurotoxins and impair animal GluR function, because GluR channels remain sensitized to non-native ligands, such as BMAA (Ross *et al.*, 1989). On the other hand, animal GluR channels become desensitized to glutamate, the native agonist. The mechanism of desensitization is essential for the proper functioning of animal GluR channels (Geoffroy *et al.*, 1991). The finding that the long-hypocotyl phenotype induced by BMAA can be reversed by the addition of glutamate is consistent with the hypothesis that BMAA affects seedling morphogenesis by competing with the natural ligand glutamate, and therefore blocking proper *GLR* function in *Arabidopsis* (Brenner *et al.*, 2000). An alternative hypothesis is that BMAA activates specific *AtGLR* channels necessary for regulating ion flow and cell expansion, therefore leading to increase in hypocotyl elongation.

The pharmacological experiments discussed here suggest that glutamate-gated ion channels may be involved in light signal transduction and photomorphogenesis. However, future experiments are necessary to link these processes to plant *GLR* genes. One of these ongoing experiments is discussed in the next section.

4.3. Forward Genetic Approaches to Study *GLR* Function in *Arabidopsis*

To ascertain whether BMAA is affecting photomorphogenesis through plant *GLR* genes or a *GLR*-mediated pathway, Brenner *et al.* (2000) have isolated *Arabidopsis* mutants (*bim*) that are insensitive to the effects of BMAA. Normally, wild-type *Arabidopsis* seedlings that are grown on BMAA have long hypocotyls and partially closed cotyledons. *Arabidopsis bim* mutants are selected as seedlings that have short hypocotyls and open cotyledons in the presence of BMAA, that is, they behave as if they are resistant to the effects of BMAA (Figure 16.6, see color insert). Different classes of *bim* mutants were isolated, and they are in the process of being characterized and mapped to locations on *Arabidopsis* chromosomes. The identification of the affected loci in *bim* mutants will help to determine if plant *GLR* genes are involved in light signal transduction and photomorphogenesis.

4.4. Reverse Genetic Approaches to Study *GLR* Function in *Arabidopsis*

With the cloning of full-length *Arabidopsis GLR* complementary cDNA clones, researchers can now take a reverse genetics approach in determining the function of *GLR* genes in plants. Kim *et al.* (2001) have constructed transgenic *Arabidopsis* plants that overexpress *AtGLR3.2*. *AtGLR3.2* overexpressing plants exhibit symptoms of calcium deficiency, which can be alleviated by supplementing the plants with calcium. The calcium deficiency phenotype is apparently not due to decrease in calcium uptake since the overall levels of calcium in tissues are not significantly different from control plants transformed with an empty vector. It was therefore concluded that the calcium deficiency phenotype of *AtGLR3.2* overexpressor plants is due to inefficient calcium utilization instead. Kim *et al.* (2001) suggested that *AtGLR3.2* may encode an ion channel protein that forms nonspecific cation channels, which are permeable to sodium, potassium, as well as calcium. Ectopic expression of *AtGLR3.2* may therefore lead to excess uptake of potassium and sodium ions, which could then affect calcium utilization. The proposition that *AtGLR3.2* overexpression may lead to increased uptake of potassium and sodium ions is also supported by the fact that *AtGLR3.2* overexpressor plants are hypersensitive to sodium and potassium ionic stresses. An alternative hypothesis is that overexpression of *AtGLR3.2* may disturb calcium transport directly (Kim *et al.*, 2001).

5. Future Prospects

The results discussed here represent encouraging first steps to elucidating the function of *GLR* genes in plants. Future experiments are necessary to establish links between plant

GLR genes and the different processes in which they are potentially involved. With the forward genetics approach, if any of the *bim* mutants are eventually mapped to *AtGLR* locus, it will mean that BMAA affects development and photomorphogenesis by acting on *AtGLR* proteins. Other *bim* mutants that do not map to any of the *AtGLR* locus may represent interacting partners of *AtGLR* proteins.

To examine the function of *GLR* genes in plants, it will be important to obtain transgenic plants in which *GLR* genes are knocked out (reverse genetics). Since transgenic plants are easily constructed in *A. thaliana*, it will be advantageous to use *Arabidopsis* as a model plant for these experiments. One potential problem with this approach is that since the *AtGLR* gene family consists of 20 members, some of the *AtGLR* genes might have redundant functions. As a result, knocking out the function of one *AtGLR* gene may not lead to any obvious phenotype. In that case, double and triple mutants may need to be constructed for defects to be uncovered. Once *AtGLR* mutants are obtained, they can be tested for any changes in glutamate-gated calcium influxes and membrane depolarization, as well as changes in other channel activities characterized in plants, for example nonselective cation channels (Demidchik and Tester, 2002). These experiments will help to answer whether *AtGLR* genes encode ion channels *in vivo*.

Alternatively, researchers have been expressing *AtGLR* genes in heterologous systems, such as *Xenopus* oocytes. Results from these experiments will not only confirm that *AtGLR* genes encode functional ion channel subunits, but will also answer questions concerning ligand selectivity and ion selectivity of these channels. This information will be instrumental in the understanding of *AtGLR* gene function in plants.

References

AGI (2000). Analysis of the genome sequence of the flowering plant *Arabidopsis thaliana*. *Nature* **408**, 796–815.

Brenner, E.D., N. Martinez-Barboza, A.P. Clark, Q.S. Liang, D.S. Stevenson, and G.M. Coruzzi (2000). Arabidopsis mutants resistant to S(+)-β-Methyl-α, β-Diaminopropionic acid, a cycad-derived glutamate receptor agonist. *Plant Physiol.* **124**, 1615–1624.

Chen, G.Q., C. Cui, M.L. Mayer, and E. Gouaux (1999). Functional characterization of a potassium-selective prokaryotic glutamate receptor. *Nature* **402**, 817–821.

Chiu, J., R. DeSalle, H.M. Lam, L. Meisel, and G.M. Coruzzi (1999). Molecular evolution of glutamate receptors: A primitive signaling mechanism that existed before plants and animals diverged. *Mol. Biol. Evol.* **16**, 826–838.

Chiu, J.C., E.D. Brenner, R. DeSalle, M.N. Nitabach, T.C. Holmes, and G.M. Coruzzi (2002). Phylogenetic and expression analysis of the glutamate receptor-like gene family in *Arabidopsis thaliana*. *Mol. Biol. Evol.*, **19**, 1066–1082.

Demidchik, V. and M. Tester (2002). Sodium fluxes through nonselective cation channels in the plasma membrane of protoplasts from Arabidopsis roots. *Plant Physiol.*, **128**, 379–387.

Dennison, K.L. and E.P. Spalding (2000). Glutamate-gated calcium fluxes in Arabidopsis. *Plant Physiol.* **124**, 1511–1514.

Geoffroy, M., B. Lambolez, E. Audinat, B. Hamon, F. Crepel, J. Rossier, and R.T. Kado (1991). Reduction of desensitization of a ionotropic glutamate receptor by antagonists. *Mol. Pharmacol.* **39**, 587–591.

Jefferson, R.A. (1989). The GUS reporter gene system. *Nature* **342**, 837–838.

Kim, S.A., J.M. Kwak, S.K. Jae, M.H. Wang, and H.G. Nam (2001). Overexpression of the *AtGluR2* gene encoding an Arabidopsis homolog of mammalian glutamate receptors impairs calcium utilization and sensitivity to ionic stress in transgenic plants. *Plant Cell Physiol.* **42**, 74–84.

Knight, M.R., A.K. Campbell, S.M. Smith, and A.J. Trewavas (1991). Transgenic plant aequorin reports the effects of touch and cold-shock and elicitors on cytoplasmic calcium. *Nature* **352**, 524–526.

Koornneef, M., E. Rolff, and C.J.P. Spruit (1980). Genetic control of light-inhibited hypocotyl elongation in *Arabidopsis thaliana*. *Z Pflanzenphysiol.* **100**, 147–160.

Lacombe, B., D. Becker, R. Hedrich *et al*. (14 co-authors). (2001). The identity of plant glutamate receptor. *Science* **292**, 1486–1487.

Lam, H.M., J. Chiu, M.H. Hsieh, L. Meisel, I.C. Oliveira, M. Shin, *et al*. (1998). Glutamate receptor genes in plants. *Nature*, **396**, 125–126.

Ross, S.M., D.N. Roy, and P.S. Spencer (1989). β-N-oxalylamino-L-alanine action on glutamate receptor. *J. Neurochem.* **53** 710–715.

Turano, F.J., G.R. Panta, M.W. Allard, and P. van Berkum (2001). The putative glutamate receptors from plants are related to two superfamilies of animal neurotransmitter receptors via distinct evolutionary mechanisms. *Mol. Biol. Evol.* **18**, 1417–1420.

Index